THE
ASTRONOMER'S
UNIVERSE

THE ASTRONOMER'S UNIVERSE

Stars, Galaxies, and Cosmos

HERBERT FRIEDMAN

W·W·NORTON & COMPANY
New York London

Revised and updated Norton paperback edition first published 1998

Printed in the United States of America.

The text of this book is composed in Linotype Walbaum,
with display type set in Inverserif Light Italic.
Composition and manufacturing by
the Maple-Vail Book Manufacturing Group.
Book design by Jacques Chazaud.

Library of Congress Cataloging-in-Publication Data

Friedman, Herbert, 1916–
The astronomer's universe : stars, galaxies, and cosmos / Herbert Friedman.—1st ed.
p. cm.—(Commonwealth Fund Book Program)
Includes index.
1. Astronomy. 2. Cosmology. I. Title. II. Series: Commonwealth Fund Book Program (Series)
QB351.F69 1990
520—dc20 89–9362

ISBN 0-393-31763-3 pbk.

W. W. Norton & Company, Inc., 500 Fifth Avenue, New York, N. Y. 10110
W. W. Norton & Company Ltd., 10 Coptic Street, London WC1A 1PU

1 2 3 4 5 6 7 8 9 0

To my brightest stars
Mali, Luke, and Jed

Contents

CONTENTS

Portraits by Jon R. Friedman

FOREWORD

The scientific enterprise, in all its aspects, has advanced a greater distance in the twentieth century than in all the previous centuries of human inquiry taken together, and the speed with which totally new information is being acquired is accelerating even more rapidly now in the latter half of the century. There has never been a time like it, looking back as far as the eye and history can take us. Among the fields in which progress has occurred on a breathtaking scale is astronomy.

The Commonwealth Fund made the decision, several years back, to launch a new series of books to be written by working scientists about their own fields of science. The hope is to meet at least part of the public need for a more general and accessible understanding of the events that are transforming our view of nature and the world around us.

This volume, by Dr. Herbert Friedman, tells where matters now stand in our comprehension of the most immense of all puzzles, the universe.

The advisory committee for the Commonwealth Fund Book Program, which recommended the sponsorship of this volume, consists of the following members: Alexander G. Bearn, M.D.; Donald S. Fredrickson, M.D.: Lynn Margulis, Ph.D.; Maclyn McCarty, M.D.; Lady Jean Medawar;

Berton Roueché; Frederick Seitz, Ph.D.; and Otto Westphal, M.D. The publisher is represented by Edwin Barber, Director of the Trade Department at W.W. Norton & Company. Suzanne H. Heyd serves as administrative assistant. Margaret E. Mahoney, president of the Commonwealth Fund, has actively supported the work of the advisory committee at every turn.

LEWIS THOMAS, M.D., Director
ALEXANDER G. BEARN, M.D., Deputy Director

PREFACE

> . . . What unites us—the ultimate ground of our claim to equality—is our common ignorance of the central questions posed for us by the universe: whence, and why, and whither.
>
> —PAUL FREUND (1908–1992)
> Professor, Harvard Law School, 1940–1992

Man's love of the stars goes back to the beginning of recorded history, but for almost all of that time, it was only an adolescent romance. The universe we perceive in the twentieth century is incomparably greater in its design and infinitely more mysterious in its ways than anyone could have predicted in earlier generations. Barely 100 years ago, Simon Newcomb, the greatest American astronomer of his time, remarked, "We are probably near the limit of all we can know about astronomy." He was, of course, referring to celestial mechanics, the preoccupation of astronomers since the ancient Greeks. Stars were thought of only as points in a cosmic tapestry that was ageless and invariant. Astrophysics, the science of the life and death of stars, and modern cosmology, which deals with the origin and fate of the universe, are developments that barely span our contemporary lifetime.

Astronomy, the most ancient science, has become a prodigy of the Space Age. As soon as small rockets began to pop their instrumented noses above the atmosphere, a few adventurous physicists became rocket astronomers. Through new windows in the ultraviolet and x-ray regions of the electromagnetic spectrum, we perceived a celestial scene of cataclysmic violence marked by massive gravitational collapse of giant stars to neutron stars and black holes, and explosion phenomena on scales that make nuclear weapons pale into insignificance. At the same time, radar engineers translated their wartime experience into radio astronomy. Interferometric radio telescopes now exceed the imaging resolution of all other modern tools of astronomy and probe to the heart of fantastic energy sources such as quasars and the nuclei of dynamic galaxies.

The present explosion of knowledge in astronomy is not a transient phenomenon. In recent decades the acquisition of new information has been growing almost exponentially, and our efforts to grasp its meaning are mind-stretching. Can we, as British historian John Bagnell Bury asks, "be sure that someday progress may not come to a dead pause, not because knowledge is exhausted, but because our resources for investigation are exhausted—because, for instance, scientific instruments have reached the limit of perfection beyond which it is demonstrably impossible to improve them?" ("The Idea of Progress: An Inquiry into Its Origins and Growth," 1932).

It would be naive to believe that we are approaching scientific closure; as long as we can look forward to exploiting the remarkable powers of next-generation observing facilities, the astronomical horizon will continue to recede. The anticipated rate of future progress can only be guessed from an extrapolation of past experience, from the time that Galileo Galilei extended human vision with a primitive telescope to the orbiting of the magnificent Hubble space telescope. Beyond lies the potential of still-greater space observatories projected after the turn of the century. We try to plan what needs to be observed and to predict what lessons we may learn, but serendipity will most likely out-perform all of our logical expectations.

When I was a graduate student in the late 1930s, one of the most eminent physicists of the time remarked that the future truths of physical science were to be sought in the sixth place of decimals. Astrophysicists and cosmologists suffer no such dreary inhibition; great exhilaration can often be derived from knowing the truth to an order of magnitude (factor of 10), and the stock of truth that remains to be discovered "in Nature's infinite book of secrecy" (Shakespeare) is limited only by the ingenuity of the human mind.

> Nothing is too wonderful to be true.
>
> —Michael Faraday

In these chapters we survey some of the highlights of astronomical progress of the past few hundred years. The lesson to be drawn is that surprise is normal and that there is no limit to the rate of growth of knowledge other than the steadily growing costs of future facilities. The

Challenger accident in 1986 was a catastrophic setback for space science as well as a great human tragedy. It brought home the enormous costs of space transportation and the growing burdens of designing to ever more demanding reliability specifications. In these circumstances the words of Sir Arthur Eddington in his 1927 book, *Stars and Atoms*, bear special meaning:

> In ancient days two aviators procured to themselves wings. Daedalus flew safely through the middle air and was duly honoured on his landing. Icarus soared upwards to the sun till the wax melted which bound his wings and his flight ended in fiasco. . . . The classical authorities tell us, of course, that he was only "doing a stunt"; but I prefer to think of him as the man who brought to light a serious constructional defect in the flying-machines of his day.
>
> So, too, in science. Cautious Daedalus will apply his theories where he feels confident they will safely go; but by his excess of caution their hidden weaknesses remain undiscovered. Icarus will strain his theories to the breaking-point till the weak joints gape. For the mere adventure? Perhaps partly, that is human nature. But if he is destined not yet to reach the sun and solve finally the riddle of its constitution, we may at least hope to learn from his journey some hints to build a better machine. (p. 41)

Acknowledgments

I am indebted to many astronomers and their scientific institutions for generous permissions to use illustrations from their collections. Credits are cited in the captions that accompany the figures. The portrait sketches by my son Jon Friedman are based on photographs borrowed from colleagues or found in the archives of the American Institute of Physics, the Astronomical Society of the Pacific, California Institute of Technology, George Washington University, Harvard University, the University of Chicago, the Lick Observatory, and the files of *Mercury* magazine and *Sky and Telescope*.

My editor early on was Antonina Buis, who served as my intermediary with the editorial board of the Commonwealth Fund Book Program and helped me to shape the structure of the book. The penultimate draft was edited by Cheryl Simon, whose gentle urging encouraged me to simplify the text and reduce the obstacles that might be encountered by lay readers. Both editors encouraged me to use anecdotal material out of my personal experience and to adopt a narrative rather than textbook style throughout.

HERBERT FRIEDMAN, June 1989

Introduction

> Lost, among bright stars on this most weary unbright cinder, lost!
>
> —THOMAS WOLFE,
> "Look Homeward, Angel"

How far are the stars? How big is the universe? When *Sputnik* was launched, public perceptions of space were generally little more sophisticated than the views of the ancient Greeks. I recall riding in a taxicab to the Hayden Planetarium in New York City to deliver a lecture on rockets and satellites. When my driver learned that I was an astronomer, he confessed great puzzlement about the placement of stars and satellites. How was it possible that *Sputnik* went around and around up there without "bumping into all them stars?" Yet Aristarchus of Samos, in the third century B.C., already appreciated that the apparent immobility of the stars meant that they were very far away indeed.

In 434 B.C. the Greek philosopher Anaxagoras thought the Sun was a ball of fire about 35 miles in diameter, hanging about 4000 miles above the Earth. His logic and mathematics were good, but were flawed by his assumption that the Earth was flat. Two centuries later Eratosthenes assumed a spherical Earth and derived a radius of 4000 miles, which is remarkably close to the correct figure. Not much later Hipparchus, who based his model on eclipse measurements, found a good value for the distance of the Moon—about 200,000 miles—and from that measure

concluded that the Sun was 37 times as far away. Although it was far short of the true distance, his figure was accepted for the next 17 centuries. The correct distance of 92 million miles became known only 2 centuries ago from observations made with telescopes.

Stellar distances are so great that astronomers describe them in units of a light-year, the distance traveled by a ray of light at a speed of 186,000 miles per second in 1 year (1 light-year = 6 trillion miles). In 1840 Wilhelm Friedrich Bessel, a German astronomer and mathematician, made the first precise determination of the distance to a star. From measurement of the trigonometric parallax of Alpha Centauri—its maximum angular displacement as seen from opposite extremes of the Earth's orbit—Bessel calculated a distance of slightly more than 4 light-years, or about 25 trillion miles. One could only guess at the distances of stars with no measurable parallax. Large telescopes now observe quasars as far away as 13 billion light-years, which are so distant that their light rays began the journey to us long before the Sun and Earth were born.

From Celestial Mechanics to Astrophysics

Beginning with Galileo's telescope, astronomers began to discover the universe in ways that could never have been predicted. In 1609, when he presented a 9-power spyglass to his patrons in the Venetian Senate, Galileo shrewdly emphasized its potential importance in time of war. He then hurried home to construct a 20-power instrument and aim it at the Moon, stars, and planets. Immediately, he discovered four of the satellites

FIGURE *1.* Cecelia Payne-Gaposhkin (1900–1979). *While on a scholarship to Cambridge University in 1919, Cecelia Payne was inspired by Sir Arthur Eddington to make her career in astronomy. She enrolled in graduate work at Harvard College Observatory and after 2 years received her doctorate. Her thesis on "Stellar Atmospheres" combined laboratory spectroscopic data and hundreds of her own photographic plates of stellar spectra, creating a new field. She deduced the temperature and elemental composition of each spectral type at a time when knowledge of stellar atmospheres was very primitive. In essence she demonstrated that stars, the Sun, and the Earth were made of the same chemical elements in similar ratios of abundance. The outstanding discrepancy between her astrophysical and terrestrial abundances was in the high stellar concentrations of hydrogen and helium. Only years later was it recognized that the solar atmosphere is almost pure hydrogen and helium with only a trace of heavier elements.*

In 1956 Cecelia Payne-Gaposhkin became the first woman appointed to a full professorship at Harvard; she held the chairmanship of the astronomy department for 12 years.

of Jupiter and the mountains and craters of the Moon; the Milky Way resolved into countless swarms of stars. His *Starry Messenger*, a treatise published in 1610, described those amazing observations and set off a revolution in our perception of the universe. But for the next 300 years, astronomy remained simply a matter of putting eyeball to telescope and slowly cataloging the celestial panorama. The great centers of astronomy in the nineteenth century, the national observatories at Paris, Pulkova, Greenwich, England, and Washington, D.C., focused their efforts on celestial navigation. Why the stars shine was beyond comprehension, and the Milky Way was thought to be the entire universe.

With the development of astrophysics, emphasis switched from stellar motions to physical mechanisms. Spectroscopy, the analysis of characteristic wavelengths emitted and absorbed by gaseous matter, enables the astrophysicist to deduce stellar temperature, pressure, density, and composition, the degree of ionization, the strength of gravity, and electric and magnetic field strengths, and to characterize turbulent heat motions. The

FIGURE 2. *Penetration of rays into the terrestrial atmosphere to a level at which the intensity is reduced to 1/2.7. Visible light penetrates to the ground through a narrow window in the atmospheric absorption spectrum. Short-wavelength radio waves reach the ground but longer waves are reflected back to space by the ionosphere. Far-ultraviolet, x-rays, and gamma rays cannot be detected below balloon altitudes. The absorbing constituents of the atmosphere are shown at their effective wavelengths.*

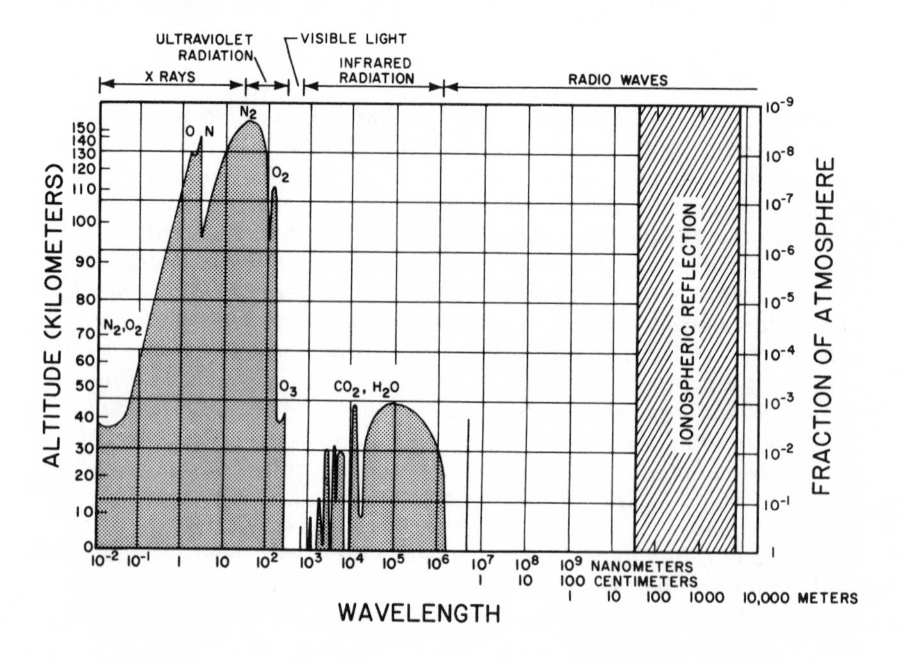

spectrum lines that appear in gaseous discharges in the laboratory also are observed in stars from thousands to billions of light-years distant.

The use of spectroscopy in understanding the constitution of stars advanced markedly in 1925 when Cecilia H. Payne, a young Radcliffe astronomy student recently arrived from England, wrote a landmark treatise. Acclaimed as the most brilliant doctoral dissertation ever written in astronomy, it enforced the concept of a cosmic abundance of the elements; that is, that the relative abundances of elements in the stars are constant and are not different from those of the Sun.

Except for the narrow rainbow band of visible light, the atmosphere is largely opaque to shorter and longer wavelengths. With balloons, rockets, and satellites, astronomers since World War II have moved their observatories beyond the atmospheric veil. Whereas classical spectroscopy was confined to the visible, with slight extensions to the ultraviolet and infrared, modern spectroscopy now spans the entire range from ultrashort gamma rays to long radio waves. An entire new subject, astrochemistry, has grown up around the radio microwave technique. More than 100 molecules, including some as large as ethyl alcohol, have been detected in interstellar gas.

The Milky Way and Beyond

In recent years spacecraft have ventured forth to explore the Solar System, and people are more aware of its vast scale. But voyages to the planets are local excursions compared to a trip to the nearest star. Pluto, at the edge of the Solar System, is about 4 billion miles from the Sun. Proxima Centauri, the Earth's nearest star, is 6000 times farther away. In 1989 the *Voyager 2* spacecraft reached Neptune, which is relatively near at 2.7 billion miles, or about 242 light-minutes, but the trip still took 11 of our calendar years.

In 1883 a Dutch astronomer, Jacobus Cornelis Kapteyn, undertook the enormous task of recording the positions of all stars that had been observed through nineteenth-century telescopes. He produced the first primitive star maps of our galaxy, the Milky Way, but the scientific picture has changed enormously since Kapteyn's time. In 1918 a major advance came from the ideas of Harlow Shapley, who in a bold stroke of inspiration mapped out the extent of the Milky Way and 100 known globular star clusters, which are dense concentrations of as many as 100,000 stars, arranged in a galactic halo. He concluded that the globular clusters were the outposts of the Galaxy and that they were spherically distributed around a center of their own; that is, the center of the Milky Way. Thus 400 years after Copernicus, Shapley removed the Sun and Earth from a privileged central position in the Milky Way and placed them in the suburbs, some 30,000 light-years from the galactic nucleus that we orbit once every 250 million years at a speed of 250 kilometers per second.

Shapley's new picture of the Milky Way created a storm of controversy. His most distinguished opponent was Heber D. Curtis, a classical scholar as well as an astronomer. They confronted each other at the National Academy of Sciences on 26 April 1929 in what came to be known in scientific circles as "The Great Debate." Curtis severely critized Shapley's distance estimates, but Shapley's broad concept won out. On the other hand, Shapley was the loser in an equally important aspect of the debate, that of the significance of the nebular patches seen against the star fields of the Milky Way. Curtis had come to realize that these nebulae were not gas clouds local to the Milky Way, but distant galaxies, the "island universes" that philosopher Immanuel Kant had suggested. Shapley persisted in arguing that the nebulae were on the fringe of the Milky Way.

In defense of Shapley's otherwise remarkable intuition, we now recognize that he was misled by incorrect observations published by a reputable Dutch-American astronomer, Adriaan van Maanen. Van Maanen was concerned with the measurement of rotation in nebulous objects. He claimed to have evidence of rotation for Andromeda and several other spiral nebulae. If that much rotation were evident, the nebulae had to be small and rather close by. Shapley accepted his results, but before long van Maanen was found wrong. It was difficult to understand his errors: Perhaps he was a victim of a not uncommon syndrome that sometimes leads scientists to "find" what they are looking for when the evidence is too marginal to trust. The dispute was finally settled by Edwin Hubble in 1934, when he determined that the distance to the Andromeda nebula was about 1 million (later revised to 2 million) light-years, far outside the boundary of the Milky Way.

The Milky Way as we perceive it today is a typical spiral galaxy like billions of others in the observable universe. Its winding arms define a visible disk approximately 500 light-years thick and 100,000 light-years across that bulges toward the center. The congregation of about 100 billion stars spans all ages. Some are newly born and some are as old as the Galaxy itself; about 15 billion years. (Our Sun is a second-generation star, born about 5 billion years ago.) Like a string of diamonds, bright young stars of high temperature stud the spiral arms.

A rift of obscuring dust that hugs the plane of the Galaxy hides the center from view but is now being penetrated by radio, microwave, infrared, x-ray, and gamma-ray telescopes. The nucleus of the Galaxy turns out to be a truly mysterious place. From recent radio, gamma-ray, and infrared observations of strong activity in the nucleus, astronomers now suspect the presence of a black hole 1 million times as massive as the Sun that is temporarily starving and ready to feed on any nearby stars that come within its grasp, supplying a great flood of new energy. With the advent of space-borne gamma-ray telescopes of much greater spatial resolution than those used thus far, such ideas might very well be confirmed or rejected before long.

Beyond the Milky Way lies a cosmic ocean of awesome extent, filled with perhaps another 100 billion galaxies, each with its complement of tens upon tens of billions of stars. Some are serenely spinning spirals like the Milky Way and Andromeda. Others are elliptically and irregularly shaped. Many are characterized by active galactic nuclei, the scenes of energetic processes so violent that they thus far defy understanding. Almost in contact with the Milky Way are its two nearest neighbors, the Magellanic Clouds. The farthest galaxies that our optical telescopes can study are almost 15 billion light-years away. Translated into miles, that distance is 9 followed by 22 zeros. When we look at those galaxies we see light that conveys an image 15 billion years old.

The Expanding Universe

In the 1920s Vesto Melvin Slipher at Lowell Observatory in Arizona noted a shift in the wavelengths of light from distant galaxies that soon became the key to understanding the expansion of the universe. He was

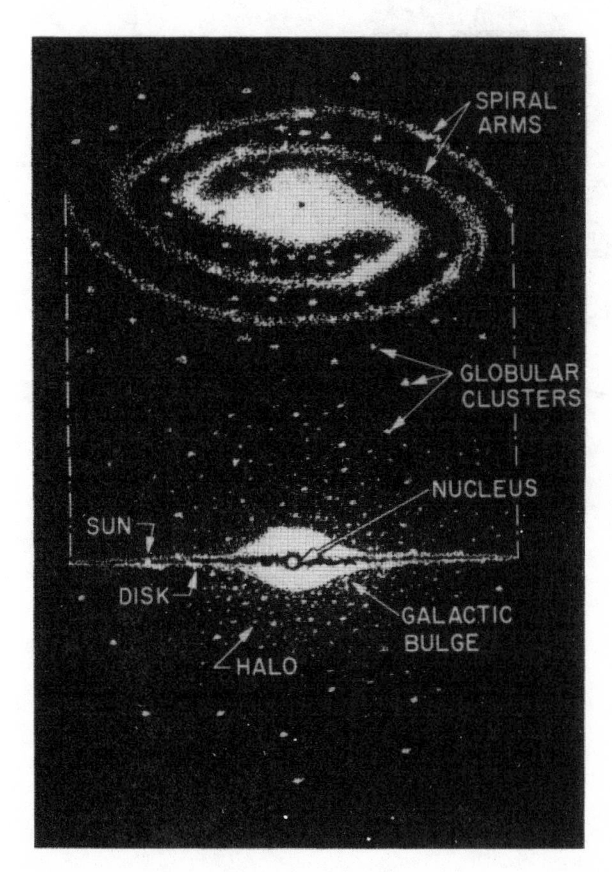

FIGURE 3. *The shape of the Milky Way galaxy viewed from above and from the side. Spiral arms unwind from a hub, a dense central bulge of stars, and flatten into a thin disk. Obscuring clouds of dust in the plane appear to divide the disk into two halves. Faint stars and globular star clusters envelope the Galaxy in a spherical halo. The diameter of the Galaxy is 100,000 light-years, and the Sun is located two-thirds of the distance from the nucleus to the edge. At the very center there is evidence of a massive black hole.*

observing a phenomenon first described by Christian Doppler in Vienna in 1842 and thenceforth known as the Doppler effect. The shift in wavelength and the corresponding frequency applies to all wave motions, to sound as well as light. As the whistle on an approaching train speeds toward you the pitch sounds higher than when the train has passed and is traveling away from you.

Christoph Buys Ballot, a Dutch meteorologist, arranged an unusual confirmation of Doppler's theory a few years later. He stationed several musicians with perfect pitch on a railroad platform while a group of tooting trumpeters were carried past the platform on a train. Although the players were in pitch, their notes sounded sour to the keen-eared bystanders.

If a source of light is moving toward an observer, the waves are compressed and the color is shifted toward the blue. Recession of the light source stretches the waves and produces a *redshift*. By 1936 Edwin Hubble had connected the amount of redshift to the distance from Earth of each galaxy; the greater the redshift, the greater the distance. It followed that the universe has been expanding from some point in time billions of years past and that the most-distant galaxies have been racing apart the fastest. This relationship is fundamental to the concept of the "Big Bang," the primordial explosion that sent all of the mass of the universe hurtling

FIGURE 4. Albert Einstein (1879–1955). *While working in the Swiss patent office, Albert Einstein in his spare time produced the special theory of relativity and published it in 1905. Perhaps the best-known result is that matter and energy are interchangeable according to the simple relationship,* $E = mc^2$, *from which it was deduced that the fusion of hydrogen to helium produces the energy of stars. Ten years later his general theory of relativity provided a profoundly different explanation of the relationship among mass, space, and time. Space is curved in the presence of mass, and gravity is a warp in spacetime. His theory describes the shape of the universe—closed, open, or flat—in terms of the mass that it contains. Although his field equations indicated an unstable condition that would cause the universe to expand or contract, Einstein was so convinced that the universe is static that he deliberately added a cosmological gravitational constant to prevent expansion. At the time Hubble had not yet discovered the expansion of the universe.*

Einstein had an aversion to the uncertainty implications of quantum theory that he expressed in his oft-quoted statement: "God does not play dice." The statistical interpretation of quantum theory implies an underlying indeterminacy in the laws of physics that he could not accept. After the age of 40, he spent his life seeking a unified theory of all the physical forces, but he never succeeded. Physicists today have unified the electromagnetic and weak nuclear forces and still seek to include the strong nuclear force and gravitation. Einstein's dream may yet come true.

apart in all directions. As galaxies condensed from the out-rushing matter, they inherited its high velocity. So firmly established is the Big Bang theory in modern cosmology that any contradiction would be an enormous shock to the scientific community.

The concept of an expanding universe developed with almost no interaction between theory and observation. Albert Einstein proposed his theory of general relativity in 1916 before it was recognized that the universe was expanding. Although his equations required a dynamic universe, Einstein clung to the concept of a static universe, eternal and unchanging. He deliberately introduced a term called the *cosmological constant* to make his theoretical universe hold still. In later years he ruefully referred to this artificial restriction on expansion as the greatest blunder of his life.

In 1922 a young Russian mathematician named Alexander Friedmann showed that Einstein's equations best fit an expanding universe. Russian science was in disarray during those early years of the Russian Revolution, and Friedmann is said to have first published his ideas in his hometown Petrograd newspaper, where they received little attention. When he sent his paper to the *Zeitschrift für Physik* it came to the attention of Einstein, who treated it with suspicion and gave it no credence. Later in the year, Einstein grudgingly reversed himself. Friedmann died of typhus in 1924, and little or no note was taken of his work for another half-dozen years. His ideas were basically reproduced in 1927 by Belgian astrophysicist Abbé George Lemaître, who envisioned all the mass of the universe condensed into a "cosmic egg" that exploded at the beginning of time.

FIGURE 5. Alexander Alexandrovitch Friedmann (1888–1925). *In 1922 Alexander Alexandrovitch Friedmann, a young mathematician and meteorologist, wrote a paper "On the Curvature of Space" for the* Zeitshrift für Physik. *At the time there were two cosmological models: Einstein's, in which matter filled a static space, and De Sitter's strange solution of Einstein's equations that proposed an expanding universe devoid of matter. Friedmann found that Einstein's field equations without his arbitrarily added cosmological constant led to one of two dynamical universes. If the average density of the universe is greater than a critical value—5×10^{-30} gram per cubic centimeter—gravity must force the universe to curve back on itself and eventually implode. Lesser density leads to a universe that expands forever. Einstein was so fixed on a static universe that he summarily criticized Friedmann's paper as being mathematically correct but without physical significance, then later had to reverse himself.*

Friedman was primarily concerned with meteorological observations from hot-air balloons. In 1914 he obesrved a solar eclipse from a dirigible balloon, and during World War I he carried out bombing surveys for the Russian Air Force. In 1925 he achieved an altitude record—23,000 feet—for manned balloon flight. Very shortly afterward he died of typhus and did not live to see his cosmological theory vindicated.

About 1 year later the American cosmologist Howard Percy Robertson independently advanced the same concept. All of this theoretical work seemed to pass unnoticed by Hubble, who had sufficient skeptical reserve that he would not declare, based on his observations, that the universe was expanding. Sir Arthur Eddington came upon Lemaître's publication and finally, in 1929, put theory and Hubble's observations together.

The Big Bang

> If the Lord Almighty had consulted me before embarking on the Creation, I would have recommended something simpler.
>
> —Alfonso X of Castile,
> a medieval patron of astronomy,
> A.D. 1221–1284

As the human mind tries to grasp the immensity of the universe, it struggles with the concept of creation. Ancient myths offer many versions: All essentially envision a transformation to order out of chaos. Modern science seeks to understand the evolution of the universe through the laws of physics, but we can probe just so close to the beginning.

Most people think of the Big Bang of creation as something akin to the explosion of a super-massive bomb, but in the concept of general relativity, it is an expansion of spacetime itself, filled with incredibly hot matter and radiation. Astrophysicists search for clues to the earliest moments in terms of the physics of elementary particles and grand unified theories (GUTs) of all the fundamental forces in nature.

Scientists have always wondered what matter was made of, but until the nineteenth century it was thought of as simply solid, liquid, or gas. Then scientists returned to the concept of an ancient Greek, Democritus, who said that matter ultimately could be subdivided into indivisible atoms. At the beginning of the twentieth century over 80 different atoms were recognized, now about 100 are recognized. Dealing with such a complicated array was much too cumbersome, and physicists sought a more fundamental breakdown of matter into a relatively few basic constituents. By the time electrons, protons, and neutrons were discovered it was recognized that all the chemical elements could be constructed of these three basic particles. But it soon turned out that they were only an entrée into a new world of elementary particle physics.

To understand radioactivity it was necessary to postulate the existence of the neutrino, which was subsequently detected experimentally. Cosmic-ray particles raining down from space collided violently with molecules of the high atmosphere to create showers of new particles: positrons, muons, pions, and others. Newly developed accelerators imitating cosmic-ray bombardment eventually expanded our knowledge of the zoo of particles to include rhos, omegas, and phis. So rapidly did the list grow that Enrico Fermi remarked, "If I could remember the names of all these particles I'd be a botanist."

Physicists began to bring ordered simplicity into particle physics in the 1960s, when they found it possible to make all of the nuclei out of a smaller number of truly elementary particles. The ultimate entity was the quark, a new particle derived from the theoretical construct of Murray Gell-Mann and George Zweig at the California Institute of Technology (Caltech). The proton and the neutron, for example, each consists of three quarks. Gell-Mann took the name from a passage in James Joyce's *Finnigan's Wake:* "Three Quarks for Muster Mark!" (The three quarks denote the three children of Mr. Finn, about whom the novel is written.) No one has ever detected a free quark, and in theory a quark should not be found outside a nucleus; however, the indirect evidence for its existence is very strong.

The atom is often described as a microscopic analog of the Solar System, its nucleus tightly packed with protons and neutrons resembling the

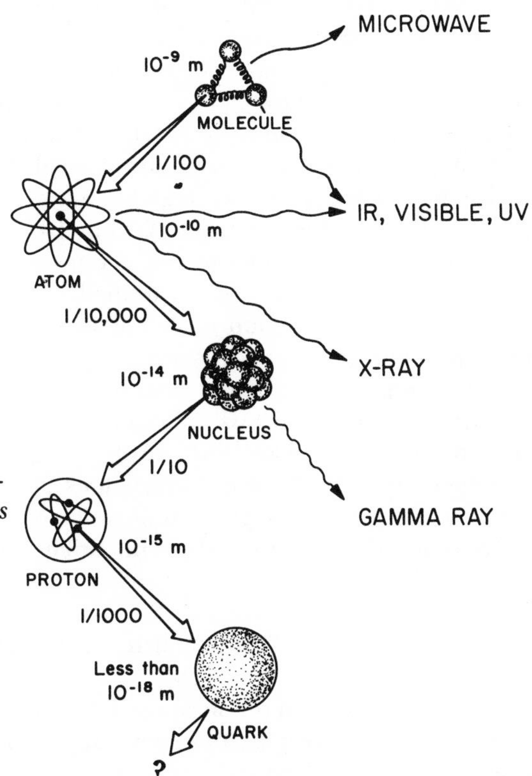

FIGURE 6. *The hierarchy of the structure of matter and related radiation processes. Typical sizes of molecules, atoms, and nucleons are shown in meters. The size of a quark is still unknown; it may have structure or it may be pointlike. Wavelengths of radiation associated with these structures correlate with the energy content of each configuration.*

Sun, and with an array of orbiting electrons resembling the planets. In the hierarchy of structure diagrammed in Fig. 6, a number of atoms combine to make a molecule that vibrates and rotates to radiate microwave and infrared radiation. Transitions of electrons in an atom lead to emission and absorption of optical photons and x-rays. Nuclear transitions produce gamma radiation. The scale of structure and the wavelength of associated radiation are roughly commensurate.

Physicists and cosmologists now construct theories that work backward to almost infinitesimal fractions of 1 second from the instant of creation. If we call that moment Time Zero, we can follow the evidence of observation and theory backward from now to then. The formation of atoms took place at approximately Time Zero plus 500,000 years. This was an epoch when the universe was bathed in visible light radiated by atoms. Earlier on, the universe was so hot that electrons were stripped off atoms as quickly as they attached, leaving a sea of nuclei and electrons in a hot bath of x-rays in a dark universe.

At Time Zero plus 3 minutes, the universe was a fireball at a temperature nearly 100 times higher than the thermonuclear furnace in the core of the Sun. Condensations of nuclei—for example, the deuteron (one proton plus one neutron); the tritium nucleus (one proton plus two neutrons); and the helium nucleus (two protons and two neutrons)—could no longer remain bound. Still closer to Time Zero, there existed a domain of such extremely high temperatures that the elementary particles in the fireball were of the kinds known to be created only in the highest-energy collisions produced by manmade accelerators. At Time Zero plus 1 millisecond (10^{-3} seconds), the temperature exceeded 1000 billion degrees kelvin and particle energies were too high for quarks to combine into the hundreds of known elementary particles. Within 10^{-35} second of the moment of creation, the fundamental forces of our familiar world no longer functioned individually; they were "unified." A super "grand unified force" dominated a cosmic soup of elementary particles that were constantly dissolving and reforming in a sea of primordial radiation within the volume of a universe no bigger than a baseball. The theoretical challenge today is to understand this realm of unified forces from 10^{-43} to 10^{-35} seconds after the clock of the universe began to run. A favorite current hypothesis is that a "phase change" took place in the primordial soup at 10^{-35} seconds when the grand unification of forces dissolved, causing the universe to inflate enormously in the briefest instant of time.

Matter and Antimatter

In 1930 the English physicist Paul Adrien Maurice Dirac advanced the concept that for every particle—electron, proton, positron, and so on—there must exist a sister particle of the same mass but of opposite electric charge or spin. These mirror particles are called *antimatter.* For example, the electron has 1 unit of negative charge and the antielectron (positron) has 1 unit of positive charge. Particles combine to form the atoms of our familiar world. Similarly, antiparticles can make antimatter. In all respects an antimatter galaxy seen from a distance would be indistinguishable from our Milky Way. Its gravitation and atomic spectra would be no different in appearance. But when a particle meets its antiparticle both annihilate instantaneously, and their mass is transformed entirely to photons of gamma radiation. Annihilation is a perfect converter of mass to radiation, nearly 100 times as efficient as nuclear fusion. Twenty-five pounds of matter combined with 25 pounds of antimatter would suffice to produce all the electrical energy consumed in New York State in 1 year.

There are scientists who believe in a fundamental symmetry to the universe that requires equal amounts of matter and antimatter. Astrophysicists question whether there could be entire galaxies made of antimatter in distant parts of the universe that would never encounter matter galaxies. In 1955, inspired by a lecture given by physicist Edward Teller, who is often referred to as the father of the hydrogen bomb, Harold Furth, a leading researcher in fusion physics, composed the following poem:

Well up above the tropostrata
There is a region stark and stellar
Where on a streak of antimatter,
Lived Dr. Edward Anti-Teller.

Remote from Fusion's origin,
He lived unguessed and unawares
With his antikith and kin,
And kept his macassars on his chairs.

One morning, idling by the sea,
He spied a tin of monstrous girth
That bore three letters: A. E. C.
Out stepped a visitor from Earth.

Then, shouting gladly o'er the sands,
Met two who in their alien ways
Were like as lentils. Their right hands
Clasped, and the rest was gamma rays.

—Harold Furth,
"Perils of Modern Living," 1955.
From *The New Yorker,* 10 November 1956.
By permission.

Instruments on balloons first discovered Dirac's positrons and, more recently, antiprotons in the cosmic rays, but thus far there is no evidence of heavier antiparticles in space.

And God Said, "Let There Be Light"

In 1966 Andrei Sakharov, the brilliant dissident Soviet physicist, offered a theoretical solution to the mystery of the matter–antimatter imbalance. He showed how all the antimatter created in the Big Bang could have been eliminated almost instantly when the newly born universe had cooled to a temperature of 1 billion billion billion degrees, leaving a cosmos created solely of matter. At the moment that the cosmic clock reached 10^{-35} second, most of the chaos of the primordial soup abruptly vanished. For every 10,000,000,000 particles of antimatter that emerged there were 10,000,000,001 particles of matter. Within one-millionth of a second the matter and antimatter particles annihilated each other and released a blaze of radiation; only a tiny fractional surplus of matter was left out of which to create the present universe of stars and galaxies. From this small residue the material universe was created. According to this scenario, the real world barely made it at birth.

When Sakharov first put forth his hypothesis of a matter–antimatter imbalance in the primordial universe it was not taken very seriously, but today it appears to be a natural consequence of GUTs (p. 295). This view requires that protons be slightly unstable and decay into other particles and radiation with an average life of about 10^{31} years. Several large, expensive experiments whose aim is to find evidence of proton decay are now being undertaken. Such evidence would strongly support the GUTs of the early universe, and nothing in the physical universe would be immortal.

A Whisper of the Big Bang

Support for the theory of the Big Bang comes from observations of a celestial background of relic radiowaves. It is almost inconceivable that an event that occurred from 10 to 20 billion years ago could have left a signature on the sky that is observable today. But like a bright, hot fire that has burned to dull coals, the residual heat can still be felt and the earlier intense flame can be inferred. The discovery of 3°K microwave background radiation is considered by most cosmologists to be the most important confirmation of the Big Bang. It is a marvelous example of serendipity in scientific research.

In fact, the theoretical stage was set for this discovery by the late 1940s works of George Gamow, Ralph Alpher, and Robert Herman, but theorists and observers failed for many years to make what, in retrospect, seemed to be an almost obvious connection. Gamow and his students started out to explain the cosmic abundance of the elements from nuclear synthesis in the epoch of the Big Bang. That theoretical effort failed, but Alpher and Herman recognized that high-temperature radiation from the hot, dense early epoch of expansion should be observable today, enormously redshifted to microwaves by the expansion of the universe. Alpher approached my colleagues in radio astronomy at the U.S. Naval Research Laboratory, but they concluded that the expected flux would be too weak for the instruments of that time to detect.

It took until 1965 to hatch Lemaître's "cosmic egg" concept and fully realize the meaning of the radio background evidence of the Big Bang. At Holmdel, New Jersey, Arno Penzias and Robert Wilson of AT&T Bell Laboratories were searching for sources of radio noise from the Milky Way. Their 20-foot receiver resembled a giant ear trumpet. It was designed to be used with the first Telstar communications satellite to relay television signals and radio conversations across the Atlantic. At the time that Penzias and Wilson took over the microwave horn it had been in disuse for about 10 years. They upgraded it to superb sensitivity by installing a maser amplifier and received approval from their supervisors to use it for radio astronomy as well as communications problems.

As Penzias and Wilson searched for sources of radio noise from the Milky Way, the receiver produced a persistent hiss at 7.35 centimeters even after they took meticulous precautions—including cooling the receiver

FIGURE 7. *Arno Penzias* (right) *and Robert W. Wilson used the horn antenna in the background to detect the cosmic 3°K microwave radiation background relic of the Big Bang fireball. (Courtesy of AT&T.)*

to liquid helium–temperature—to eliminate all known sources of interference. Suspicious that some source of noise till persisted in the antenna, Penzias and Wilson recalled that a pair of pigeons had taken shelter in the throat of the horn. The pigeons were caught and removed to distant parts, but they soon returned. When more effective measures finally vindicated the pigeons of any guilt, Penzias was still suspicious of what he delicately referred to as "white dielectric material" they had left behind. The antenna throat was dismantled and cleaned thoroughly, but the hiss persisted. They concluded that the noise was real. It amounted to about 400 microwave photons per cubic centimeter of the universe and was characterized by a temperature between 2.5° and 4.5°K. (Stand outdoors and 1000 trillion of these relic photons will strike your head each second.) Thus transpired the most important cosmological discovery of this half-century. Almost immediately, a group of Princeton University physicists led by Robert H. Dicke made the connection between the 3°K radiation of Penzias and Wilson and the light of the hot early universe, redshifted from 5000°K by the expansion of the universe.

Penzias and Wilson recall that they hardly realized the impact of their discovery until it ran as front-page news in *The New York Times.* In 1978 they received the Nobel Prize. This story illustrates as well as any example in the history of science how difficult it is to progress on theory alone, how insulated observers and theoreticians often are from each other's concepts and discoveries, and how large a part is played by chance in scientific discovery. But, as Louis Pasteur once said, "Chance favors a prepared mind."

Supernovae and Burned-Out Stars

Lying on his back and gazing up at the night sky, Huckleberry Finn pondered "whether [the stars] were made or just happened." How stars are made and what then happens over millions and billions of years is of paramount interest to astrophysicists. Ten billion years is a typical lifetime for a star like the Sun; smaller stars, which are more abundant, live longer. Bright blue giants are profligate energy burners and can exhaust their power in a matter of mere millions of years.

The death and transfiguration of a massive star comes in a catastrophic convulsion. In a matter of seconds the burned-out core comes crashing down and a rebounding shock wave turns the body of the star inside out. For several days the titanic explosion—a supernova—shines with the combined brilliance of all the stars in the Milky Way. What remains at the heart of the explosion may stabilize as a neutron star or disappear from the universe in a black hole. In February 1987 a supernova flashed in the Large Magellanic Cloud, closer to Earth than any similar event of the previous 400 years. It has presented astrophysicists with remarkable new insights into the physics of stellar collapse and explosion.

Theoretical modeling of the death of stars en route to becoming white dwarfs, neutron stars, and black holes has been spurred especially by recent improvements in observational and computing power. In the course of stellar evolution, those stars that have an initial mass of less than 1.4 solar masses are destined to become white dwarfs. When their hydrogen fuel is exhausted after billions of years, they collapse until they stabilize at a density of about 1000 tons per cubic inch (10^9 grams per cubic centimeter), a density beyond which electrons can be compressed no further. Over tens of billions of years, the dwarf radiates away its store of heat to become a dead clinker in space. For stars between 1.4 solar masses and somewhat less than 3 solar masses, electron pressure theoretically does not suffice to halt collapse, which then proceeds until a density of 1 billion tons per cubic inch (10^{15} grams per cubic centimeter) is reached. Electrons are pressed onto protons to form neutrons until a giant nucleus of about 10^{57} neutrons—a neutron star some 10 kilometers in radius—is created. When first formed, the core is a 1-billion degree K neutron superfluid of no internal friction (zero viscosity) and its surface may exceed a few hundred million degrees K. Over several thousand years it should be a brighter source of x-rays than the sun is of visible light.

A neutron star strains all our physical concepts of the most extreme conditions of matter. Its solid crust may be almost pure iron, trillions of times as stiff as steel. Beneath the crust is a thin layer of superconducting protons that maintain a surface magnetic field of about 10^{12} gauss. (Earth's magnetic field is about 0.3 gauss). In a young neutron star, a high spin rate coupled with the powerful magnetic field can produce an external electric field of the order of 1 trillion volts per centimeter. The force of gravity is so great that the surface is as smooth as a billiard ball. Starquakes involving movements as small as a one-thousandth of 1 millimeter can produce irregularities in the rotation frequency that are easily detected at distances of several thousand light-years.

If the stellar mass exceeds that of three Suns, even the neutrons are crushed, and collapse theoretically proceeds to a point of infinite density and zero volume. The gravitational field becomes so strong that light itself cannot escape, and a black hole forms inside a sphere of radius proportional to the collapsed mass, forever isolated from the rest of the universe and manifested only by its gravitational field, angular momentum, and electric charge. In theory, black holes come in all sizes. A black hole that contains the mass of the Sun is 3 kilometers in radius. If the mass is 10^8 suns, as we believe may be true of the nuclei of some galaxies, the radius is 3×10^8 kilometers. Baby black holes may be no bigger than an atom, but even these could contain a mass of 10^{20} grams (about the weight of Mount Everest).

How these almost science-fiction concepts of collapsed stars have become reality is part of the story of pulsars and x-ray stars (see Chapter 5).

Pulsars:
Lighthouse Beacons in the Galaxy

One of the great surprises in astronomy was the discovery of pulsars. In 1967 Antony Hewish and his collaborators at Cambridge University were observing the twinkling of quasars with an antenna "farm" of 2040 interconnected dipoles. As Jocelyn Bell, a young Irish graduate student, carefully examined the reams of strip chart, she noticed a series of precise pulsations about 1.3 seconds apart. The sharpness of the signals implied a source no larger than a small planet. Half seriously, the group labeled the source LGM, for Little Green Men. In short order, two more sets of pulsations were discovered: It was clearly a natural phenomenon. So much for the little green men!

Scientists came up with a variety of ideas to explain the pulsations and finally settled on one that seemed plausible. A neutron star, said Tommy Gold, was spinning, and as it did so, it beamed synchrotron radio waves from electrons whipped around at nearly the speed of light in a powerful magnetic field. Because the radiation was highly directed, when viewed from Earth it pulsed like a lighthouse beacon. Observations in space reveal that the synchrotron spectrum is continuous and, for two pulsars in the Crab and Vela, spans the entire known range, from radio to billion electron-volt gamma rays.

Theoretically, there could be tens of thousands of pulsars in the Galaxy. Their lifetime is believed to be a few million years, and a new one is formed on average about once every 30 years, when the core of an exploding star collapses, becoming a spinning neutron star—a pulsar. The slowest pulsars have periods of between 3 and 4 seconds; the fastest spins nearly 1000 times a second and at its periphery is moving at a speed so close to the speed of light that it is on the verge of tearing apart. Per unit volume, the power of the pulsar is billions of times greater than the rate of thermonuclear energy generation in the core of the Sun.

Despite the great power generated in the pulsar energy spectrum by non-thermal processes, theorists thought until recently that it would be impossible to observe a neutron star in visible light because (1) it is so small, and (2) it radiates only a tiny fraction of its luminescence in the optical portion of the spectrum. Typically, the average galactic neutron star would be 100 million times dimmer than the faintest stars seen by the unaided eye. The Faint Object Camera (FOC) on the *Hubble Space Telescope* has now brought stars 20 times fainter into view, and the field of optical neutron star observations is developing rapidly. The surface temperature of radio pulsar 1055 − 52 is estimated to be about a million degrees K. Perhaps the most efficient gamma ray generator in the universe, efforts to observe it from the ground with powerful optical telescopes in the southern hemisphere over the past eight years did not succeed. The neutron star was hidden in the glare of a normal star, 100,000 times brighter, in the line of sight, and separated by only one thousandth of a degree. When the FOC was pointed accurately at the source of radio pulsations, the neutron star lay exactly in the same direction, separated from the normal star. Although its magnitude is a very dim 24.9, it is comfortably within reach of the FOC.

Pulsar 1055 – 52 is only one of 8 neutron stars detected so far at optical wavelengths compared to about 21 that have been observed in x-rays and 760 in radio wavelengths. Fifty percent of the total radiant energy of pulsar 1055 – 52 is concentrated in billion-electron volt gamma rays. For several years, Geminga (see p. 00) has been a mystery pulsar that radiates 15 percent of its power in gamma rays. The famous Crab pulsar, first to be detected in visible light and x-rays, produces 0.1 percent of its radiation in gamma rays. Giovanni Bignami and his team of Milanese astronomers are carrying on the neutron star search with the FOC. They call Geminga, pulsar 1055 – 52, and another optical pulsar, 0656 + 14, their Three Musketeers and estimate their ages in hundreds of thousands of years, compared to less than a thousand years for the Crab.

Quasars at the Edge of the Universe

Quasars may be the most powerful objects in the universe. Their discovery is a fine example of the interplay of the old and new astronomies that has since become the norm for studying the most interesting celestial objects. Many quasi-stellar objects appeared on old photographic plates, but there was no obvious way to distinguish them from ordinary stars in the Milky Way. By 1960 radio astronomers had found a number of sources that scintillated like twinkling stars. They called them *radio stars* and believed they were stars in our own galaxy that for mysterious reasons produced far more radio emission than normal stars such as the Sun.

In 1962 Cyril Hazard pointed the 210-foot radio telescope of the Parkes Observatory in Australia at 3C-273 and observed its disappearance as the Moon passed in front and briefly masked it from view (an occultation). The position was obtained with sufficient precision to identify 3C-273 as a faint, blue, starlike object with a protruding luminous jet of gas. Certainly no known star had ever exhibited such a jet; it had to be extragalactic. Maarten Schmidt at the Hale Telescope on Mount Palomar obtained a spectrum of the blue star but was baffled by the pattern of spectrum lines until it suddenly flashed on him that the lines were spaced like the hydrogen spectrum but were shifted far to the red. The redshift placed 3C-273 at a distance of 2 billion light-years, where it radiates 100 times as much power as an ordinary bright galaxy from a volume 100,000 times smaller in diameter. Furthermore, we now know that as much power is radiated in x-rays and gamma rays as in the optical range.

Observations thus far fail to reveal the energy source of quasars. Theorists talk of chain reactions of supernova explosions, of a massive plasma cloud stabilized by spin, of all-consuming black holes of 100 million solar masses that cannibalize stars that come within capture range or swallow a steady diet of gas from an inward-spiraling accretion disk.

Although the consensus for the cosmological distances of quasars is now very strong, there were until recently a substantial number of dissidents. Their concerns centered on the interpretation of redshifts as an effect of the velocity of recession. Since quasars appear only as point sources of light, there is little information independent of redshift to establish distance.

Halton Arp of Mount Wilson Observatory provided most of the ammunition for dissident views on quasar redshifts (G. A. Field, Halton Arp, & John N. Bahcall, *The Redshift Controversy*, 1973). He is a dedicated cataloger of peculiar associations of quasars and other background galaxies. From Arp's studies, it appeared that halos of faint stars, gas, and dust extend far beyond the apparent outlines of some galaxies. Within these extended envelopes, he found quasars in several dozen cases. Was the association by chance, or real? If real, the observations would place the quasars at comparatively modest distances from the parent galaxies. The requirement that quasars radiate the power of from 100 to 1000 Milky Ways would disappear, and theorists could forget the need to involve such bizarre energy sources as black holes of 100 million solar masses in the nuclei of quasars. Although relatively few in number, the opponents of this theory were very outspoken and eloquent in questioning the cosmological distances of quasars. The cosmological distances of quasars are now so widely accepted, however, that the orthodox community no longer debates the issue, and any surviving opponents carry on guerrilla sniping on the fringe of credibility. The number of quasars detected thus far exceeds 5000, and the pace of new sightings keeps accelerating. Below and above a redshift of about 2.5 (250% change in wavelength), the number of quasars approaches zero. Such evidence places the epoch of birth of quasars at just a few billion years after the Big Bang.

Cosmic Rays Shower the Earth

Early in this century, with balloon-borne instruments, Victor Franz Hess discovered cosmic rays. His research was of profound significance to astrophysics, even though it took some 40 years for physicists to recognize the meaning of cosmic rays for energetic processes in the Galaxy. Hess, who was then associated with the Institute of Radium Research in Vienna, was trying to observe the variation of atmospheric electricity with altitude. He believed that the discharge of his electroscopes on the ground was attributable to radioactivity in the Earth, and he conjectured that if he could carry his instruments to a height of 5 kilometers, the air would shield them from ground-level radiation. With two other men, one in command of the balloon, the other a meteorological observer, on 7 August 1912, Hess flew the hydrogen-filled balloon to a height of 5 kilometers. Three months later Hess reported in the *Physikalische Zeitschrift* that his observations could best be explained by a radiation of great penetrating power arriving from outside the atmosphere. He had discovered cosmic rays and was awarded a Nobel Prize 24 years later.

The nature of cosmic rays puzzled Robert A. Millikan of the California Institute of Technology. Were they high-energy photons like the gamma rays of radioactive decay, or were they very energetic charged particles, perhaps nuclei of hydrogen and helium? Millikan devised a test. If cosmic rays were charged particles they would be deflected by the Earth's magnetic field. Near the north and south poles, where the magnetic field dips toward the Earth, particles could enter rather freely; over the equator the magnetic field would turn charged particles away.

Although simple in concept, the experiment to detect a latitude effect did not succeed until the 1930s, when Millikan, Arthur H. Compton, and a generation of cosmic-ray physicists took their instruments to mountaintops around the world or lofted them on balloons from pole to pole. Eventually, balloon-borne experiments proved that cosmic rays are charged particles consisting mostly of hydrogen nuclei (protons), helium nuclei (alpha particles), a few heavier nuclei, and electrons. All told, about 1–2 pounds per day arrive at the Earth.

Cosmic rays span an enormous range of energies, from 10^{12} to at least 10^{20} electron volts. The flux of primary particles falls rapidly, as the inverse square of the energy. Only 3 or 4 per square kilometer per century reach energies above 10 million trillion (10^{19}) electron volts, roughly the energy of a fast-pitched baseball. Observations suggest that the highest energy cosmic rays are protons. If so, they should propagate almost rectilinearly and reveal their sites of origin on the celestial sphere. But thus far there are no clear directional clues at the very highest energies. Furthermore, interactions with the 2.7-degree cosmic background radiation would rapidly degrade the energy so that the distance of origin could not exceed 20 megaparsecs.

Although relatively few in number, the average energy density of cosmic rays in the Galaxy is about 1 electron volt per cubic centimeter, which is roughly the same as the energy density of light from all the stars in the Milky Way. We also have evidence of cosmic gamma rays of energy as great as 10^{16} electron volts being produced by a binary star system deep in the Galaxy. These energies are thousands of times higher than the most energetic particles produced by man-made accelerators. The gamma rays light up tracks of fluorescence in the atmosphere, so-called Cerenkov radiation (see p. 99), that can be imaged with optical telescopes.

The modern era of cosmic ray research follows the lead of French physicist Pierre Auger. In 1938, he discovered widely spread air showers of cosmic rays by counting coincidences with two Geiger counters separated by a few hundred meters in the Alps. There is no known mechanism for accelerating cosmic rays to 10^{20} electron volts. Fewer than a dozen such high-energy particles have been recorded in more than 40 years of observation. To detect greater numbers of showers triggered by the highest energy cosmic rays, physicists seek to implement large arrays of detectors. The Pierre Auger project, now initiated by scientists from 19 countries, plans to have in place by 2005 a pair of 3000 square kilometer air shower detector arrays, one in the northern hemisphere in Utah, the other in the high desert of Argentina. With 3200 water Cerenkov detector modules spread over 6000 square kilometers, the Auger arrays may detect as many as 30 events per year above 10^{20} electron volts in each hemisphere.

Early Misperceptions of the Space Age

In the decade before World War II, there was little premonition in the astronomical community that rocket and satellite technology would revolutionize our ability to conduct observations from space. Robert Goddard's early rockets were toys compared to the V-2s developed by Wernher von Braun in Hitler's Germany. Regarding travel to the Moon, the following, by a distinguished astronomer, Professor Forest Ray Moulton of the University of Chicago, is typical of the conservative view that prevailed:

> It must be stated that there is not the slightest possibility of such a journey. There is not in sight any source of energy that would be a fair start toward that which would be necessary to get us beyond the gravitative control of the earth; there is no theory that would guide us through interplanetary space to another world even if we could control our departure from the earth; there is no means of carrying the large amount of oxygen, water, and food that would be necessary for such a long journey; and there is no known way of easing our ether ship down onto the surface of another world, if we could get there at low enough speed to avoid destruction. (*Consider the Heavens*, 1935, p. 107)

Immediately after World War II, about 100 V-2s were brought to the White Sands Missile Range in New Mexico. The warhead space was offered for high-altitude research, while the engineers studied the propulsion characteristics of the rockets. For the most part, those scientists who grasped the opportunity for rocket astronomy were physicists rather than astronomers. The astronomers, with a few visionary exceptions such as Lyman Spitzer, who foretold the era of space telescopes, seemed numb to the enormous potential of observations above the atmosphere. When *Sputnik* flew in 1957, President Eisenhower commented, "The Russians have only put one small ball in the air," and the British Astronomer Royal, Richard von der Riet Wooley, brushed it off with equal indifference.

The Violent Universe of X-Ray and Gamma-Ray Astronomy

Whenever the opportunity arises to observe the cosmos in hitherto unexplored ranges of the spectrum, totally unexpected revelations can be expected. X-ray astronomy has held center stage for the past two decades; gamma-ray and infrared astronomy are emerging from the wings. The celestial sphere has been mapped for over 50,000 x-ray sources: stars, galaxies, and clusters of galaxies.

Strong x-ray stars occur primarily in binary systems consisting of a giant star from which gas is drawn by tidal forces onto a collapsed companion, either a neutron star or a black hole (see Fig. 61). If the gravitational sink is a neutron star, the gas funnels down the magnetic field to the poles, where it piles up in a dense plasma that is hot enough to radiate x-rays. But how can a black hole radiate x-rays when theoretically no radiation can escape? If a hot blue-giant companion constantly evaporates a tenuous wind of gas that blows toward the black hole, the intense gravitational pull will draw the gas into orbit around the hole, where it will accumulate in a rotating accretion disk. As gas spirals in closer to the hole, it orbits more rapidly and "rubs" against slower-moving gas farther out. Friction causes the gas to heat to tens or hundreds of millions of degrees before it vanishes into the black hole. In this hot, accreting gas circulating the black hole, intense x-rays are generated.

Observations of the x-ray star Cygnus X-1 fulfill the theoretical predictions for a black hole of 10 solar masses so well that they almost dispel any doubt that black holes exist. If Cygnus X-1 is not a black hole, it must be something equally bizarre. Brave extrapolations lead to models of black holes of 1 million to 100 million solar masses at the cores of powerful quasars and radio galaxies.

Gamma-Ray Bursts—Apocalyptic Explosions from the Depth of Space

After the previous decade of tentative feelers in the range of gamma-ray astronomy, the 1990s have seen astounding advances. Most effective has been the NASA *Compton Gamma-Ray Observatory (CGRO)*; its greatest surprise has come from the study of gamma-ray bursts with the Burst and Transient Source Experiment (BATSE). Their discovery was a fall-out of U.S. military interest in detecting clandestine tests of Soviet nuclear bombs in space after the atmospheric test ban treaty was signed. It was suspected that secretive tests were possible in deep space. An atomic bomb detonation in the near-vacuum at very high altitude would not create a great fireball. Instead, its signature would be an intense x-ray/gamma-ray flash. To conceal it, the weapon could be triggered behind the moon. Only when the cloud of radioactive debris spread beyond the lunar mask would gamma rays be detectable. The military VELA project was established in 1960 to provide continuous monitoring of space for x-ray/gamma-ray flashes. In 1967 VELA satellites made their first detection of a gamma-ray burst but its time variation and energy range did not fit a nuclear explosion.

Whereas visible light photons have energies between 2 and 3 electron volts that encompass all the colors of the rainbow, the highest energy gamma rays in space exceed a thousand trillion electron volts. BATSE now detects an average of one burst per day with energies up to 10 million electron volts. The bursts rank with the most tantalizing and baffling mysteries of astrophysics. Over 2000 bursts were recorded in the first seven years of CGRO. They have no preferred direction of arrival or concentration toward the galactic center or galactic plane, implying that they come from well beyond the Milky Way, possibly from distances of billions of light years. The greater the distance the greater must be the intensity, making extragalactic bursts the strongest radiation flashes in the cosmos. The energy released in a burst at billions of light years, lasting only a few seconds, may exceed the total radiation from the sun in a thousand years, yet rapid variations in the sharp burst profiles imply a size smaller than the sun. Before 1997 no burst was identified with an optical or x-ray source, but now there is some evidence for an extragalactic origin.

A recent Italian-Dutch satellite, Beppo Sax, named after famed Italian cosmic ray physicist Giuseppi (Beppo) Occhialini, equipped with instruments adequate to position a burst to a few minutes of arc within hours after it first appears, succeeded in directing optical and radio telescopes at a gamma-ray burst on 11 January 1997. The results were disappointing—several radio and x-ray sources were found nearby but outside the permitted error circle. Bursts in February and April 1997 were targeted with the Hubble Space Telescope and the Keck telescope in Hawaii, but the observations were unsuccessful in identifying the gamma-ray bursts with an optical object. Finally, on 8 May 1997, the Keck telescope obtained an optical spectrum of a gamma-ray burst that provided a red-shift corresponding to a distance of approximately 4 gigaparsecs. Thus the distance was very great, but little clue to the nature of the burst was revealed. All that could be deduced was that gamma-ray bursts emerge from gigantic fireballs, billions of light years from earth.

The discovery of gamma-ray bursts has brought out a rash of theories; as many as 140 models were hypothesized by 1993. What might set off the kind of cosmic fireball hinted at in recent observations? A collision of two neutron stars or a neutron star and a black hole could generate the energy of a burst, and one observed burst per day would be plausible if the source region included the depths of extragalactic space. Joseph H. Taylor and Russell Hulse of Princeton University were awarded a Nobel Prize for their study of a binary pair of neutron stars, one a pulsar, that are slowly dissipating their energy of revolution in the form of gravitational waves. As their orbit shrinks, the speed of revolution increases. After perhaps 100 million years the two solar mass objects may speed up to 1000 revolutions per second and make contact in a catastrophic crash, merging to form a black hole, and the reverberations would reach us in a brilliant flash of x-rays and gamma rays.

Is the Universe Open or Closed?

Inevitably all advances in our understanding of the origin of the universe lead us back to the compelling question: "How will the universe

FIGURE 8. *"I think the universe expands one day, shrinks the next, remains static for a week, and then it begins all over again." Drawing by Dedini; © 1981 The New Yorker Magazine, Inc.*

end?" Numerous tests confirm the expansion of the universe from a primordial fire ball, but the rate of expansion is very difficult to determine with precision. Were all galaxies swept along with the Big Bang, and are they now subject to the expansion of space in an open universe in which space will expand forever? Or is universal gravitation braking the expansion toward a halt some tens of eons in the future? Is the universe open or closed?

A closed universe is bound by the self-gravitation of all its matter and energy. The theoretical critical mass above which expansion will eventually revert to collapse is one order of magnitude greater than has been estimated from all the luminous matter of all the galaxies. Astronomers are searching for evidence of invisible, unknown matter. If galaxies formed out of huge gas clouds, the condensation process must have been more-or-less inefficient. Only a fraction of the total mass of the universe could have condensed into visible stars and nebulae. Vestiges of protogalaxies of dark matter might pervade the universe as wisps of gas drifting in an intergalactic medium (IGM).

The IGM has been searched in all ranges of the spectrum—radio, optical, ultraviolet, x-ray, and gamma ray—but without conclusive results. If the IGM exists, it must be highly ionized and hotter than 10^6 °K. A hot IGM would radiate x-rays; thus, the early discovery of a diffuse background x-ray emission seemed to confirm its existence. Recent observations with the NASA *Einstein X-Ray Observatory* revealed that x-ray quasars account for more than 50% of the background emission. At the same time, hot intergalactic matter has been detected through x-ray observations of clusters of galaxies. But such evidence of intergalactic matter in localized regions does not provide clear clues to the external stretches of the IGM.

So where might hidden matter be? Some astronomers argue that it is in extended halos of galaxies made up of very faint dwarf stars, and physicists postulate exotic theoretical particles freed from the epoch of grand unification but as yet undetected. Cosmologists also speculate on the reported evidence of neutrino oscillations, which mean that neutrinos have a small but finite mass. They theorize that there are as many neutrinos as light photons in the universe. If each neutrino has only one–ten-thousandth of the electron mass, neutrinos would provide enough mass to close the universe. In addition, gravitational wave physics calls, in theory, for the generation of an enormous density of gravitational waves from the collapse to black holes of massive stars born before the formation of galaxies, and from the Big Bang itself. The equivalent mass of such background radiation would also contribute to closure.

INTRODUCTION: THE ASTRONOMER'S UNIVERSE

To the End of Time

What we call the beginning is often the end
And to make an end is to make a beginning.

—T. S. Eliot, "East Coker," 1940

If the universe is closed, the future is spectacular. In another 40 to 50 billion years, expansion will come to a halt and the universe will fall in on itself. Instead of redshifted light, the radiation of distant galaxies will be blueshifted. Galaxies will rush toward each other, the cosmic background microwave radiation will be compressed, and the light will shift toward the visible. In time, the sky will blaze with light. Then stars and planets will melt into a universal soup of hot particles. The "Big Crunch" will continue until there is nothing but an empty universe: a spacetime singularity at infinite temperature. Will collapse go all the way, or will it bounce? Pressure may grow so great that collapse will revert to expansion; this cycle may repeat forever in an oscillating universe. Will each successive reincarnation be the same? John Wheeler, formerly of Princeton University, suggests that the lengths of successive cycles may vary. Our cycle could be about 100 billion years; the next may be much shorter or longer. A universe that cycles in only 1 million years could hardly produce life as we know it. Wheeler asks, "What good is a universe without life?"

If the universe is flat, as required by inflation theory, or open, the future is a deep freeze. The stars will eventually burn out. Stellar evolution throughout the Galaxy will inevitably lead to a gravitationally bound system of black holes, neutron stars, black dwarfs, and the cold debris of planets, asteroids, meteorites, and dust. As the system radiates energy in gravitational waves—in about 10^{14} years—all of its galactic contents will eventually spiral into a black hole of 10^{11} solar masses with a radius of about one-fiftieth of a light-year: the ultimate galactic "disposal."

As time continues, clusters of galaxies will most likely remain gravitationally bound. Each galaxy will become a black hole, and the larger system of clusters will become a system of bound black holes. Eventually, after as long as 10^{30} years, there will remain only one supergalactic black hole of from 10^{14} to 10^{15} solar masses. In the end, all matter will decay to radiation.

Now entertain conjecture of a time
When creeping murmur and the poring dark
Fills the wide vessel of the universe.

—William Shakespeare,
King Henry V, Act IV, Prologue

PART I

THE TOOLS OF ASTRONOMY

Prior to the twentieth century, the tools of astronomy were most often provided by pharaohs and sultans, emperors and kings, scions of great private fortunes and industrial tycoons. All were intrigued with the mysticism of the cosmos and inspired by the beauty of the night sky. In our generation such princely largesse is no longer sufficient to finance the construction of the major observatories. Building the new generation of telescopes ranks with the most remarkable technological endeavors of our time, and cost projections have risen to heights that few of even the wealthiest private foundations can provide. When only national treasure can meet the price tag, the new astronomies will truly belong to the common man.

If the public pays the bill, politicians ask: "What is it good for?" Over the ages value judgments on astronomy have changed very little. Socrates' dialog is as apt today as in his time:

SOCRATES: Shall we make astronomy our next study? What do you say?

GLAUCON: Certainly, a working knowledge of the seasons, months, and years is beneficial to everyone, to commanders as well as to farmers and sailors.

SOCRATES: You make me smile Glaucon. You are so afraid that the public will accuse you of recommending unprofitable studies.

PLATO,
Republic VII, c. 370 B.C.

In the laboratory, physicists can control and examine their experiments from every angle. Astronomers have only one perspective, but their degrees of freedom lie in the full range of the electromagnetic spectrum. Astronomy is no longer simply a science of visual exploration. Observatories fly in Earth orbit to receive all the radiations of the electromagnetic spectrum from the ultrahigh frequencies of gamma rays to the very low frequency infrared rays. Radio telescopes as large as the Earth penetrate the innermost cores of quasars to study the most incredible energy machines in nature. One mile underground, an Olympic-sized pool filled with cleaning fluid served for 10 years as a detector of solar neutrinos. To catch higher-energy neutrinos from the depths of the Milky Way, physicists hope to string a lattice of nearly 1000 detectors in a cubic kilometer of ocean water. It has been demonstrated that galaxies themselves can serve as gravitational lenses to image the internal structures of even more distant galaxies and quasars. And gravitational waves predicted by Albert Einstein's general relativity may soon provide an entirely new astronomy of violent gravitational catastrophes deep in the universe.

Part I traces the development of astronomical observatories from Galileo's primitive telescope to the great observatories of our time. These astounding modern tools have created an explosion of knowledge that far outstrips all past imagination, and the best is yet to come.

CHAPTER 1

From the Naked Eye to the Space Telescope

Cataloging Stars with Quadrant and Sextant

Born in 1546 in Denmark, the land of Hamlet, 18 years before Shakespeare, Tycho Brahe, without benefit of a telescope, became the most renowned astronomer of his time. For Tycho, astronomy was a lifetime effort that required continuous observation of the stars over long periods of time by means of simple but very accurate instruments: a quadrant for measuring the altitudes of stars and a sextant to determine the angular distances between them. His instruments were unusually large and of the utmost precision that skilled European instrument makers could achieve. Tycho himself crafted a remarkably fine sextant of walnut, with bronze hinges, arms 5½ feet long, and very accurately inscribed gradations. Over many years of observation, he produced a catalog of 777 stars positioned to a fraction of a degree. Four centuries after Tycho, European astronomers sent into orbit an astrometric telescope, Hipparcos (High Precision Parallax Collecting Satellite). The mission got off to a disappointing start after a perfect launch on a French Ariane rocket in the summer of 1989.

A booster that was intended to transfer the satellite into a stable geosynchronous orbit failed, and Hipparcos was left in a highly eccentric 11-hour orbit, with apogee at 36,000 kilometers and perigee at 500 kilometers. However, frantic rescheduling of support from ground stations around the world saved the mission. Hipparcos more than doubled the number of known variable stars and discovered thousands of new double or multiple star systems.

The telescope's mirror was only 29 centimeters in diameter, but the resolution was sharp enough that one could have read a *New York Times* headline across the Atlantic from Paris. By the end of Hipparcos's mission in 1993, *Hipparcos*, the primary catalogue, listed 118,000 stars with positional accuracies of milli-arc seconds and brightness precision of 0.2 percent. A secondary catalogue, *Tycho*, contains about a million somewhat less precisely positioned stars, with 2.5 milli-arc seconds resolution and 6 percent brightness accuracy. Before Hipparcos only about 100 nearby stars had distances determined from parallax to within 5 percent. Among the important astrophysical results derived from Hipparcos was an estimated 10 percent reduction in the Hubble constant from which a 13-billion-year age of the universe was derived.

Tycho Brahe's laborious, careful mapping of the starry sky was punctuated by two great discoveries that forced the transition from pre-Copernican, Aristotelian concepts to modern perceptions of the universe. In 1572 he observed a supernova, irrefutable proof that the stars on the celestial vault were not eternal. Five years later he noticed the appearance, shortly before sunset, of a very bright star in the west. After dusk, a splendid tail became evident, and Tycho immediately recognized it as a comet. With sextant and quadrant he and an assistant measured its position from two sites, miles apart, but could detect no angular displacement. By his calculation, Tycho determined that the comet lay beyond Venus and orbited the Sun in a direction opposite to the planets.

At the time, most astronomers still held to Aristotle's view that planets were attached to transparent crystalline spheres nesting one inside another. Tycho Brahe realized that view was untenable, for how could the comet pass through the spheres? He bravely enunciated his conclusion, "There are no solid spheres in the heavens" (*Second Book about Recent Appearances in the Celestial World*, 1588). Beyond the planets there was only space, and the stars lay as far above our heads as it was possible to see. Thus did he set the stage for the revolution that was to come when Galileo turned his telescope to the sky.

The telescope was invented by chance in Holland in 1608. As the story goes, Hans Lippershey, a spectacle maker, was holding a lens in each hand while looking toward the village church steeple. To his surprise, when he chanced to peer through both lenses in line, he noticed that the weathercock at the steeple-top seemed closer. To hold the two lenses conveniently, he mounted them in a tube and thus constructed a telescope. Rumors of this invention reached Galileo, then aged 45, in Venice in 1609. Upon his return to Padua, he immediately set about his own investigations.

Although spectacles had been in use 300 years before Galileo's time, lens-makers had only just learned how to grind concave lenses. Galileo himself was skilled in the making of lenses. He put a convex magnifier at the front end of a lead pipe, a concave lens at the viewing end, and achieved a magnifying power of 9. When he revisited Venice on 25 August 1609, he brought a gift of a telescope to the Doge and shrewdly expounded upon its potential usefulness in warfare. The senate rewarded him with a lifetime lectureship at the University of Padua and doubled his salary.

Galileo did not apply his telescope seriously to astronomy for several months, during which time he devised a stable mount and improved the magnifying power to 36 diameters. When he tested the improved telescope by observing the Moon, he could hardly believe what he saw. Night after night, the patterns of light and shadow changed as though they were indeed cast by mountainous elevations. The Moon was "rough and uneven, covered everywhere just like the earth's surface with huge prominences, deep valleys and chasms," he wrote (*The Siderial Messenger*, 1610). Next he observed that Venus exhibited phases like the Moon and wrote that while pointing his glass at Jupiter, "I noticed three little stars, small but very bright, were near the planet." As he continued his nightly vigils, Jupiter's stars rotated around the planet and Galileo realized that "there are not only three but four erratic sidereal bodies performing their revolutions around Jupiter like the moon about the Earth, while the whole system travels over a mighty orbit about the sun in the space of twelve years." Beyond the planets, the constellations of the starry heaven were multiplied into myriad stars far outnumbering what he could see with the naked eye. His amazement was unbounded; within a few months after he had made his first telescope, he knew more about the universe than any other human being.

Johannes Kepler, Tycho's protégé and collaborator, and discoverer of the ellipticity of planetary orbits, was enthralled by the first reports of Galileo's telescope and observations. He referred to the telescope as "more precious than any scepter! Is not he who holds thee in his hand made king and lord of the works of God?" (Preface, *Dioptrice*, 1611). But the Church called Galileo's revelations artificial. Galileo responded that God had given him the telescope.

Although Galileo was aware that his observations and ideas ran counter to Roman Catholic doctrine, which still rejected Copernican theory and held the Earth to be the center of the universe, he did not foresee that he would soon become the object of severe censure by the Holy Office of the Vatican. All his efforts to publish and defend his concept of a heliocentric planetary system were obstructed. As a sick old man he despaired, "My life wastes away and my work is condemned to rot." In the end, under dire threat of punishment, he was forced to recant his planetary theory. Reluctantly, he uttered the acknowledgment of bitter defeat: "I do not hold and have not held this opinion since the command was given me that I must abandon it" (Giorgio de Santillana, *The Crime of Galileo,* 1955, p. 162).

Galileo lived out the last 9 years of his life in Florence under house arrest imposed by Pope Urban VIII. As he grew blind and ill, he cursed his hateful and intolerant judges for their intellectual fraud and deceit. He died in 1642. More than a century passed before the Vatican conceded the truth of his works, but his discoveries triggered a proliferation of scientific instrument designs, and astronomy flourished. To surpass Galileo, however, it was necessary to build bigger and better telescopes, not an easy task. The principles that guide the design of great telescopes, even today, were set forth by Sir Isaac Newton, who was born the year of Galileo's death.

Newton, a sickly ninth child, grew up in the quiet English countryside in the town of Woolsthorpe to become one of the greatest figures in the history of science. But his towering intellect did not express itself until he was 23. When the Great Plague spread through Europe, Newton was studying at Trinity College, Cambridge. The university was shut down from autumn 1665 to spring 1667, and Newton returned to seclusion at Woolsthorpe. During those 18 months, Newton's genius brought forth his most remarkable developments in mathematics and physical science, including the discovery of the gravitational force that binds the Moon in orbit of the Earth, the composition of white light, and the nature of color. His scientific achievements brought him widespread renown and the admiration of the intellectual world.

Immediately after returning to Cambridge in 1667, Newton became intrigued with the design of telescopes. He concluded that refracting lenses presented too many difficulties if one wished to build a large telescope. For instance, he was aware that the refraction of glass varies with color and that different colors cannot be brought to the same focus by a simple lens: a fault known as chromatic aberration. A telescopic mirror would be free of the color defect because all wavelengths obey the simple law that the angle of reflection equals the angle of incidence. Newton replaced the glass lenses with a curved mirror at the back end of the telescope tube, and an eyepiece magnifier at the front end was used for viewing the focused image. For the mirror material, he chose an alloy of copper and tin known as speculum metal, a hard, brittle alloy that gave a reflectivity of 60%. He ground and polished the surface until he could resolve the moons of Jupiter. He presented a small instrument to the Royal Society in 1671, but although the images were somewhat sharper than those afforded by Galileo's telescope, it was not markedly superior.

Newton's small mirror telescope produced no new discoveries of note, and almost a century passed before substantially larger telescopes were built by William Herschel. The greatest astronomer of his time, Herschel was in fact trained as a musician; as a teenager, he was sufficiently skilled to play oboe in a regimental band. At age 19, he fled to England ahead of the French occupation of his native city of Hanover. In 1766, when he was 28, he had achieved a distinguished musical reputation and was appointed organist at the prestigous Octagon Chapel in the city of Bath. Now afforded the leisure of a secure position, he turned his attention to building telescopes and indulged his intellectual fascination with astronomy. At the mature age of 39, Herschel produced a reflecting telescope of 20-foot focal length and 12-inch diameter. A few years later he completed a telescope that he called the "large" 20-foot reflector, which had a mirror 18 inches in diameter. It served as a model for Herschel's ultimate achievement, a 40-foot telescope of 48-inch aperture.

Herschel's interest quickly moved from telescope-building to astronomy, and he devoted his efforts to cataloging and classifying far more objects than all that had been previously studied by astronomers. For example, he observed more than 800 double stars and verified, for many of them, the motion of one star about another. In these observations he extended Newton's concepts of gravitational attraction beyond the Solar System to remote galactic objects.

In 1781 Herschel inaugurated a new era of planetary astronomy. On the evening of March 31 he spied an object he knew was too large to be a star. At first he believed it was a comet. Soon he established that it was a planet, Uranus, the first to be discovered since ancient times. He became an instant international celebrity and was elected to the Royal Society of London. In keeping with his new status, he was granted a pension by King George III, with no obligations other than to be on call as the royal family's astronomer-in-residence. At last, Herschel could devote all his energy to discovering the universe. With his sister Caroline, for the next 20 years Herschel searched the entire sky, using his 20-foot telescope. Out of this intensive effort came two catalogs of 1000 galactic nebulae each, and a third compilation of 500 more. At the time, the study of nebulae was one of the most fascinating aspects of astronomy. Herschel was especially interested in those mysterious glowing disks with an embedded star that he called *planetary nebulae.* They are now known to be the expanding shells blown off the outer portions of evolving stars.

The English astronomer William Parsons, the Third Earl of Rosse, built the largest telescope in the world in 1845, on his estate. The 72-inch mirror came to be known as the "Leviathan of Parsonstown." For 70 years it remained the largest telescope ever built. It was almost useless because of the persistently heavy overcast, but on occasion he was able to distinguish spiral structure in what were then thought to be nebulae in the Milky Way and were later recognized as external galaxies.

Silvered glass mirrors with 90% reflectivity soon replaced the speculum mirrors. In the 1930s American physicist John Strong developed the technique of aluminizing mirrors by metal evaporation in a vacuum. When exposed to air, the thin aluminum film oxidizes to a hard permanent surface that is equal to or superior to silver in reflectivity over the entire spectral range. Strong was eventually placed in charge of coating the great 200-inch mirror for the telescope at the Palomar Observatory.

Astronomy in the New World

Astronomy in America harks back to the earliest settlements in the New World. In 1660 John Winthrop, Jr. brought his telescope from England to Connecticut and studied the fuzzy image of Saturn. For 2 centuries astronomy remained, for the most part, a pasttime of wealthy amateurs. In 1825 President John Quincy Adams urged the establishment of a national observatory but failed to persuade Congress, even though Europe already had well over 100 observatories. A few years later, however, observatories began to appear, first at Williams College and then at locations including Western Reserve College in Ohio and West Point.

A renewed effort to create a national observatory led to the establishment of the Naval Observatory on a Foggy Bottom knoll overlooking the Potomac River. The ground was originally selected by George Washington as the site of a national university that never came into being. About 100 years ago, after much suffering of personnel from swamp fever and malaria, the observatory was transferred to high ground at its present site on Massachusetts Avenue. Great Congressional debate ensued at the time over whether it should be a national or a naval observatory. One legislator sneered that "a naval observatory assignment was for naval officers who are made seasick by ocean voyages." Another countered that civilian scientists were too extravagant whenever they got hold of public funds. In the end, the Navy retained the observatory, which took its place among the European national observatories at Greenwich, Pulkova, and Paris. These institutions were primarily concerned with stellar astronomy as the foundation of navigation. The science of astronomy was celestial mechanics; astrophysics did not emerge until the end of the nineteenth century.

In the second half of the nineteenth century, the Lick Observatory in California became the most powerful in America. Its benefactor was James Lick, a cabinetmaker who made a fortune in South America manufacturing piano cases and returned to California in 1848 with $30,000 in Peruvian gold doubloons. When he landed in San Francisco, he found a village of some 800 rowdy, impoverished inhabitants living in shanties or worse. Seventeen days later a gold nugget was found at a site near what is now Sacramento. Within weeks the Gold Rush was on in full force. As people sold off their land to get to the gold fields, Lick, in a matter of months, bought roughly one-half the future city of San Francisco. Inflation soared, especially the price of land. Lick became by far the wealthiest man in the area, eventually becoming a multimillionaire.

Lick had a passion for growing flowers and for horticulture in general. Intellectually, he was fascinated by the mystery of the universe. He had a mystical belief in the cosmos as the handiwork of God and was impelled to do something great to further the understanding of the universe. He was also an adventurer in spirit and is said to have forecast, "One day men will go to the moon and back!" At the age of 77, he published a will specifying the disposition of his fortune. It was to go entirely to the people of California in various forms, with a stipulation that the greatest sum was to be spent for the world's largest telescope. At one time he considered establishing an observatory as a memorial to himself, even seriously considering a site at the corner of 4th and Market Streets in San Fran-

FIGURE 9. *Affirming his love for the stars, James Lick specified in his will that his casket be placed in the concrete base of the telescope that he gave to the University of California. (Mary Lee Shane Archives of the Lick Observatory.)*

cisco. He was dissuaded from building a reflector and agreed instead to the construction of the largest refractor in the world, 36 inches in diameter, atop Mount Hamilton. He died in 1876; in 1877, per instructions in his will, his remains were removed from a cemetery in San Francisco and taken to the as-yet-uncompleted observatory, where he was interred in the concrete base of his telescope.

In arriving at his decision about the telescope, Lick consulted two of the leading American scientists of the time, Joseph Henry and Simon Newcomb. Henry told him of the bequest of the Englishman, James Smithson, out of which the Smithsonian Institution was born. Commenting on Lick's motivation to build a great telescope, Newcomb remarked that:

> [astronomy] seems to have the strongest hold on minds which are not intimately acquainted with its work. The view taken by such minds is not distracted by the technical details which trouble the investigator, and its great outlines are seen through an atmosphere of sentiment, which softens out the algebraic formulae with which the astronomer is concerned into those magnificent conceptions of creation which are the delight of all minds, trained or untrained. (*Harper's* magazine, February 1885)

Many of the great observatories of the nineteenth and early twentieth centuries were privately endowed by wealthy men whose inspiration was properly described by Newcomb.

The man most responsible for the great American telescopes that set the stage for modern astrophysics was George Ellery Hale, after whom the 200-inch telescope on Mount Palomar is named. Four times in his career he initiated the construction of new telescopes, each the world's largest telescope of its time. Hale was born into a wealthy family. As a teenager and as a student at the Massachusetts Institute of Technology, he used his resources to satisfy a passion for astronomy. Through his comfortable association with them, he gained the support of rich and powerful men for a series of great new observatories. In Chicago he cultivated Charles Yerkes, a wealthy streetcar and real estate tycoon with a reputation for somewhat shady business dealings. Yerkes gave him carte blanche: "Build the observatory," he told Hale. "Let it be the biggest in the world and send the bill to me" (Helen Wright, *Explorer of the Universe,* 1966, p. 98). Hale held him to his word, although his dealings with Yerkes grew difficult and often acrimonious. Yerkes never did come up with $1 million that he originally promised, but he provided enough to see the project to completion. At age 29 Hale established the Yerkes Observatory at Williams Bay, Wisconsin, and equipped it with a 40-inch refractor. That telescope remains the largest refractor in the world; all subsequent larger telescopes have been reflectors.

Hale's appetite for larger telescopes was fed with money donated by his father. He obtained a 60-inch mirror and chose Mount Wilson in California as the site on which to erect a telescope. There he established a solar observatory with $30,000 of his own money and the backing of Andrew Carnegie, the steel magnate who had founded the Carnegie Institution. Even before the 60-inch mirror was in operation, Hale had set his sights on a 100-inch mirror. Carnegie was so persuaded by Hale's enthusiasm that he doubled the Carnegie Institution endowment to cover the added cost of operating the 100-inch telescope, named the Hooker telescope. The site for the 100-inch mirror, 1 mile above Los Angeles, could be reached only by climbing a tortuous 8-mile mule trail from Pasadena. Upon completion of the Mount Wilson Observatory, there began a remarkable period of discovery, from the 1920s to the early 1930s, which encompassed Edwin Hubble's observations of the expansion of the universe.

Hubble was a wonderfully gifted astronomer and quite an athlete to boot. At the University of Chicago he starred in boxing, track, and basketball. Upon graduation, he had to choose between accepting a Rhodes scholarship or the lucrative enticements of a fight promoter. Fortunately he took the Rhodes, but while abroad, he did fight an exhibition bout with the world light-heavyweight champion, Georges Carpentier. Hubble studied law during his Rhodes scholarship, and upon his return to the United States in 1913, he set up practice in Louisville, Kentucky. A lawyer's life was not for him, however, and he abandoned his practice after about 1 year. He later remarked that "astronomy is like the ministry—I got the calling" and returned to the University of Chicago.

By 1916 Hale had already brought Harlow Shapley to Mount Wilson, but the promising young Hubble refused Hale's inducements. Hubble went off to France to serve in the Army, instead. At the end of the war he belatedly accepted Hale's invitation, and in 1919 he arrived at Mount Wilson to become the master of the 100-inch telescope.

Although Hale was in poor health and was afflicted by repeated nervous breakdowns and unbearable headaches, he was obsessed with the idea of building telescopes much larger than the Hooker telescope, perhaps as large as 200 or 300 inches in diameter. With his characteristic power of persuasion, he obtained the enthusiastic support of the Rockefeller International Education Board and, subsequently, the General Education Board, who pledged $6 million for the construction of a 200-inch telescope.

The saga of the casting of the 200-inch disk is replete with episodes of near disaster and brilliant solutions to unexpected engineering problems. The Pyrex mirror was poured from 42,000 pounds of glass at the Corning glassworks on 2 December 1934. It was placed in an annealing oven for 2 months and then was slow-cooled for 8 months, after which the disk

was ready for figuring. Encased in a steel shell and protected by steel plates, the mirror was stowed upright on a flatbed car and drawn to California by a steam locomotive. It took 2 weeks to cross the country at slow speed. During the day huge crowds gathered along the route to cheer its passage. At night the disk was parked at sidings and illuminated with floodlights, and guards armed with rifles cordoned off a no-man's zone to prevent vandalism.

Hale died in 1938 while the grinding was in progress in the optical shops of the California Institute of Technology. World War II interrupted the work, which would not be resumed until 1946. During the grinding, 5.25 tons of glass were removed with 31 tons of fine abrasives. The finished weight was 14.5 tons. By 1948 the disk was mounted inside an aluminum dome, 7 stories high, that matches the great Roman Pantheon in size and shape. Its parabolic surface was polished to an accuracy of one-half of one-millionth of 1 inch and, lastly, was coated with a thin layer of aluminum only 1000 atoms thick. On 3 June 1948 the instrument was dedicated as the Hale telescope. "First light," the astronomers' term for the inaugural use of a telescope, took place on 26 January 1949 when Edwin Hubble took a photograph of nebula NGC2261 and pronounced the telescope a complete success, to enthusiastic nationwide acclaim. Its light-gathering power was superior to any other telescope in the world.

FIGURE *10.* George Ellery Hale (1868–1938) *was the greatest telescope builder of his time. Four times in his career he led the construction of the largest telescopes in the world. The Yerkes 40-inch refractor on the shore of Lake Geneva in Wisconsin was completed in 1897. Even before the Yerkes instrument went into operation, Hale dreamt of bigger reflectors for stellar astronomy. He purchased a 60-inch blank with family money and persuaded private philanthropies to finish the mirror and erect an observatory 14 years later atop Mount Wilson in Southern California. He next acquired support for his third great telescope, the 100-inch Hooker mirror. Edwin Hubble was lured to Mount Wilson to become the master of the 100-inch and soon discovered the expansion of the universe. By then Hale's appetite was whetted for a much grander instrument. Although stricken with fierce headaches and depressed to the point of nervous breakdown, Hale was still able to gain support for creating a 200-inch telescope on Mount Palomar. Sadly, he did not live to see it through to completion, but the great telescope carries his name.*

Robert Browning's lines ("A Grammarian's Funeral," 1855) make a fitting epitaph:

That low man seeks a little thing to do,
Sees it and does it.
This high man, with a great thing to pursue,
Dies ere he knows it. . . .

FIGURE 11. *Andrew Carnegie, with great coat misbuttoned, stands with George Ellery Hale before the Mount Wilson Observatory on the occasion of its dedication in 1908.*

Carnegie remarked, "I should like to be satisfied before I depart that we are going to repay the old land some part of the debt that we owe them by revealing more clearly than ever to them the new heavens." (Hale Observatory.)

Built at a time when radio and talking movies were the height of advanced technology, when television, jet planes, and computers were still unknown, the Hale Telescope was a marvel of optical and mechanical engineering. Even though the massive supporting tube and yoke weigh more than 500 tons, it rides with almost no friction on pressure-oil-pads and requires only a small 1/12-horsepower motor drive to follow the stars. American astronomers proudly called it the "Big Eye."

The Hale telescope mirror has been exceeded by the 6-meter mirror at the Zelenchukskaya Astrophysical Observatory in the Caucasus Mountains of the Soviet Union. The Soviet telescope is flawed, however, and has never performed up to expectations. To preserve the mirror figure from the stress of gravity, as it varied in different positions, the engineers of both the Hale and the Soviet mirrors resorted to making them massively rigid. Shaped from a 42-ton block of glass, the Soviet mirror is monolithic, without the weight-saving feature of honeycomb structure, and suffers from excessive thermal lag. The mirror follows the day–night temperature cycles so slowly that it never comes to equilibrium with the ambient environment. Even small temperature changes appear to cause significant reductions in image quality. The mirror now in use by the Soviets is their third effort; the first cracked; the second was marred by bad spots that had to be blacked out, with attendant loss of resolution and image intensity; the third is still subject to thermal problems. All of the world's largest telescopes are close to the limits of what can be accomplished with massive cast blocks of glass.

New-Technology Telescopes

There are about 28 telescopes worldwide that have apertures of 2 meters or more, and about two-thirds of them have come into operation since 1970. But astronomers are now turning to much larger multiple-mirror reflectors that can be made thin and lightweight, and comparatively cheaply. Although they are placed on independent moveable mounts, they can be controlled by a computer with such precision that the combined mirrors act as a single perfect surface. The concept can be traced back to Erasmus Darwin (Charles Darwin's grandfather). In 1779 he wrote: "Suppose 20 glasses or concave specula, are so placed as to throw all their images of a certain object onto the focus—there will be one image with twenty times the brightness that one lense or speculum could produce" (Sandra Blakeslee, "Astronomy from the Ground Up," *Mosaic 17* (2), 1986, p. 20). For that era the mechanical problems of translating that idea into a large telescope were insurmountable. In the 1950s the United States Navy began to build the largest fully steerable radio dish in the world, 600 feet in diameter, out of articulated mirror segments. The project was dropped when the military requirement was

solved by other means, since radio astronomy alone was not sufficient to justify the enormous cost.

A multiple-mirror telescope went into operation in 1980 atop Mount Hopkins in Arizona, just a short distance from Tucson. Each of its six mirrors was 1.8 meters (72 inches) in diameter. In concert, they mimicked the performance of a single high-resolution 4.5-meter mirror. At present it has been reconfigured to carry a single 6.5-meter mirror of lightweight design, but the multiple-mirror feasibility has been so well demonstrated that astronomers were encouraged to plan several new-technology multiple-mirror telescopes of very large aperture. Among those already operational are the two 10-meter Keck telescopes atop Mauna Kea in Hawaii and a European Southern Observatory Very Large Telescope (VLT) comprised of four primary mirrors, each larger than the 200-inch telescope on Mount Palomar. The Keck telescopes were built jointly by the California Institute of Technology and the University of California, with the help of a $75 million grant from the Keck Foundation. A segmented mirror is a mosaic of 36 hexagons, precisely positioned and continuously adjusted by computer (Appendix C).

In the past the glass disk for a telescope mirror was always cast with a flat surface and then laboriously ground to parabolic shape. Manufacture of the Hale telescope required 11½ years. Early in this century, physicist Robert W. Wood at The Johns Hopkins University demonstrated that when one spins liquid mercury in a beaker, the mirrorlike surface assumes a paraboloidal shape. It recently occurred to astronomer James Roger Angel at the University of Arizona that spinning molten glass in a container on a turntable would allow centrifugal forces to shape a glass mirror in similar fashion. He tested his idea with Pyrex glass in a home-made furnace and found that it worked. Angel and his colleagues then

FIGURE *12.* Edwin P. Hubble (1889–1953). *Edwin P. Hubble considered a career as a professional boxer or as a lawyer before settling into a lifetime commitment to extragalactic astronomy and observational cosmology. In 1926, using the 100-inch telescope on Mount Wilson, he developed a classification scheme for galaxies that sorted them into spirals, ellipticals, and irregulars. Hubble then concentrated on distance measurements, making especially good use of the Cepheid variable stars as his standard candles. In 1924 he identified 1 dozen Cepheid variables in the Andromeda nebula and found a distance of about 1 million light-years, a distance so large that it finally settled the long-standing disputes over whether the spiral nebulae lay inside the Milky Way or were indeed island universes far beyond its bounds.*

By 1929 Hubble had obtained distances for 22 galaxies out to the Virgo cluster and could show the linear relationship between distance and redshift, since known as Hubble's law, that revealed the universe to be expanding.

FIGURE 13(a). *Rear view of the 200-inch mirror blank being inspected by engineers at the Corning Glass Works in New York State. The world's largest piece of glass was cast in a waffle mold to lighten the mirror weight. (Corning Glass Works.)*

FIGURE 13(b). *After grinding, polishing, and coating in the Caltech optical shop, the 200-inch mirror is hauled up Mount Palomar. (California Institute of Technology.)*

set out to spin-cast a large mirror in a rotating oven. Their prototype kiln, capable of casting mirror blanks as large as 6 feet across, was supported by a ball-bearing 1.2 meters in diameter and was rotated at up to 15 revolutions per minute. As the molten glass spun, it flowed outward from the center and climbed the wall of the mold to assume a paraboloidal

FIGURE *13(c). Audience assembled under the dome of the Hale 200-inch telescope at the dedication ceremony in 1948. (Los Angeles Times, Photograph by Gordon Wallace.)*

shape. Once the glass was cool and the shape frozen in, only a modest amount of final polishing was required.

Angel went a step further to reduce the weight without sacrificing rigidity. Hexagonal refactory plugs of water-soluble aluminum silicate fiber were anchored to the bottom of the mold to displace the bulk of the glass. The molten glass filled a honeycomb pattern between the plugs and flowed over them to produce a thin mirror surface. When the glass solidified and cooled, the refractory plugs were easily flushed out under a stream of water at high pressure. The end product was a lightweight, high-strength mirror of surprisingly good figure. In their first test of this spin-casting method, Angel's team produced a 0.8-meter mirror, figured to an accuracy of 1 millimeter.

The process of monitoring the casting from close up required that an "oven pilot" ride on the turntable. Through a window in the oven, the pilot kept watch on the temperature, keeping it within a range of 1 or 2 degrees, and adjusted the melting and casting program to the required

FIGURE 14. *The Multiple Mirror Telescope (MMT) atop 8550-foot Mount Hopkins, south of Tucson, Arizona, is a sextet of identical telescopes operated as a single mirror. The individual mirrors are 72 inches in diameter. In combination they equal the light-gathering power of a 176-inch telescope. The entire telescope and all of its peripheral computers are housed in a squat building, five stories tall and weighing 500 tons, that rotates with the telescope. (Smithsonian Institution.)*

FIGURE 15. *Rotating furnace with walls and hearth plates assembled in preparation for spin-casting a mirror at the Stewart mirror laboratory, University of Arizona. The 4.5-meter lid hangs on the wall. (Photograph by Carl Matter.)*

schedule until the mirror solidified. Fortunately one of the Arizona group, Dan Watson, was found to have the stomach for the pilot job and became the expert oven pilot. He was the only member of the group who could ride the carousel without becoming nauseated and disoriented. For the future, fortunately, the "oven pilot" has been automated.

A radically different technique is the "thin meniscus" approach adopted by the European Southern Observatory for a set of 8-meter telescopes at La Silla in Chile. Each mirror is a sheet of glass only a few centimeters thick, making it lightweight and flexible. To keep the surface figure precise to a fraction of a micrometer, the mirror is backed by a set of mechanical pistonlike drivers that will deflect the mirror under computer control by exactly the right amount to compensate for the distorting effects of gravity.

From Photographic Film to Silicon Pixels

During most of astronomy's history, watchers of the skies drew only the minimum information from starlight, which was only enough to qualify them as astrologers. After the invention of the telescope, astronomers were content to position the stars accurately, draw pictures of their impressions of nebulae, and sketch the canals of Mars. Then, during the midnineteenth century, photography emerged at last and revolutionized astronomy. Astronomers could now expose their photographic plates for hours to build up images of faint sources and their spectra.

When the Hale telescope went into operation in 1948, photographic emulsions registered about 1 out of every 300 photons. In spite of this low efficiency, photography had served as the mainstay of astronomical imaging for more than 100 years. Now the new miracle of electronic imaging is displacing photography with electronic detectors roughly 200 times as sensitive as the photographic plates of the 1940s. Hale's obsession with building larger and larger telescopes was to gather more light and reach fainter objects, yet his photographic film was throwing away more than 99% of the photons that reached it. Making the Hale telescope twice the diameter of the Mount Wilson telescope improved its light-gathering power by a factor of 4, but if Hale could have used a modest electronic detector such as those currently available, he would have achieved the same performance with a telescope only 50 inches in diameter.

Silicon, next to oxygen the most common element on Earth, has become astronomer's "gold." The miracle of the silicon chip that appeared in the early 1970s can make all telescopes at least 10 times better. On the light-sensitive surface of silicon, it is possible to create hundreds of thousands of microscopic islands, called *pixels* like the retinal cells of the eye. Photons of light falling on the pixels are detected with efficiency as high as 80% when the chip is cooled to low temperature.

The incident photons release electrons in the silicon that are then read off electronically and digitally with a dynamic range of 1 million, as compared to a useful range of between 10 and 100 in a photographic emulsion. Doubling the exposure of a photographic plate does not make it twice as black upon development; in other words, the density is nonlinear with respect to exposure. To reveal in full a celestial object with detail covering widely different levels of intensity—for example, a galaxy with a low brightness halo of stars, a bright nucleus, and a faint jet—several exposures of widely different duration are necessary. The electronic cameras take in all the information at once, and the image can be constructed over a very wide range of brightness.

Around 1978 the so-called CCD, for Charge Coupled Device, was introduced to astronomy. As many as 640,000 pixels in an 800-by-800 array were created on a chip the size of a postage stamp. The charges induced by the photoelectric effect of the light are transferred off the chip row by row, in a manner somewhat analogous to buckets of water being passed down the line of a fire brigade. When the image is reconstructed electronically, it has the dotted pattern of a newspaper picture or magazine photoengraving, but is far superior to a commercial television image. Each pixel stores a gradation of grayness that provides final image contrast, just as the density of each developed grain in a photographic emulsion determines the gray scale of a photo.

With electronic imaging, astronomers no longer put their eyes to the telescope. More likely they sit at a console in an air-conditioned room remote from the telescope. Over computer–telephone line link-ups, the image might just as well arrive 1000 miles away. The old-fashioned astronomer fighting the cold of the night in an electrically heated suit while seated at the focus of the telescope has become an anachronism.

The miracle of electronics has had an enormous impact on the sheer volume of data processed by astronomical research programs today. An astronomical image made by a CCD requires about 0.5 megabyte (1 megabyte = 1 million characters). The floppy disk commonly used with personal computers can store about 0.3 megabyte, barely enough to hold one CCD image. In a typical night of observing, the astronomers may wish to record as many as 50 images, or about 25 megabytes of data, which would require as many as 75 floppy disks for storage. It becomes impractical to store the enormous flood of data coming out of telescope observations on floppy disks. Instead most of these results are stored on magnetic tapes, thousands of them per year. The future offers promise of laser disks for data storage as an alternative to magnetic tape. It would be possible to store 1000 megabytes of information on a single laser disk so that a comparatively small number of such disks would satisfy the requirements of almost the entire world output of astronomical observatories.

THE TOOLS OF ASTRONOMY

Toward the Ultimate Optical Sensitivity and Image Sharpness

Astronomers rate the brightness of stars on a logarithmic scale similar to that used by Hipparchus of Samos in ancient Greece. In 134 B.C. Hipparchus prepared a catalog of the brightest stars. Each star was mapped in a system of longitude and latitude and assigned a "magnitude" corresponding to one of six levels of brightness: the higher the magnitude, the dimmer the object. (Like golfers, the brightest stars have the lowest scores.) First magnitude included the 20 brightest stars in the sky; sixth magnitude included 2000 or so of the dimmest stars that are just barely visible to the naked eye on a moonless night. The second, third, and fourth magnitudes lay between these extremes. Each step on the modern magnitude scale corresponds to a brightness factor of about 2.5; a difference of 5 magnitudes is approximately a factor of 100.

As late as the 1960s, astronomers thought that the faintest object that could be detected with the largest telescopes was about twenty-first magnitude, roughly 1 million times fainter than what could be seen with the naked eye. The limit is set by the background brightness of the sky; under ideal conditions, perhaps twenty-fourth magnitude could be reached. With the CCD, twenty-fifth magnitude is well within reach. Further improvement in sensitivity can be achieved only by moving telescopes into space. No matter how large the telescope, its resolution is limited on the ground by the shimmering turbulence of the atmosphere. Sir Isaac Newton stated the problem well:

> If the theory of making telescopes could at length be fully brought into practice, yet there would be certain bounds beyond which telescopes could not perform, for the air through which we look upon the stars is in perpetual tremor, as may be seen by the tremulous motion of shadows cast from high towers and by the twinkling of the fixed stars. The only remedy is a serene and quiet air such as may perhaps be found on the tops of the highest mountains above the gossamer clouds. (*Opticks*, 4th ed., p. 110)

Newton was only partly right. The best remedy is to exceed the air itself and place the telescope beyond the atmosphere. Henry Norris Russell, Princeton University's famous astronomer, jested that all good astronomers go to the Moon when they die so that they can observe the universe without looking through a dirty atmosphere.

The resolution of the human eye is only a couple of minutes of arc, which is sufficient to distinguish a penny at a distance of about 100 feet. The Moon, 250,000 miles away, subtends an angle of 0.5 degree, and the eye can easily resolve its disk and detailed surface features. If the Moon

FIGURE 16. *"It says it's sick of doing things like inventories and payrolls, and it wants to make some breakthroughs in astrophysics." (Copyright © 1981 by Sidney Harris—American Scientist Magazine.)*

were as far away as the Sun, it would subtend an angle of only 6 arc seconds and would appear fuzzy rather than sharp edged. The Sun at 92 million miles also subtends 0.5 degree, and the eye perceives clearly the edge of its disk and even small sunspot markings. However, if the Sun were moved to 4.5 light-years, the distance of the nearest star, it would subtend an angle of only 0.003 arc second and appear as no more than a twinkling point of light, even when observed with the 200-inch telescope.

Resolution improves in direct proportion to the diameter of a telescope and inversely as the wavelength of the light being detected. In principle the larger telescopes should give finer detail, but the turbulence of the atmosphere effectively degrades them all to about the resolution of a 12-inch mirror. Telescope sites on mountain tops improve the "seeing," but even at the best sites a resolution of 0.5 arc second is rarely obtained. Consequently the 200-inch mirror provides no sharper resolution than a telescope one-tenth its diameter. With the largest Earth-based telescopes, our penny could be resolved no farther than 4 to 5 miles away.

The Hubble Space Telescope

The earliest savants of rocketry and space travel, beginning with Russian visionary Konstantin Tsiolkovsky late in the nineteenth century, foresaw astronomical telescopes in orbit and on the moon. In 1946, Lyman Spitzer of Princeton University advanced persuasive arguments for a space telescope with a primary mirror diameter of 400 inches, twice as large as the world's greatest telescope at that time on Mount Palomar in California. After NASA was created Spitzer's proposal received serious attention. To fit into the space shuttle that would place it in orbit, the mirror diameter could not exceed 94 inches, yet it would be the most sensitive optical telescope ever to search the sky. Astronauts would have to develop the skills for handling massive objects on space walks and for exchanging large replacement instruments from time to time, skills that would be useful eventually in the assembly of a space station.

Newton's concern for atmospheric turbulence was not the only problem confronting astronomers. Ground-based telescopes cannot escape man-made light pollution from street lamps and illuminated billboards, so that locating observatories on mountaintops in regions remote from cities is a necessity. But moonlight scattered from dust and aerosols remains a persistent interference. Even on moonless nights the sky is not jet-black. Airglow from the interactions of electrons and ionospheric ions colors the night sky. On the darkest nights the sky shines as though there were a star of twenty-fifth magnitude in every square arc second. In space these forms of interference would be eliminated or greatly minimized. Named after the leader of American astronomy in the first half of the century, Edwin Hubble, the *Hubble Space Telescope (HST)* would be the most ambitious telescope project ever attempted. Its cost of more than $1.5 billion would exceed the cost of all the world's major optical telescopes combined. It would be the answer to Henry Norris Russell's prayers, a great telescope on an Earth-orbiting satellite as free of the atmosphere as the Moon. No astronomers would be aboard because the marvels of electronic imagery and radio telemetry would bring the telescope's pictures directly to them on the ground. It was expected to be the finest achievement of NASA technology and the fulfillment of astronomers' dreams. In orbit it would attain 10 to 20 times the best resolution possible on the ground, and over a broader spectrum, from ultraviolet to near infrared.

"First Light" at the *HST* in orbit was expected to climax the scientific and technological revolution of the twentieth century and to inaugurate an era of "The Great Observatories" in space. The entire scientific world

FIGURE 17. *Astronauts Story Musgrave (on remote manipulator arm) and Jeff Hoffman (in cargo bay) repair the* Hubble Space Telescope *on the space shuttle* Endeavor *in December 1993. Solar arrays were replaced and a new camera and corrective optics were installed.*

held its breath as the telescope lifted off on 24 April 1990 with a crew of five astronauts. Expectations immediately came to grief. The first images were blurred and an incredible error in the main mirror was discovered. It had been ground to the wrong shape—too flat by 2 microns—just a few percent of the thickness of a sheet of writing paper, but intolerable for the desired performance of the telescope. The flawed mirror was derided as the most scandalous mistake in the history of telescopes; NASA and astronomers everywhere were stunned. While it could carry out journeyman work about as well as telescopes on the surface of the earth, its performance fell an order of magnitude short of the design goals.

Various salvage options were frantically considered and astronaut repair in orbit was chosen at a cost of $700 million. Three years were required to mount a mission to repair the telescope optics, while the program struggled with its compromised performance. Furthermore, the mirror was not the only problem. The extended solar panels that provided power to the spacecraft shook whenever the spacecraft orbit carried it from day to night and from night to day. Shortly after launch two of the gyroscopes that stabilized the *Hubble* in its orbit malfunctioned. It appeared that the *Hubble* was a star-crossed, disgraceful, $1.5-billion dud.

Improvised fixes were rushed to the rescue. Ground controllers worked around periods when the solar panels jittered and other on-board gyros were diverted to the stabilization. At the same time, all of the optical expertise available in the country was called upon to cure the blurred imaging. To correct the *Hubble*'s flawed mirror, a set of small mirrors had to be inserted into the optical train. A device named COSTAR (for Corrective Optics Space Telescope Axial Replacement), with more than 5000 parts, was the solution. COSTAR was an optical bench that carried an array of mirrors the size of small coins with motors to adjust them. Because of the modular design of the *Hubble*'s instruments, the COSTAR shell and its integral components could be substituted as a unit for one of the least used of the original instruments, the High Speed Photometer. In December 1993, astronauts fitted the telescope with COSTAR "spectacles" to correct its vision. The operation was a great success and the telescope recovered near perfect sight. The *Hubble* images are now remarkably detailed and almost magical in their beauty.

The performance of the *HST* far surpasses the capabilities of Hale's great 200-inch mirror on Mt. Palomar. Resolution is 0.1 arc second or better, between 5 and 10 times as good as the best ever achieved by Earth-bound telescopes. Because the image of a star would be so highly concentrated at this resolution, the superior contrast to background makes it possible to detect objects 50 times fainter and 7 times as distant as any that have been seen with ground-based instruments. The limiting visual magnitude is about 29, about 1 billion times fainter than can be seen by the unaided eye. To stabilize the telescope during exposures it locks on to guide stars as faint as 14.5 magnitude and holds steady to 0.007 arc second. Besides obtaining an unparalleled sharpness of image, the *HST* reaches the far ultraviolet, where observation is impossible from the ground.

The original management strategy for the *Hubble Space Telescope* called for repeated "house calls" by astronauts every two or three years. Instruments would be replaced with superior versions as they were developed, and damages to various elements such as solar-power panels could be repaired. In the February 1997 service mission, an extended robotic arm captured the telescope and secured it in the shuttle's cargo bay where the team of astronauts removed two instruments, the Goddard High Resolution Spectograph (GHRS) and the Faint Object Spectrograph (FOS). These were replaced with a $125 million Space Telescope Imaging Spectrograph (STIS) and a $105 million Near Infrared Camera and Multi-Object Spectrometer (NICMOS), each roughly the size of a household refrigerator. STIS can gather spectra simultaneously from dozens of celestial objects. These instruments will attempt to answer questions about how stars and galaxies age and how many galaxies have black holes in their cores.

To advance cosmological understanding, the *Hubble* provides a tenfold greater power in searching the early universe, analyzing its composition, gauging its rate of expansion, and estimating its age. In hunting for evidence of planets outside the solar system, it has focused on the Orion Nebula, a nursery of newborn stars, and found disks of dust clouds enveloping young stars much as astronomers believe our solar system's planets were formed. There is a strong consensus among astronomers that quasars derive their almost incredible luminosities from supermassive black holes—millions to billions of suns collapsed into infinitesimal volumes of space in the nuclei of galaxies. *Hubble* observations already provide diagnostics from which to select among current astrophysical models.

The Ultraviolet Sky

The *HST* has great power in the far-ultraviolet as well as in visible wavelengths, and remarkable progress has already been achieved in ultraviolet astronomy with short-lived rocket-borne instruments and a modest but highly versatile satellite, the *International Ultraviolet Explorer.* The story of galactic ultraviolet astronomy begins in the mid-1950s, when my colleagues and I attempted galactic astronomy observations in the extreme ultraviolet (one-fourth the wavelength of visible light) with Aerobee rockets. We were amazed to discover that the entire night sky was flooded with Lyman alpha radiation, the strongest spectrum line of hydrogen, at 1216 angstroms in the far ultraviolet. The bright glow was soon explained as a backwash of solar Lyman alpha radiation scattered into the shadow cone of the night sky by a far-reaching cloud of atomic hydrogen, a geocorona extending to about 50,000 miles above the Earth.

At wavelengths slightly longer than Lyman alpha, it was possible to see into the Galaxy and carry out photometric surveys of the hottest stars. Collections of 4- and 6-inch mirror telescopes with photon counters at the foci were mounted to look outward through the skin of the rocket at the passing celestial scene as the rocket spun in flight. These early flights showed, as expected, that the ultraviolet sky was dominated by young, hot blue stars at temperatures as high as 50,000°K and ordinary stars like our 6000°K Sun were invisible.

A signal achievement of the early years of rocket astronomy was the discovery of molecular hydrogen in interstellar space. George Carruthers of the United States Naval Research Laboratory succeeded in building a sophisticated electronic image converter spectrograph for an Aerobee rocket; with this he recorded the extreme ultraviolet spectra of stars seen through dusty regions. He found evidence that molecular hydrogen was the most abundant constituent of dark interstellar clouds. In the 1960s radio astronomers found little evidence of atomic hydrogen's radio waves when they searched these clouds, but theorists predicted that molecular hydrogen would be found to be abundant. They proposed that dust grains would act as surface catalysts for the combination of hydrogen atoms to molecules. Carruthers' observations proved them right.

The crowning success of ultraviolet astronomy has been the *International Ultraviolet Explorer (IUE)*, launched 1978 January 26 and now in its second decade of successful operation. The design was conceived by British physicist Robert Wilson at the Culham Laboratory, and its development and operation was a joint effort of NASA, the European Space Agency (ESA), and the British Science and Engineering Research Council. The satellite is in a geosynchronous orbit 22,000 miles above the Atlantic. As it orbits the Earth with a 24-hour period, it appears to astronomers based in the United States and Europe to hang stationary. Unlike other satellites in low Earth orbit, which need to preprogram detailed observation schedules because they are out of contact with ground stations for much of the time, *IUE* is operated by direct command in real time, like typical ground-based telescopes. Not very long ago it would have seemed sheer fantasy to imagine an astronomer sitting at a console on the ground, commanding a telescope 22,000 miles in space, and with delicate touch controlling every detail of its operation.

Most of the *IUE* observing time has been allocated to the study of hot stars; those with temperatures higher than 10,000°K. In contrast to the Sun, which expels a gentle wind of hot, tenuous gas at some tens of kilometers per second, fierce winds of thousands of kilometers per second blow from the hot stars. *IUE* has found that the more intense the radiation from the brilliant surface of a hot star, the stronger the wind driven by that radiation pressure. A few one-millionths of 1 solar mass is lost each year, which may seem small; however, on the scale of stellar lifetimes it is sufficient to have a drastic effect on the star's evolution.

The *IUE* also followed the evolution of Supernova 1987A's ultraviolet light curve. Many unexpected features of the supernova can be explained by theorists in terms of the different chemical abundances in the Magellanic Clouds, as revealed by the *IUE*.

The success of the *IUE* has given strong impetus to the preparation of a follow-on observatory designed to study wavelengths shorter than 1150 angstroms, which the *HST* will not be able to reach. The new observatory will also be placed in geosynchronous orbit and operated in the same direct manner as the *IUE*.

CHAPTER 2

The Invisible Universe of Radiowaves and Infrared

The opening of a new astronomical frontier, the invisible universe, began in 1933 in a most unpretentious way. In a New Jersey potato field Karl Jansky listened to the noisy radio hiss of the cosmos. From the time of Galileo to the 1930s, astronomy had been limited to the narrow wavelength range of light seen by the human eye or photographic film. Then a new cosmos was suddenly unveiled in radio waves, to be quickly followed in the Space Age by the opening of all the wide spectrum of electromagnetic waves, from the short-wave gamma rays to x-rays and ultraviolet light, and to the long-wave infrared. The blindfold of the atmosphere was fully removed when rockets began to carry astronomical detectors to the near-vacuum of space, and astronomers discovered there is far more to the universe than meets the eye.

Following Heinrich Hertz's laboratory demonstrations of radio waves in 1887, several unsuccessful attempts were made in England, Germany, and France to detect such waves from the Sun. In the United States, the first serious thoughts of detecting solar radio noise have been attributed to Thomas Alva Edison. His concept is succinctly described in a letter written by his associate, Arthur E. Kennelly, in 1890. The entire short note reprinted here (from C. D. Shane, "Radioastronomy in 1890? A Proposed Experiment," *P.A.S.P. 70,* 1958, p. 303) is revealing of the scientific aspect of Edison's genius that is not generally appreciated:

> Professor E. S. Holden, Principal
> Lick Observatory
> Mount Hamilton
>
> Dear Sir:
>
> . . . I may mention that Mr. Edison who does not confine himself to any single line of thought or action, has lately decided on turning a mass of iron ore in New Jersey that is mined commercially, to account in the direction of research in solar physics. Our time is of course occupied at the Laboratory in practical work, but on this instance the experiment will be a purely scientific one. The ore is magnetic, and is magnetic not so much on its own account like a separate steel magnet but rather by induction under the early's polarity. It is only isolated blocks of the ore that acquire permanent magnetism in any degree. Along with the electromagnetic disturbances we receive from the sun which of course you know we recognize as light and heat (I must apologize for stating facts you are so conversant with), it is not unreasonable to suppose that there will be disturbances of much longer wavelength. If so we might translate them into sound. Mr. Edison's plan is to erect on poles round the bulk of the ore, a cable of seven carefully insulated wires, whose final terminals will be brought to a telephone or other apparatus. It is then possible that violent disturbances in the sun's atmosphere, might so disturb either the normal distribution of magnetic force on this planet, as to bring about an appreciably great change in the flow of magnetic induction embraced by the cable loop, enhanced and magnified as this should be by the magnetic condensation and conductivity of the ore body, which must comprise millions of tons. . . .
>
> Yours faithfully,
> A. E. Kennelly

Edison apparently never carried out his plan. If he had, the experiment would have been too insensitive to the short radio waves that penetrate the ionosphere. More than 40 years were to pass before Karl Jansky made the seminal discovery of cosmic radio waves.

Jansky's serendipitous discovery occurred shortly after he came to the Crawford Hill facility of Bell Labs in Holmdel, New Jersey, in 1928 at age 22. Overseas radio telephone communications had been inaugurated by AT&T 1 year earlier. Jansky's task was to investigate the cacophony of hissing, crackling noises that interfered with transatlantic radio telephone communications. He constructed a turntable that rode around a circular cinderblock track on four wheels from a junked model-T Ford. On it he erected a 100-foot-long antenna made by mounting 400 feet of brass pipes on wooden posts. With this merry-go-round contraption turning at 3 revolutions per hour, Jansky hoped to sort out the directions from which the radio-noise interference was coming. As the antenna swept the sky, noise seemed to come from all directions. Jansky, although he suffered from a chronic kidney ailment and was very frail, was also highly dedicated. After 4 years of patient, persistent effort, he felt sure enough of his conclusion that he published a paper entitled "Directional Studies of Atmospherics at High Frequencies" in the *Proceedings of the Institute of Radio Engineers* (1932). (*Atmospherics* is a term that refers to the strong contributions of lightning and thunderstorms in the radio-noise background.)

When Jansky subtracted the thunderstorm contribution, there still remained a residue of hiss that he couldn't pin down to any specific source. The hiss varied in intensity with the time of day, and the maximum noise appeared 4 minutes earlier each succeeding day. In those 4 minutes lay his clue to the source: If the radio hiss came from the stars, the diurnal variation and the 4 minutes advance were exactly what one would expect. Sidereal time is measured with respect to the fixed stars. Twenty-three hours and 56 minutes is a sidereal day, as distinct from the solar day, which is 24 hours long. Even if the Earth did not spin on its axis, it would still make one full turn a year relative to the stars as it circled the Sun. One turn a year amounts to about 1 degree per day, or 4 minutes of time.

Jansky had detected the "music of the spheres," radio waves generated among the distant stars of the Milky Way and, in particular, from Sagittarius, in the direction of the center of the Galaxy. He realized the significance of his finding but could not interest astronomers and engineers in following it up. Perhaps scientists were then so preoccupied with the explosion of knowledge in their own disciplines that they paid little attention to other fields. (In that year, 1932, deuterium, the neutron, and the positron were discovered as well.)

The publication of Jansky's results did arouse considerable popular interest. Headlines in *The New York Times* of 5 May 1933 read "New Radio Waves Traced to Center of the Milky Way." On May 15 the NBC Blue Network broadcast Jansky's hissing signal via a direct link-up from Holmdel to their New York studio and transmitted it around the country. According to press reports the following day, it sounded like steam escaping from a radiator. *The New Yorker* magazine commented: "It has been demonstrated that a receiving set of great delicacy in New Jersey will get a new kind of static from the Milky Way. This is believed to be the longest distance anybody ever went to look for trouble" (17 June 1933).

Jansky's superiors at Bell Labs saw little potential scientific or practical

FIGURE *18. With this crude instrument, Karl Jansky discovered in 1932 that the heart of the Galaxy is a strong source of radio waves. Note the mounting on Ford Model-T wheels riding on a circle of cinder blocks.* Insert: *photo of Karl Jansky. (Courtesy of AT&T Bell Laboratories.)*

value in pursuing these investigations, and he was discouraged from doing further research. During World War II he worked on methods of detecting directions of radio signals. His illness steadily worsened, and he succumbed to a stroke at the age of 44.

One radio engineer's curiosity was sparked by Jansky's discovery. Grote Reber began a hobby of the study of galactic radio noise. With $2000 of his own hard-earned Depression wages and no helper, he built a bowl-shaped antenna 31 feet in diameter in his backyard in the suburb of Wheaton, Illinois. For a decade he carried on his efforts in radioastronomy, rushing home each night from his job in Chicago, 30 miles away, to moonlight at astronomical research. Upon reaching his home in the evening he would sleep a few hours, then awaken and become a backyard radio astronomer from midnight to 6 A.M., carrying on his lonely vigils when radio interference was at a minimum. Reber not only confirmed Jansky's detection of radio noise from the direction of Sagittarius, he also discovered new sources in the constellations of Cygnus (the swan), Canus Major (the large dog), and Cassiopeia (the mother of Andromeda). Most puzzling, some of the signals came from patches of the sky without bright stars; suggesting celestial objects that radiated more power in radio waves than in light.

Reber's dish and his surprising discoveries attracted some attention, but no financial support. Eventually a committee from a research foundation came to see his operation on a day that was overcast and rainy. Assuming that nothing could be observed on such a foul day, they were so surprised when Reber showed them that the weather had little effect on radio transmission that they approved a grant. In 1940 Reber submitted his first scientific paper to *The Astrophysical Journal.* Otto Struve, a distinguished astronomer, was then the editor. He and other astronomers at Yerkes Observatory were not able to judge the meaning of the data that Reber submitted, data such as they had never seen before. Struve decided to send a delegation from the University of Chicago. On the day they arrived at Wheaton, Reber had to delay the demonstration of the dish because his mother was using it as one end of her clothesline, but he soon convinced the experts that he knew what he was doing experimentally, although they doubted his theoretical interpretations. When Reber's paper was published, it was analyzed by two expert astrophysicists, Louis G. Henye and Philip C. Keenan, who correctly explained the observed radio signals.

The Radio Pioneers of World War II

World War II brought radar into wide use, with important consequences for radio astronomy. On 12 February 1942 British radar experts learned that the German warships *Scharnhorst* and *Gneisenau* had passed through the English Channel from Brest to Kiel almost totally unbeknown to the Allies. British radars that should have detected the ships were jammed. J. Stanley Hey, the chief trouble-shooter for such matters, was asked to investigate the jamming. Within 2 weeks he was surprised by another severe jamming incident, this time of antiaircraft fire control radars. The expected bombing never did materialize, and Hey's inquiries revealed that the radar operators reported such jamming only in daytime and only when the antenna pointed to the sun. Jamming was most intense when a large sunspot group reached the central meridan of the solar disk on February 28. Hey made the correct association: The jamming was the result of solar radio noise. He reported his conclusion in a secret memo, and most of the scientific community remained ignorant of it for almost 4 years.

In the years after World War II, radio astronomy grew apace. The mysterious radio sky inspired Albrecht Unsold, a German astronomer, to remark poetically in 1946: "The old dream of wireless communication through space has now been realized in an entirely different manner than many had expected. The cosmos' short waves bring us neither the stock market nor jazz from distant worlds. With soft noises they rather tell the physicist of the endless love play between electrons and protons" (*In* W. T. Sullivan, III, "Preface," *Classics in Radio Astronomy*, 1982). Antennas sprouted in many countries, but the practitioners of radio astronomy were regarded as amateurs by the optical astronomers. Otto Struve still referred in 1949 to "astronomers" and "radiotechnicians" in a typically patronizing way. It is worth noting that in the last decade of his life, Struve became director of the National Radio Astronomy Observatory.

The skies seemed filled with patches of radio emission, but it was not possible to identify what they were. The radio view was like a badly out-of-focus optical picture. There were hundreds of visible objects in the neighborhood of each radio patch, but no way to decide which optical image matched up with a particular radio source. Radio resolution was just too poor.

The discovery of the first discrete radio source beyond the Sun was an accident. While conducting a survey of the broad distribution of galactic radiation, Hey and his co-workers found a region in Cygnus that was extremely bright, but they could not localize the source with the crude resolution of his telescope. The brightness fluctuated in a random fashion, however, on a time scale of seconds. Later it was recognized that the fluctuations were not intrinsic to the source itself but were a twinkling due to varying electrical patches in the high ionosphere. The phenomenon of radio scintillation was an analog of optical twinkling and indicated that the source was starlike. Cygnus A was soon recognized as the brightest discrete radio source in the sky.

The catalog of radio sources built up very slowly. By 1955, after 8 years of study, there were only eight sources detected that were reliable, but they suggested very interesting connections with spectacular optical sources. Among them were the Crab nebula, the exploded remains of a 1000-year-old supernova; Virgo A, a giant elliptical galaxy with an extended, skinny, knotted jet; Casseopeia A, another supernova remnant made of a ball-shaped mass of rapidly moving filaments; and Cygnus A, then thought to be a pair of colliding galaxies, 1 billion light-years away. These sources are now known to produce their radio intensity as a result of fantastically violent events, but the energy that ultimately arrives at Earth is only the faintest whisper. From the Crab, the total power over the Earth is barely 100 watts, enough to light an ordinary lightbulb. Quasars that radiate the power of hundreds of galaxies are billions of light-years distant and feed the Earth with barely 1 watt. After 2 decades of radioastronomy, scientists estimated that all the telescopes in the world had collected the energy equivalent of the impact of a single snowflake. For such small signals, radio astronomers have sought to build ever-larger telescopes and ultra-sensitive receivers.

Toward Bigger Dishes and Sharper Radio Images

The resolving power of a telescope—how much detail it can distinguish—is proportional to the ratio of its diameter to wavelength. Visible light has a wavelength of a few hundred-thousandths of an inch, which is microscopic when compared to the diameters of optical telescope mirrors. That is why they can reveal structures of 1 arc second or less in angular diameter, the equivalent of a small crater on the Moon or the stitches on a baseball several miles away.

Radio waves are about 1 million times longer than visible light waves. To match the resolution of an optical telescope, the radio telescope must be 1 million times larger. The earliest radio telescopes were hardly able to distinguish one constellation in the sky from another. The development of radio astronomy has been a history of continually increasing angular resolution by enlarging the dimensions of radio dishes and perfecting the surface to permit working at the highest frequencies. The great 1000-foot reflector at Arecibo, Puerto Rico, is the world's largest fully surfaced (filled) reflector, but it is built into a natural bowl and is not steerable. At 21 centimeters, the strongest radio-emission wavelength of the neutral hydrogen atom, it has an angular resolution of about 3 arc minutes; only one-half as good as your eyes. To compete with large optical telescopes working at the wavelength of green light, the radio astronomer would need a dish from 10 to 20 miles across. Making telescopes bigger and going to shorter wavelengths quickly becomes impractical, but there is a large agenda of observations that can best be made with large, fully steerable dishes.

The most prominent exponent of the big dish in the 1950s was the British astronomer, Sir Bernard Lovell. From the time he brought his ex-army radar equipment to Jodrell Bank in December 1945, Lovell was the leading promoter of ever-larger telescopes. For many years his great 250-foot Mark I telescope was the largest steerable dish in the world. But while he struggled in Britain to gain support for a still-larger telescope, the Mark V, scientists and engineers in Germany were developing an ingenious new design for a 100-meter (330-foot) telescope that could retain a paraboloidal shape accurate to 1 millimeter, for any position of elevation.

The German engineers recognized that a structure of that size would sag and droop out of tolerable shape as it was tilted, no matter how much steel bracing was added to increase its stiffness. At that time Sebastian von Hoerner, a German scientist working at the National Radio Astronomy Observatory in Green Bank, West Virginia, had demonstrated with computer models that it was possible to design a structure that could tolerate severe gravitational stresses and still remain a near-perfect paraboloid. His so-called homology principle created self-adjustments in the surface that allowed the focal length to change but preserved the figure.

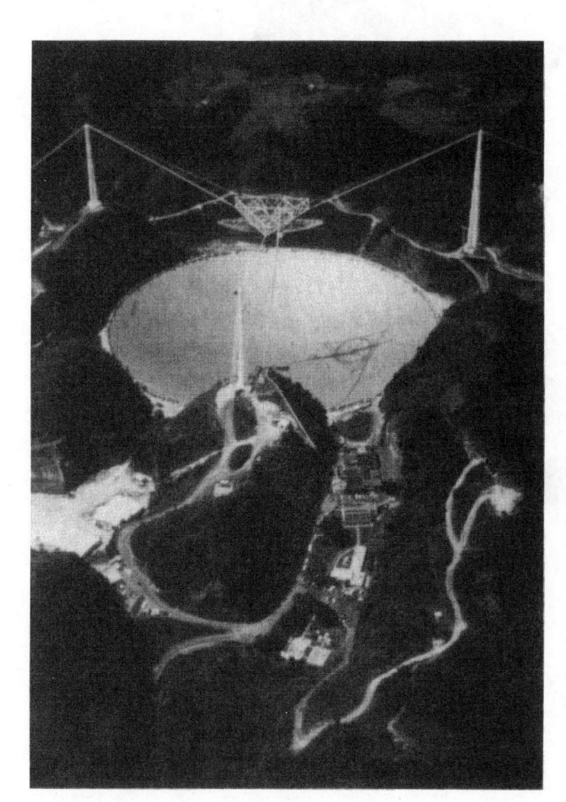

FIGURE *19. The Arecibo radio telescope dish in Puerto Rico is the largest focusing mirror in the world. Set in a natural karst sinkhole, it measures 1000 feet in diameter and is 167 feet deep. A 600-ton platform carrying various feed antennas is suspended 426 feet above the mirror surface. (National Astronomy and Ionosphere Center)*

By 1968 the Germans were committed to building the 100-meter telescope on the homology principle at Effelsberg, 40 kilometers southwest of Bonn. When completed in 1970, its surface matched a true paraboloid to 0.22 millimeter in all positions, even though deflections of 10 centimeters occurred in the surface. The success of the Bonn dish persuaded Lovell not to compete any further for construction of the Mark V, and British astronomers turned to the design of interferometers.

Splitting and Recombining the Radio Image

The success of radio interferometers has been phenomenal; they now reveal details 1000 times finer than the best that optical telescopes can achieve. In principle a radio telescope could be sawed in half and the sections moved far apart. If the signals could be properly combined, the split telescope would have the resolution of a single dish as large in diameter as the separation of the parts. The simplest radio interferometer consists of a pair of antennas spaced a considerable distance apart and connected to a receiving system. After amplification and filtering, the signals from the two antennas are combined in a correlator. As the two antennas track the target, the signals form a fringe pattern that contains information about the position and structure of the radio source being studied. Measurements must be taken at many position angles if the shape of the radio source is to be determined.

The concept of radio interferometry was pioneered and developed most

FIGURE *20.* Martin Ryle (1918–1986). *Martin Ryle's career was distinguished by his development of radio interferometry to a remarkably fine resolution and ultimately to detailed imaging capability by means of aperture synthesis. Using his first modest two-element interferometer in 1948, Ryle showed that the very distant, powerful radio source in Cygnus had a very small angular extent. With a team of students he soon built a more sensitive interferometer and located some 50 discrete radio "stars" that were not obviously coincident with known optical objects.*

While most radio astronomers were intent on building larger dishes, Ryle continued to exploit radio interferometry with comparatively modest resources. He constructed a four-element interferometer in 1955 of parabolic trough reflectors made up of 36 sections that required manual adjustments by three persons—pushing, pulling, and locking into position—to steer it in declination. With this simply engineered instrument, the impressive 3C (3rd Cambridge) survey was carried out. In 1971 Ryle completed his most powerful telescope, a 5-kilometer interferometer that used eight parabolic dishes, of which four were movable on rail mounts. Its resolution was 0.6 arc second, and a wealth of fine detail on the structure of radio galaxies was revealed. In 1974 he shared the Nobel Prize with Antony Hewish, who was recognized for his discovery of pulsars.

effectively by Martin Ryle in England. He realized that if he had a super telescope with an aperture as large as 1 mile across and fully steerable, as the Earth rotated he could orient the telescope to observe a radio source from all angles until he finally derived a detailed map of the structure. Since a 1-mile-wide single radio dish is, of course, impossible to build, Ryle's idea for aperture synthesis was to use two smaller telescopes to make large numbers of observations over many days while they were being moved in all possible positions over an area 1 mile in diameter. As the telescopes are jockeyed about, always pointing to the same radio object, they gradually accumulate the signals that would be received by one telescope of aperture equal to the area of operation.

Early on Ryle received a grant to produce a crude cylindrical paraboloidal dish formed of wires. By placing a smaller instrument in different positions relative to that one, he demonstrated that a series of combined observations could simulate the performance of a telescope of much larger aperture. Next he received funding for two 52-foot-diameter paraboloids placed in fixed positions, and for a third dish moveable on a railway track. He predicted that he would obtain a resolution some 200 times better than had ever before been achieved. The instrument that was finally completed became known as the "One Mile Telescope." In its final form, two telescopes were placed 5000 feet apart on the east–west line, and the third was pulled by diesel tractor along a railway track between them. Of course the telescopes in line simulated a narrow synthetic aperture, but for every 24-hour rotation of the Earth, the axes of the telescopes made a complete revolution. Thus, every half-day an oval ring was synthesized; to fill in the full aperture the intermediate telescope was set to a different position each day for 64 days.

Three years after the start of construction in 1962, Ryle was able to demonstrate that his predictions were correct. His three-element interferometer was a brilliant success. In 1974 Ryle received the Nobel Prize for his achievement.

Today the most successful interferometer is the Very Large Array (VLA) in the New Mexico desert, perhaps the only expanse of flat real estate in the United States able to accommodate it. A few-dozen miles west of the town of Socorro, one comes upon a surreal sight: 27 dishlike antennas lined up for miles over the flat desert plains of San Augustin. Airline pilots who fly over the VLA have dubbed it "the mushroom patch." For the first time, radio astronomers can obtain high angular resolution and image detail comparable to the best that can be obtained with optical telescopes. When the telescopes are set in the largest configuration, the resolution is better than 1 arc second. In the more compact arrangements, the telescope is used like a wide-angle lens to observe more extended objects, such as nebulae, without loss of sensitivity.

The antennas of the VLA rapidly fill in the aperture to simulate one big antenna. Looking down on the VLA from high in space, each dish is

FIGURE *21. The Very Large Array (VLA) is composed of 27 identical radio dishes, each 82 feet in diameter. Together they simulate the resolution of a telescope 20 miles in diameter. Each telescope weighs 215 tons and is towed on railroad tracks laid in the form of a "Y" over a dried-up lake bed on the plains of San Augustin, New Mexico (see* insert). *High-resolution radio images of strong sources are generated in a matter of minutes, compared to as much as 10 hours by earlier radio interferometers that used only a few dishes. (National Radio Astronomy Observatory.)*

one of many pairs. The 27 dishes make 351 pairs [$(27 \times 26) / 2$]. As time passes and the VLA rotates with the Earth, the baselines turn and fill in more and more of the area inside the boundaries of the array. Within a few hours, thousands of different baseline orientations synthesize an antenna with the collection area of a 430-foot-diameter dish and with the resolution of a telescope 21 miles across.

Advances in computer technology are what make the VLA work. Every second the central computer must receive 1 million bits of data from the network along a 40-mile-long system of wave guides buried beneath the desert plane. During a typical 12-hour observation period, three computer tapes record enough digital information to fill 1 dozen big-city telephone books.

Telescopes as Large as the Earth

The angle subtended by the nuclei of radio galaxies is less than one-thousandth of 1 arc second, from 100 to 1000 times beyond the resolution limit of the VLA. To produce radio images on a scale of 10^{-3} arc second requires dimensions comparable to the size of the Earth. But when the separations are made much larger than those of the VLA, the interconnection of antennas by transmission lines becomes too costly; there are also the practical problems of how to carry those lines across rivers and hills.

Interconnection with radio links has no limit in principle, but it requires the installation of repeaters every 50 kilometers or so; the cost escalates prohibitively if the array is to spread across a full continent. Satellite repeaters could be used, but to operate a broad bandwidth array would absorb the full capacity of a modern communications satellite. The problem is solved by dispensing with the real-time connection between interferometer elements.

In the late 1960s groups in the United States and Canada began experimenting with signals recorded simultaneously on magnetic tape at widely separated telescopes. The tapes were then sent to a central computer, where they were combined. It required an extremely accurate atomic clock stationed at each antenna to guarantee that the recordings would be synchronized to within 1 microsecond. Such clocks now lose less than 1 second in millions of years. Any less-precise match would make it impossible to derive a useful picture. This approach is the basis of very long baseline interferometry (VLBI).

Kenneth Kellermann of the National Radio Astronomy Observatory at Green Bank, West Virginia, tells of overcoming technical and bureaucratic obstacles to establishing a link early on with the Crimea Astrophysical Observatory on the shores of the Black Sea. Due to heroic efforts on the part of the scientists, they succeeded despite all difficulties. En route to the Crimea, the battery of the atomic clock brought from the United

States failed, and the clock stopped. Swedish radio astronomers quickly responded by flying a clock from Stockholm to Leningrad, where Kellermann retrieved it from suspicious border officials and raced away by plane and car to the Crimea, as the battery grew steadily weaker. At all times Kellermann had to watch for any opportunity to plug the clock into a power outlet and rejuvenate it for the next lap of the journey. It was a replay of the *Perils of Pauline*, but Kellermann won the race and he and his Soviet colleagues obtained their data. Carrying out their experiment involved telephone coupling of the Crimea Astrophysical Observatory with Green Bank via Moscow, entailing the problems of language barriers as well as the generally poor service of the Soviet telephone system. Lastly, the tapes had to be shepherded past custom officials in both countries who were naturally suspicious of anyone carrying tapes claimed to be very precious but containing nothing but radio noise.

Many measurements have since been made utilizing borrowed time on perhaps a half-dozen dishes scattered over North America and Europe. The very wide spacing simulates an antenna that is almost global in size. But coordinating different telescopes (in one attempt as many as 18) can become very complicated, and processing the trillions of bits of data that need to be correlated can take weeks.

The VLA, with its well-automated setup, produces images in about 1 hour and is best used to map the overall structure of a radio galaxy. VLBI is much slower but can look further and probe the finer details inside the cores of radio galaxies and quasars, regions no more than a few light-years across. The resolution is comparable to being able to read a newspaper halfway across the country. Astronomers in the United States have constructed a Very Long Baseline Array (VLBA) of 10 radio dishes spread across the entire width of the continent. The VLBA is a dedicated network based on 25-meter telescopes like those of the VLA at sites in Hawaii, Washington, California, Arizona, New Mexico, Iowa, Texas, Massachusetts, and Puerto Rico. Every movement of every dish is controlled from a master computer in Socorro. The aim is to make this system work at wavelengths as short as 7 millimeters because the shorter the wavelength, the deeper one can probe into the nuclei of galaxies. Eventually the designers hope to get down to 3 millimeters.

When the Earth Is Too Small

After ground-based antennas have been combined across the ends of the Earth, more resolving power can be had only by moving VLBI into space. A test demonstration was carried out successfully a few years ago by deploying an antenna on the space shuttle and combining its signals with those picked up on NASA tracking antennas on the ground. The Japanese have launched a deployable 8-meter antenna in an elliptical orbit extending up to 15,000 kilometer above the Earth, and radio astronomers around the world will couple a network of telescopes to it interferometrically. The antenna unfurled in space employs a novel tension truss design with several dozen booms and cables for structural rigidity. The VLBI Space Observatory Program (VSOP) should improve resolution over ground-based VLBI by at least a factor of 5, to about 60-millionths of 1 arc second. (A resolution of 1-millionth of 1 arc second would be needed to probe the centers of active galaxies). As the orbiting telescope circles the globe, it will act as a multitude of antennas positioned along the orbit and will thereby produce a very detailed synthesis of the image.

Radio astronomers dream of an ASTRO–ARRAY several decades in the future, with all its antennas in space. Thirty radio antennas would parade around the Earth in 60,000-mile orbits, about one-fourth the distance to the Moon. All 30 dishes would combine their information and mimic a single telescope with a baseline of about 120,000 miles. As fantastic as such an array would appear to be, the technologies required are in principle natural extensions of present-day capabilities.

If the description of radio telescopes in deep space seems like a distant future vision, it is modest compared to projections for radio astronomy on the surface of the Moon. Ideas of manned bases on the Moon are inspired largely by those who see exploration by astronauts as the form of space activity most appealing to the general public and to politicians, who view it as an arena of prestige competition between the big powers. But radio astronomers would be happy to grasp any opportunity to establish lunar observatories by astronaut construction crews for purely scientific purposes. The observatories would be operated unmanned, except for occasional maintenance.

Whereas atmospheric absorption by oxygen and water vapor sets limits on any ground-based observations of millimeter and submillimeter waves, the vacuum at the Moon eliminates all absorption obstacles. On the back side of the Moon a radio telescope would be shielded from all the radio noise that floods the terrestrial environment. There are many features of the Moon's terrain that could be advantageous. Large telescopes could be emplaced in craters near the lunar poles, where there would be little need for sun shades to reduce thermal distortion. A lunar crater could support a dish like the Arecibo telescope. It could be built in large segments and its shape controlled by computer. Material for the surface of the telescope, in the form of plastic-reinforced carbon fibers, could be mined and manufactured on the Moon. Proponents of these ideas believe that it would be possible to achieve a surface accuracy of a few microns with active optics and to attain a resolution of better than 0.1 arc second.

For the present, astronomy on the moon is more a fantasy than a realistic prospect. Enthusiasts for settlement of the Moon are not deterred, but their sights must be set well beyond the turn of the century.

The Cold Infrared Universe

Infrared radiation was discovered and named by Sir William Herschel in 1800. He placed a thermometer in the different colors of the spectrum of sunlight and measured the rise in temperature. To his surprise, beyond the red limit of the visible spectrum, the thermometer continued to receive heat; thus infrared rays were discovered. Thomas Alva Edison's interest in astronomy was evident even before his aborted attempt to observe solar radio noise that we described earlier. He invented a sensitive infrared detector and set it up in a chicken coop in Wyoming to observe the solar eclipse of 1878. As darkness came on, the chickens came in and so upset his plans that totality passed without his gaining more than a fleeting glimpse.

Today astronomers study the infrared spectrum all the way from the visible red to radio microwaves: from 7000 angstroms (0.7 micron) to 1 millimeter. As compared to the range of the visible spectrum, the infrared covers a one-thousandfold spread of wavelengths and is marked by highly variable spectral transparency of the atmosphere. The "smog" of atmospheric water vapor and carbon dioxide is the bane of infrared astronomers, but there are moderately transparent "windows" at various wavelengths through which to observe.

There is no fundamental difference between visible and infrared astronomy. It is only that the eye is blind beyond 0.7 microns. All objects, at whatever temperature, radiate: Hot stars radiate in the visible part of the spectrum, cool bodies radiate in the infrared. People radiate an average of 50 watts of heat (long-wave infrared), but we see each other only by reflected visible light, as we do the planets. In the photographic infrared, from 0.7 to about 1.1 microns, the sky looks much the same as in the visible spectrum, but the brightest stars are comparatively dim.

Why the special interest in the infrared? Interstellar space contains a fine soot, about 2 dust grains per million cubic meters. That might not seem like much, but visible light traveling from the nucleus of the Galaxy to the Sun (30,000 light-years) is weakened by a factor of 1 trillion, and the central region of the Galaxy is completely hidden from our view. At longer infrared wavelengths, the radiation can penetrate the entire distance from the center of the Galaxy to the Sun. Even at the relatively short infrared wavelength of 2.2 microns, the dimming is only 16 times.

The Orion nebula is a nursery of newly born stars that are wrapped in cocoons of dust and gas impenetrable to visible light but readily observable in the infrared. The infrared is especially sensitive to the most-distant galaxies and quasars because their intrinsically brilliant optical and ultraviolet light is Doppler-shifted into the infrared. Improved visibility is also an advantage of the infrared because the atmospheric distortion of starlight decreases at longer wavelengths.

Most of the frontier research now being carried out in infrared astronomy cannot be done photographically and owes its success to new technologies in infrared sensing. Bolometers were among the earliest detectors used in the near infrared, and modern bolometers are 1 trillion times as sensitive as common thermometers. In essence, a *bolometer* is a piece of material whose electrical resistance varies with its temperature. Because it requires very low temperature for operation, it is placed in a vacuum bottle (a cryogenic Dewar flask) filled with liquefied helium. Starlight enters through a small window in the side of the Dewar flask and raises the temperature of the bolometer. The star signal must be distinguished from the sky background and from the radiation of the structural assembly of the telescope. To determine the background contribution, the telescope is pointed slightly off the star and again measures the signal, which now represents empty sky, the assembly surrounding the detector, and anything else that contributed pollution to the first observation. After the second measurement is subtracted from the first, the remaining signal is a measure of the star itself. Unfortunately it isn't as simple as it sounds. Every speck of dust that passes in front of the telescope radiates in the infrared. Insects and moths, for example, can contribute very strong spurious signals. In an infrared mission aboard the space shuttle, a small stray scrap of Mylar plastic film drifted into the field of the telescope and overwhelmed the detectors with its infrared emission.

Modern infrared detector and imaging technology was until recently a proprietary province of the military, but has now been largely declassified; this has been an enormous boon to astronomy. Semiconductor microchips the size of a fingernail are constructed from various substances that give unique wavelength sensitivities (see Fig. 22). Arrays of chips have been fashioned into imaging formats as large as 64-by-64 pixels. Sensitivity to the far infrared is provided by crystals of pure germanium infused with a trace of copper, mercury, or gallium. Each dopant gives a characteristic wavelength signature. Cooling to liquid helium temperature greatly reduces the thermal background and yields a tremendous increase in the signal-to-noise ratio.

The recent, very successful *Infra-Red Astronomical Satellite (IRAS)* mission was prepared by a United States–Dutch–British consortium. The 2249-pound *IRAS*, which scanned the infrared sky for 10 months in 1983, identified as many as 250,000 celestial sources. On each of two daily passes over a ground station in Chilton, England, the satellite dumped 350 million bits of data. Gerry Neugebauer, one of the principal scientists for the telescope, estimated that if he could throw a baseball high enough above New York, its infrared emission could be detected by *IRAS* in California.

When the *IRAS* mirror focused on the young star Vega, 27 light-years from Earth, it detected what appeared to be a protoplanetary system extending outward about 15 million miles from the star, in the form of a ring of "pebbly" particles of miscellaneous sizes. Altogether *IRAS* found

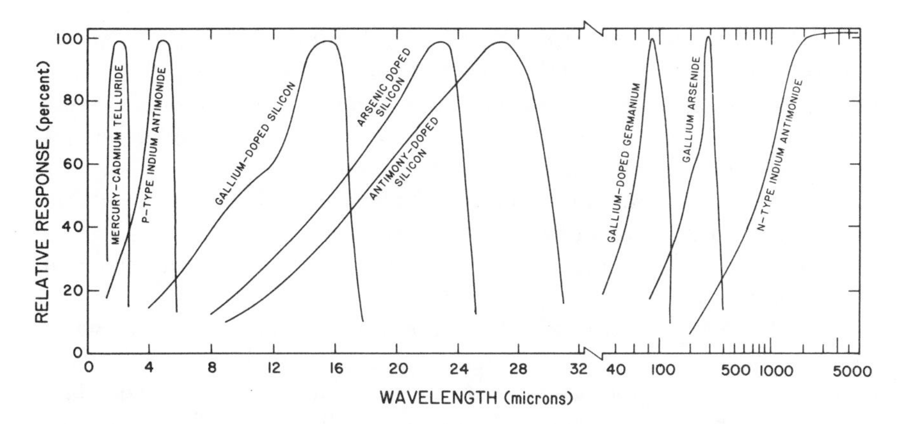

FIGURE 22. *The materials used in infrared detector arrays. Silicon that has been "doped" with various impurities responds to near and mid-infrared. Doped germanium covers the submillimeter wavelengths.*

FIGURE 23. *The* Infrared Astronomy Satellite (IRAS). *To shield the detectors from the radiation environment of the satellite, the entire telescope was refrigerated at close to absolute zero temperature in a bulky dewar of liquid helium. A gold-plated sunshade kept the heat of the Sun off the detectors. (Jet Propulsion Laboratory.)*

some 40 stars within 75 light-years that exhibited rings of protoplanetary material similar to Vega. The satellite telescope also observed infrared tails on comets and picked up signals from about 20,000 asteroids. Among other surprises were "infrared cirrus" forms, stringy clouds of interstellar dust straddling the galactic plane, and miscellaneous objects characterized as "wisps, fingers, knots, and globs." Starburst galaxies emerged that were one or more orders of magnitude brighter in infrared wavelengths than in the optical domain. These *IRAS* galaxies are more numerous than quasars and have similar energy outputs.

One of the highest priorities for future missions of the United States astronomy community is the large deployable reflector (LDR) being designed for work in the far-infrared and submillimeter regions. The giant telescope, from 20 to 30 meters in diameter, will be built of 96 structural sections and a mosaic of 49 glass mirror sections. To transport all the parts of the telescope to orbit would require five space shuttle flights and 100 days of assembly at a space station.

The LDR will increase the light-gathering power by between 400 and 900 times compared to most 1-meter class telescopes that are now in operation. It would also be an advanced instrument for the study of the far-infrared parts of the spectrum that contain some of the most powerful emission features of the celestial sky. Among its anticipated capabilities is the study of discontinuities in the background radiation that are predicted by new models of an "inflationary universe" (this will be discussed in Chapter 7). These discontinuities occur along narrow strings or along domain walls that stretch for enormous distances across the universe. Other missions for the LDR would include the study of the dark matter that some scientists believe may account for a large component of invisible mass in our galaxy and in extragalactic systems. Quasars and galaxies formed at the earliest epochs would appear brightest in the infrared sensitivity range of the LDR.

CHAPTER 3

The Violent Universe of High-Energy Astronomy

To a classical astronomer prior to 1945, the prospect that x-rays would provide many of the most dramatic developments in astronomy would have seemed very remote indeed. The surprising x-ray discoveries of the 1960s were startling even to the most optimistic pioneers of rocket astronomy. Since x-rays are a highly energetic form of electromagnetic radiation, we might naturally expect them to come from violent processes in the environments of superstars, collapsed stars, and black holes. But the discovery of an invisible "hot" universe of x-ray sources often radiating at millions of times the power of their visible radiation was totally unpredicted. Our sun, the only known x-ray star prior to 1962, is normally 1 million times weaker in x-rays than in visible light. Unlike the permanent stable sky of visible stars, the x-ray sky flickered, flashed, and pulsed with great variability on all time scales from microseconds to years. Temperatures from tens to hundreds of millions of degrees were indicated in whirling accretions of particles funneling onto neutron stars and black holes, and in the vast pools of gas that fill the interspaces of great clusters of galaxies.

The discovery of x-rays in the laboratory by German physicist Wilhelm Roentgen has just passed its centennial anniversary. Appropriately, a great x-ray observatory named *ROSAT*, for *Roentgen Satellite*, was launched close to the anniversary date.

Roentgen turned his interest to cathode-ray tube experimentation in 1894 and 1 year later discovered x-rays. For several decades cathode rays had been a popular subject of study with the Crookes tube, a simple arrangement of two electrodes in an evacuated glass envelope to which a high voltage was applied. It was the primitive antecedent of modern cathode-ray and television tubes. Electrons were accelerated from the negative plate to the positive plate, causing the glass walls to fluoresce wherever struck by the electrons. When Roentgen began his experiments it had already been noted that a screen painted with barium platinocyanide fluoresced if placed close to a Crookes tube that was equipped with a thin aluminum window, but the phenomenon was not observed through the heavier walls of glass tubes. Roentgen suspected that the external fluorescence could be detected without benefit of a thin window if care were taken to mask the strong glow of the glass wall itself. The following is an account of Roentgen's experiment (Otto Glasser, *Dr. W. C. Roentgen*, 1958, p. 35):

> Selecting a pear-shaped tube from the rack, he covered it with pieces of black cardboard, carefully cut and pasted together to make a jacket . . . and then hooked the tube onto the electrodes of the Ruhmkorff coil. After darkening the room in order to test the opacity of the black paper cover, he started the induction coil and passed a high tension discharge through the tube. To his satisfaction no light penetrated the cardboard cover.
>
> He was prepared to interrupt the current to set up the screen for the crucial experiment when suddenly, about a yard from the tube, he saw a weak light that shimmered on a little bench he knew was located nearby. It was as though a ray of light or a faint spark from the induction coil had been reflected by a mirror. Not believing this possible, he passed another series of discharges through the tube and again the same fluorescence appeared, this time looking like faint green clouds moving in unison with the fluctuating discharges of the coil. Excited, Roentgen lit a match, and to his great surprise discovered that the source of the mysterious light was the little barium platinocyanide screen lying on the bench.

The mysterious rays readily passed through a book that he interposed. Since it was well known that cathode rays did not penetrate the walls of a Crookes tube, let alone an external obstacle. Roentgen realized that he had discovered a new form of highly penetrating radiation and named them *x-rays* to emphasize their mysterious nature.

Roentgen was a superb investigator. He determined that x-rays were produced where the cathode rays struck the wall or the target of the tube. The rays were not deflected by a magnet and were attenuated more strongly when passing through denser materials. A thick sheet of lead was opaque, but a block of wood was relatively transparent. He failed to focus the x-rays with a lens but proved by means of pinhole pictures that the rays travel in straight lines. When a heavy metal target such as platinum was used as the anode, the x-ray intensity was much stronger than with an aluminum target. In these simple diagnostic tests, Roentgen identified many of the essential properties of x-rays.

A few days before Christmas, 1895, Roentgen aimed his x-rays across his wife's outspread hand and onto a photographic plate. The rays passed through her flesh but were blocked by the bones. His x-ray photographs of the bony skeleton created both a scientific and a public sensation. For the medical profession, x-ray radiography clearly held great promise. In the popular press, lurid stories appeared of voyeurs who used x-rays to see through clothing; a fiction akin to Superman's x-ray vision. Laws were passed in the United States to forbid the use of x-rays in opera glasses.

Although the first Nobel Prize was awarded to Roentgen not many years after his discovery, physicists remained baffled by the mysterious radiation. All attempts to observe interference effects typical of electromagnetic waves failed. The answer came 17 years after Roentgen's discovery, when German physicist Max von Laue discovered that x-rays are diffracted by the regularly arranged planes of atoms in a crystal. In effect, the wavelength of x-rays is comparable to the spacing of atoms in a crystal lattice and is at least 1000 times shorter than visible light.

After hearing Roentgen's report of his cathode-ray experiments to the French Academy of Sciences, Henri Becquerel speculated that fluorescent material itself could be a source of x-rays. Becquerel specialized in the study of absorption of light in crystals. At the time of Roentgen's exhibit of the x-ray photograph of his wife's hand, Becquerel had been working with a highly phosphorescent salt of potassium-uranium sulphate. Recalling that Roentgen's rays seemed to emerge from a greenish fluorescent spot on the glass wall of his Crooke's tube, Becquerel guessed that the fluorescent material was the source of the x-rays.

Becquerel's uranium salt crystals became phosphorescent when exposed to sunlight. He covered his luminous crystals with metal sheets inside a wrapping of black paper and placed them next to a heavily packaged photographic plate in the hope that a new kind of ray, like Roentgen's x-ray, would escape the phosphorescent salt, penetrate the metal sheets and paper, and blacken the photographic plate. Becquerel's suspicion seemed fully vindicated, for when he developed the plate, it was black. On 24 February 1896 he informed the French Academy that he had reproduced Roentgen rays without benefit of an electrical discharge, merely by allowing his salts to absorb sunlight and then fluoresce invisible light as well as visible.

A few days later Becquerel set out to repeat his experiment with boxes of uranium salt crystals, but the weather was gloomy and continuously overcast, and he could not expose the salts to sunlight to excite phosphorescence. He lay the boxes in a drawer that happened to contain a package of photographic plates and waited a few days for the Sun to shine forth again. Here Becquerel demonstrated his scientific objectivity. Even though it went against the theory he had expounded to his fellow academicians only the week before, he decided to develop the photographic plates before exposing them to phosphorescent crystals, on the outside chance that they might have been affected even though the crystals were not luminous. To his astonishment, he found clear images of the crystals on the photographic plates that had lain underneath them. Even more penetrating than x-rays, the uranium radiation blackened film after passing through several sheets of metal. Becquerel had discovered the gamma rays of radioactivity, which are 1000 times shorter in wavelength than Roentgen's x-rays.

Mapping the X-Ray Sky from Rockets and Satellites

The physics of x-rays and gamma rays and their practical applications were studied intensively throughout the first half of this century, and many Nobel prizes were awarded for seminal discoveries. But there was no hint of the enormous portent for the future of astronomy. Astronomers remained oblivious to the potential of x-ray astronomy, and physicists concerned with the behavior of the electrified upper atmosphere, the ionosphere, puzzled over what form of solar radiation could be responsible for the ionization. Although the ionosphere waxed and waned from dawn to nightfall with the overhead passage of the Sun, there was no plausible connection with visible light or even with near-ultraviolet radiation. A suspicion was expressed that solar x-rays could be the source, but since it has a temperature of 6000°K, the solar disk cannot produce x-rays. The first clues came from studies of the very faint extended solar corona. The production of the highly ionized plasma of the corona requires temperatures of 1 million degrees or more; such superheated gas would generate x-rays. When rocket astronomy burst on the scene after World War II, the early discovery of solar x-rays (see Chapter 3) opened the way to the high-energy universe of galactic and extragalactic x-ray astronomy.

The 1950s were marked by an intensive effort to study solar x-rays and ultraviolet radiation. Successes in solar observations inspired thoughts of searching beyond the Sun into the far reaches of galactic space. A breakthrough came in 1962 when a small Aerobee rocket instrumented with Geiger counters rose above the atmosphere and discovered x-rays from beyond the Solar System. The early discoveries in x-ray astronomy were made with comparatively small Geiger counters and proportional counters (window areas of from 10 to 1000 square centimeters) that operated without the benefit of telescopes to focus the view of the sources. The detectors faced the open sky behind mechanical sights in the form of slats or of "egg crate" honeycomb structures that shuttered the field of view to a few degrees. They mapped the sky by allowing the spin of the rocket to sweep the collimator across the stars. As pointing controls were developed for the rockets, it became possible to concentrate the observations on individual sources while the collimator reduced the background signal. With these simple techniques, the pioneers of x-ray astronomy discovered intense discrete x-ray sources in the galaxy that are thousands to millions of times as powerful in their x-ray emission as ordinary stars are over the entire visible spectrum. Seyfert galaxies and quasars with intensely active nuclei were found to be even brighter in x-rays. The pulsar in the Crab nebula is a far more powerful source of x-rays than radio waves. Its x-ray emission is 10,000 times as strong as the radio emission.

The first x-ray astronomy satellite was launched from Kenya on 12 December 1970, the anniversary of that country's independence. The new observatory was named *Uhuru*, Swahili for *freedom*, in honor of the occasion. *Uhuru* was the first of the series of small NASA satellites named *Explorers*. It weighed 150 pounds. The satellite spun at a slow 12 minutes per revolution and could be made to scan the entire sky. By the end of 1972 *Uhuru* had built a catalog of almost 200 sources. Now x-ray stars could be monitored for extended periods instead of for the few minutes of a rocket flight. Several sources were found to be neutron stars in close orbits about giant blue stars, drawing streams of superheated gas into swirling disks that reached temperatures in the tens of millions of degrees. One object, Cygnus X-1, gave strong evidence of being a black hole. Beyond the Milky Way, great clouds of superheated gas filled the spaces within clusters of galaxies.

Uhuru was followed by a series of much larger satellites, the *High Energy Astronomical Observatories (HEAO)*. These were carried into orbit by giant Atlas-Centaur rockets 131 feet tall, and weighing as much as 165 tons. The *HEAO-1* spacecraft itself, the largest and heaviest before the era of space shuttle launches, was 19 feet long, about the length of a small truck, and weighed 3.5 tons. The detectors were banks of proportional counters that combined to about 50 times the sensitivity of the *Uhuru* sensors. Months of technical problems preceded the launch of *HEAO-1*, and the last few hours were filled with special frustrations. As launch time approached on the night of 11 August 1977, an electrical storm moved into the Cape Kennedy area, delaying the launch for 1 hour. Then the spacecraft computer failed, and the program had to be loaded through Goddard Space Flight Center on an emergency basis. While that went on, several fishing boats entered the rocket impact area. Attempts to reach them by radio and request that they clear the area failed. Helicopters flew out to drop messages on the decks, but the crews had gone below because of the storm. The launch was delayed another hour. Finally the weather cleared

FIGURE *24.* HEAO-1, *launched in 1977, was the largest and heaviest spacecraft of the pre–space shuttle era. Banks of proportional counters added up to a total sensitive area of about 3 square meters, roughly 50 times as great as the sensors of its predecessor, Uhuru. (NASA.)*

and the fishing boats left the area. At last the rocket was launched and everything functioned well.

As the first x-ray observatory with high-resolution imaging capability, *HEAO-2* was a spectacular success. Small x-ray telescopes for solar imaging had been used briefly on some early rocket flights and on the *Skylab* mission. But the step to the large *HEAO-2* telescope mirror was a superb technological feat that brought x-ray imaging of galactic sources to a level of sophistication comparable to optical imaging. Not only was the mirror technology unique, it also required a new class of imaging x-ray detectors.

When x-rays encounter a mirror at a large angle, they are not reflected but enter into the surface to be absorbed in much the same way that a stream of bullets penetrates a wooden wall. However, bullets hitting a wall at a glancing angle ricochet, and x-rays that impinge on polished-glass mirror surfaces at grazing angles also reflect with high efficiency.

FIGURE 25. *Map of the x-ray sky in galactic coordinates obtained with large area proportional counters aboard* HEAO-1. *Larger dots indicate stronger sources. The collection of sources includes active galactic nuclei, supernovae remnants, x-ray binaries, clusters of galaxies, stellar coronae, and a majority of objects that are still unidentified. (U.S. Naval Research Laboratory.)*

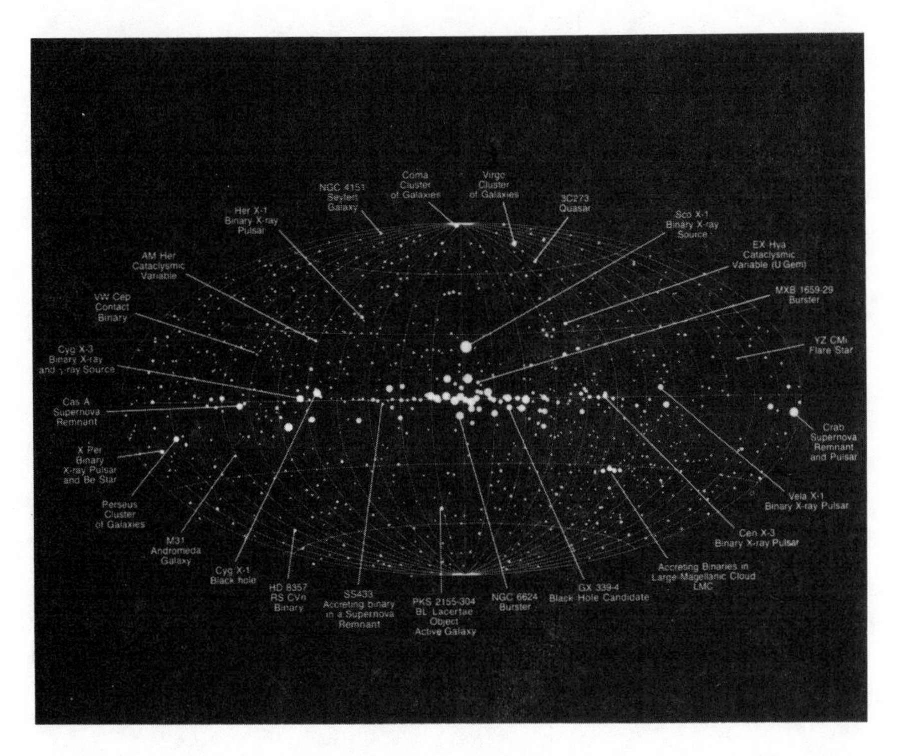

German physicist Hans Wolter investigated x-ray mirrors in an effort to construct an x-ray microscope in the 1950s. The basic mirror design was substantially adapted to x-ray telescopes.

The x-ray telescope bears little resemblance to optical telescopes. Overall, it looks like a long, slightly tapered cylinder. X-rays entering the telescope along the axial direction strike the inner surface at a glancing angle and deflect slightly toward the axis. Far down the axis, well beyond the cylinder itself, the deflected x-rays reach the focal plane. The Wolter telescope can produce a sharp image, free of distortion. However, it has a major deficiency: The grazing incidence leaves only a thin ring of x-rays near the wall of the cylinder to be collected. Most of the incident x-rays must be blocked by a central aperture stop to prevent their passing straight down the tube without being reflected, thus washing out the focused image.

The x-ray telescope flown aboard NASAs *Einstein Observatory (HEAO-2)* in 1979 had a diameter of 0.6 meters, and its spatial resolution was better than 3 arc seconds. The glass mirrors were first diamond-ground to their approximate shape and then polished, coated with chrome and

FIGURE *26(a). The* Einstein Observatory *undergoing final integration before launch. The dimensions accomodate the 3.44-meter focal length and 58-centimeter aperture of the nested mirrors. (NASA.)*

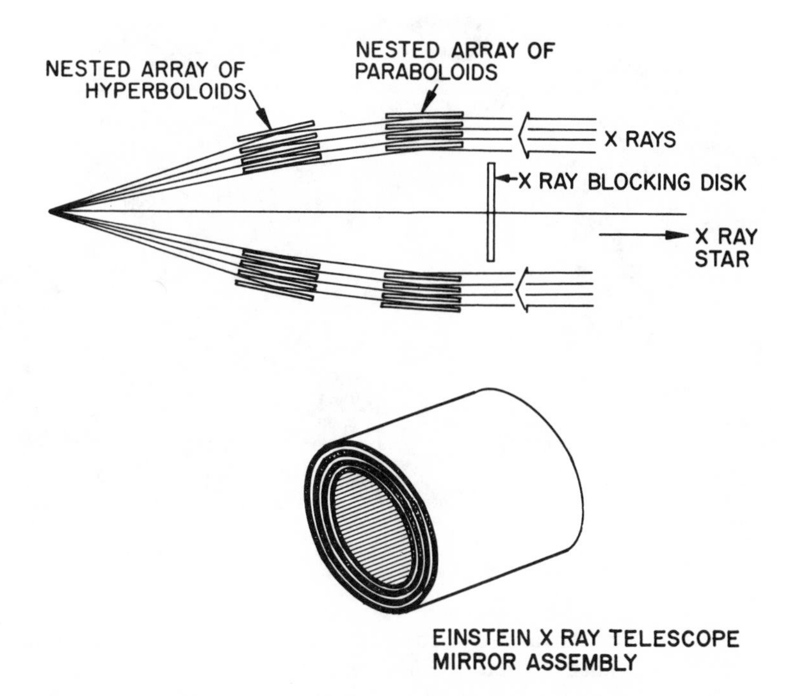

FIGURE *26(b). The* Einstein Observatory (HEAO-1), *which operated in orbit from 1978–1981, employed successive reflections, first from paraboloidal and next from hyperboloidal mirror sections, to a common focus. The double reflection eliminates image aberrations. Because the x-rays can be reflected only at a very shallow angle, the overall telescope is shaped like a slightly tapered cone. Four concentric sets of mirrors with the same focus were nested to increase the total reflecting surface.*

nickel, aligned, and bonded to a support structure. The surfaces had to be polished internally, which is far more difficult to do than figuring the typical, nearly flat optical mirror. Nor was it possible to use conventional methods to measure the smoothness of the surface. No rough spots greater than about one–ten-millionth of 1 inch could be tolerated. A rougher surface would scatter radiation and smear the image. It was even difficult to measure the roundness of the mirrors because they were very flexible and deformed under their own weight. The final measurement of roundness was made while the mirror floated in a vat of mercury.

Although small x-ray telescopes that were returned by parachute from Aerobee rockets or brought back by astronauts from the *Skylab* space station had produced images of good quality on photographic film, the unmanned *HEAO-2* required a fully electronic system of image transmission by radio telemetry. Two imaging detectors were developed for the

FIGURE 26 *(c).* The Advanced X-ray Astronomy Facility (AXAF), *under construction and being tested at the Marshall Space Flight Center, is scheduled for launch before the end of 1998. It is expected to be 100 times as sensitive as the* Einstein Observatory *and will reach 10 times as far into the cosmos. In the photograph we see only the outermost of four concentric mirror assemblies.*

Einstein Observatory mission: the Imaging Proportional Counter (IPC) and the High Resolution Imager (HRI). The former was designed for modest resolution (arc minutes) and wide field of view; the latter has to match the from 1– to 2–arc second resolution of the telescope. This was a challenge well beyond the state of the art when its development was undertaken.

The combination of mirror and IPC was highly sensitive; sources thousands of times fainter than the weakest in the *Uhuru* catalog were easily reached, but its imaging capability was poorly matched to the high resolution of the telescope. At the cost of some fivefold sacrifice in sensitivity, the HRI was developed to image with the full resolution of the mirror, some 30 times sharper than the IPC pictures. No traditional gas-discharge detector could satisfy this requirement. The HRI borrowed from electronic imaging technology developed earlier for the ultraviolet, the microchannel plate, which was a collection of millions of closely packed fine glass capillary tubes. Each tube acts as an x-ray converter to electric current, and its outgoing signal is 1 pixel of the full x-ray image.

The process of HRI development was one of endless trial and tribulation. valuable assistance came from british scientists Kenton Evans and Ken Pounds, who were involved early on with the development of x-ray microchannel plates. They worked closely with Riccardo Giacconi's team at the Smithsonian Astrophysical Observatory as time became very tight and it appeared that the HRI might impose a serious delay of the launch date. When a prototype was completed, it of course had to be flight-tested on a rocket. What followed was an incredible series of rocket failures for which the HRI was not at all to blame. On the first try a high-voltage relay failed; next time the pointing control for the rocket payload misbehaved; a third try was useless because the rocket doors stuck and did not open to expose the detector; and then another switch failure stalled the prototype HRI test once again. Finally, a single successful test flight was achieved just months short of the launch date.

When the excellent mirror and a perfectly functioning HRI eventually worked in the *HEAO-2* Mission, x-ray astronomy had truly come of age. The image quality was a close match to ground-based optical pictures. In the short space of only 16 years, x-ray astronomy had made more-rapid progress than any of the other new astronomies of the twentieth century. The *Einstein Observatory* could not only study the spectacularly powerful sources picked up with simple rocket payloads in the 1960s, but it could even detect x-rays from coronas of faint dwarf stars, as common in the Galaxy as the Sun.

Success with the *Einstein Observatory* pointed the way to the next generation of x-ray mirrors. *ROSAT (German Roentgen Satellite)* has given startling confirmation of the ubiquity of x-ray stars. Equipped with a 30-centimeter mirror for soft x-rays, it has mapped 50,000 x-ray stars. *ROSAT* has demonstrated that a large percentage of all normal stars are detectable x-ray sources. Next to dominate x-ray astronomy will be NASA's *AXAF (Advanced X-ray Astronomy Facility)*, scheduled for launch before the end of 1998. The high-resolution mirror assembly consists of four pairs of concentric grazing incidence paraboloid-hyperboloid mirror elements fabricated out of Zerodur, a glass ceramic with a very low coefficient of thermal expansion. For high reflection efficiency the mirrors are coated with iridium. The outer mirror is 1.2 meters in diameter, and its focal length is 10 meters. Tests of the mirrors indicate a spatial resolution of 0.2 arc seconds. The combined surface area of the mirrors is 6.7 times that of the primary *Hubble Space Telescope* mirror and it is polished to a smoothness better than 5 times the size of a hydrogen atom. *AXAF* is expected to be 100 times as sensitive as *Einstein* and to reach ten times as far into the universe. It will be launched into a highly elliptical orbit with an apogee of 140,000 kilometers, the perigee of 10,000 kilometers, and an orbital period of 64.3 hours. Unlike the *Hubble* it will not be serviceable in orbit, but its design lifetime is 5 years.

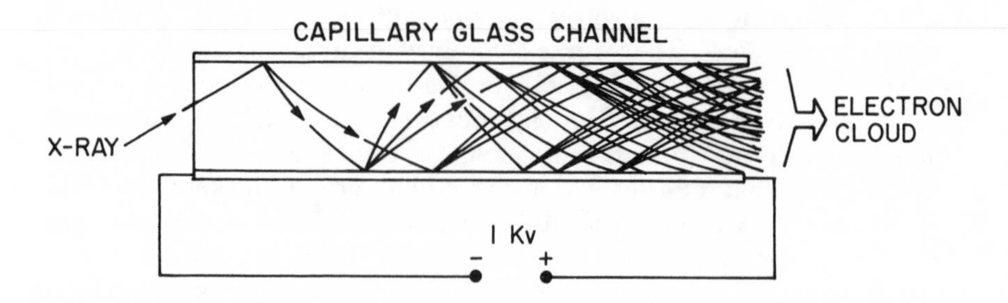

FIGURE 27(a). *The Channeltron. A single glass capillary converts an incoming x-ray to an avalanche of electrons that emerges from the opposite end.*

FIGURE 27(b). *The High Resolution Imager (HRI). A microchannel plate consists of millions of fine glass capillaries. The plate thickness is about 1 millimeter and the diameter can exceed 100 millimeters. The emerging electrons reproduce a greatly intensified image of the x-ray pattern.*

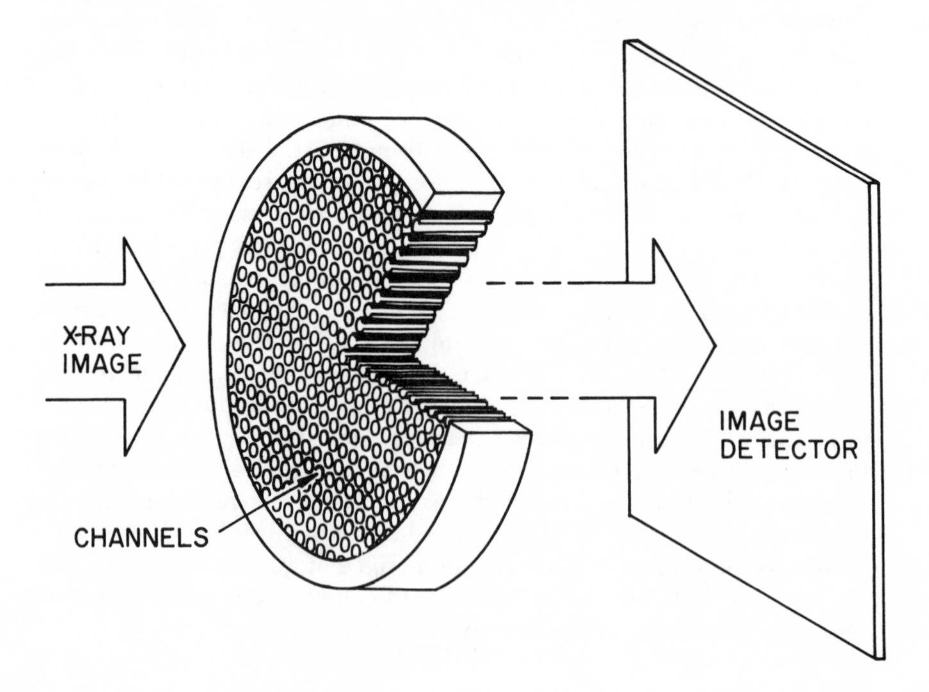

Gamma-Ray Astronomy

Nuclear processes play an important role in astrophysics, from the comparatively low-energy reactions that generate stellar energy and those that accompany the explosion of stars to the very high energy interactions of cosmic rays with the interstellar medium. Since gamma rays provide a unique signature of nuclear processes, it was naturally expected that gamma-ray astronomy would eventually assume special importance in the new era of space observatories. Before the first successes of galactic x-ray astronomy, Eugene Feenberg and Henry Primakoff of the University of Washington in St. Louis speculated that high-energy electrons speeding through interstellar space would collide with photons of light and boost the photon energies into the gamma-ray range. The process is known as *Compton scattering*.

In 1958 Philip Morrison of the Massachusetts Institute of Technology (MIT) spurred observers to search for gamma rays from a variety of sources. Besides Compton scattering, he pointed out that whenever an electron meets its antimatter mate, the positron, the two will annihilate and produce a pair of 0.5-million electron volt gamma rays. Nobel laureate Hideki Yukawa had predicted even earlier that neutral pi-mesons (pions), short-lived particles produced in cosmic-ray collisions with interstellar hydrogen, would disintegrate with the emission of gamma rays in a broad energy range centered at about 70 million electron volts. The entire galactic disk should shine with gamma rays.

Energies of gamma-ray photons range upward from x-rays to millions and billions of electron volts. One discrete source, Cygnus X-3, is known to radiate photons with energies as high as 1 quadrillion electron volts. We characterize gamma radiation as *hard*, by which we mean very penetrating. Typically, in laboratory experiments with gamma-ray sources the experimenter must be protected by a wall of lead bricks. In one sense, however, the label "penetrating" is misleading because the gamma rays from space never reach the ground as do sunlight and starlight. Most celestial gamma rays barely penetrate the thin stratospheric air to a level at which balloons can still float and U-2 aircraft can fly. The earliest attempts to observe gamma rays from space were made from balloons flying at altitudes of from 20 to 30 kilometers.

Gamma rays are absorbed indiscriminately by all materials. There are no relatively opaque or transparent materials as, for example, a thin metallic foil versus a thick plate of glass for visible light. It is only a matter of the number of electrons encountered. The gamma rays see the sheet of lead as no different from wood or earth except for the density of electrons.

Those of us who are senior citizens remember the common practice of medical x-ray fluoroscopy with the aid of zinc sulphide phosphors. The scrutinizing eyes of the doctor are now largely supplanted by electronic imaging tubes. In similar fashion, combinations of photoelectric tubes and fluorescent crystals provide the means of detecting and measuring hard x-rays and gamma radiation in the range up to about 10 million electron volts.

A thin fluorescent screen is fine for x-rays of moderate energy but does not efficiently absorb higher-energy x-rays and gamma rays. Crystals of thallium-activated alkali-halides such as sodium iodide and cesium iodide fluoresce strongly under irradiation and can be grown large enough to absorb even high-energy gamma rays. When a gamma-ray photon enters a crystal of sodium iodide, it interacts with atoms to release energetic electrons, which in turn dissipate their energy very quickly in the excitation of ordinary light via a host of secondary collisions. The resulting flash of light can be detected by a sensitive photomultiplier tube and converted to an electrical pulse. A high-energy gamma-ray photon produces a brighter pulse than one of lower energy. By counting pulses and measuring their brightness, we obtain a measure of the photon's intensity and spectral energy distribution. In the 1950s and 1960s many scintillation detectors were flown on balloons, and successful observations were made, first of solar flares and later of the pulsars in the Crab and Vela. The spatial resolution of these instruments was very poor.

Gamma-Ray Observatories in Space

The promise of gamma-ray astronomy with imaging capability began in 1968 with the flight of a modest instrument, weighing just a few pounds, on board the NASA *Orbiting Solar Observatory (OSO-3)* satellite. It just barely perceived the background gamma radiation of the Milky Way. All told, it collected only 621 gamma photons over its lifetime. The *Small Astronomy Satellite (SAS-2),* launched in 1972, collected 8000 photons in a flight that lasted 8 months. *COS-B,* a satellite mission of the ESA launched in 1975, carried about 200 pounds of instruments and functioned perfectly for almost 7 years, over which time it collected more than 100,000 photons. For each of these missions, the scientific knowledge obtained was roughly commensurate with the number of photons collected.

The spatial resolution of *COS-B* was about 2 degrees. Its mission was to measure the diffuse radiation of the galaxy, and it was not expected that it would find strong discrete sources. Indeed, it did show that the plane of the Milky Way shines brightly with gamma rays in the 100 million electron volt range. These rays originate in collisions between energetic cosmic rays that roam the Galaxy and the dilute hydrogen gas that concentrates toward the galactic disk. The energy conversion mechanism involves the production of intermediate energy pi-mesons, ephemeral particles that decay rapidly with the emission of gamma rays. *COS-B* provided clear evidence of a direct relationship between the density of interstellar matter and the intensity of cosmic gamma rays. But these results were not the most exciting returns from *COS-B;* rather, this was the surprising discovery of many apparently discrete sources of gamma rays; that is, of new types of celestial objects that produce most of their radiation in the gamma-ray range.

Since 1989, a number of successful gamma ray observatories have been launched, most notably the NASA *Compton Gamma-Ray Observatory* in April 1991, with more than 10 times the sensitivity of *COS–B.* Among the sources that have been studied for almost a decade, the most intense have been active galactic nuclei (AGN). *COS–B* found only the quasar 3C-273 at energy greater than 100 megavolts. *Compton* has observed 3C273, 3C279, and 14 other AGN. It is thought that all AGN

FIGURE *28. The gamma-ray universe as seen from the* COS–B *satellite. The region of the sky not covered by* COS–B *scans is shaded, and dots mark positions of discrete sources. The Crab and Vela pulsars are uniquely identified by the timing of their pulsations. High above the galactic plane, at a distance of 2500 million light-years, lies the brightest quasar, 3C-273. Its gamma-ray luminosity exceeds its x-ray and optical power. Rho Ophiuchi is a giant molecular cloud in which cosmic-ray collisions with interstellar gas produce the gamma radiation. Most of the other sources hug the galactic plane, which suggests that they are associated with younger objects, but until recently none had been identified with any known visible or x-ray source. Geminga was known for twenty years as a mystery source that could not be identified with any object in other parts of the spectrum. (In the Italian Milanese dialect,* Geminga *means "It is not there.") Now there is positive identification with an old x-ray pulsar of period 0.237 second and a twenty-fifth magnitude star, the faintest optical counterpart to a high-energy source yet found. Virtually all its energy is radiated as x-rays and gamma rays. Other unidentified sources may also be pulsars whose beams miss the earth. (European Space Agency.)*

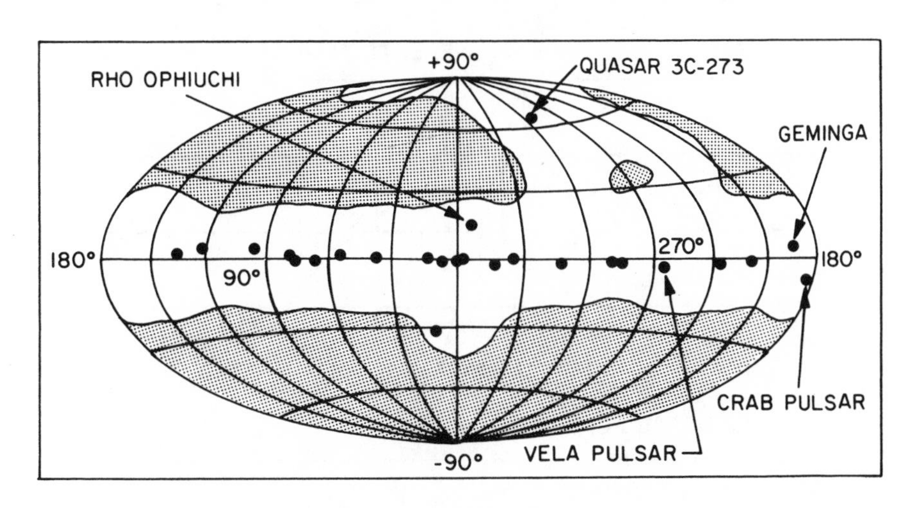

are powered by massive rotating black holes and characterized by outward-extending plasma jets. When the jets are aimed at the earth the gamma-ray intensity is believed to maximize.

Gamma Rays Light-Up the Air

With increasing energy, the number of celestial gamma-ray photons that strikes the atmosphere diminishes. To detect them requires larger and larger instruments, until the size becomes impractical. But the atmosphere itself can serve as a detector, and as for size, the sky is the limit.

High-energy gamma rays that strike the upper atmosphere initiate cascades of secondary particles and lower-energy gamma rays. These showers begin at an altitude of about 20 kilometers. At the first interaction, the gamma ray usually converts to an electron–positron pair, with each particle carrying off one-half the energy of the incoming gamma ray. A

FIGURE 29(a). *A spark chamber consists of a set of metal plates in a container filled with neon gas. When a high voltage is applied to alternate plates, the passage of a charged particle causes visible sparks to jump between successive pairs of plates along the ionization trail.*

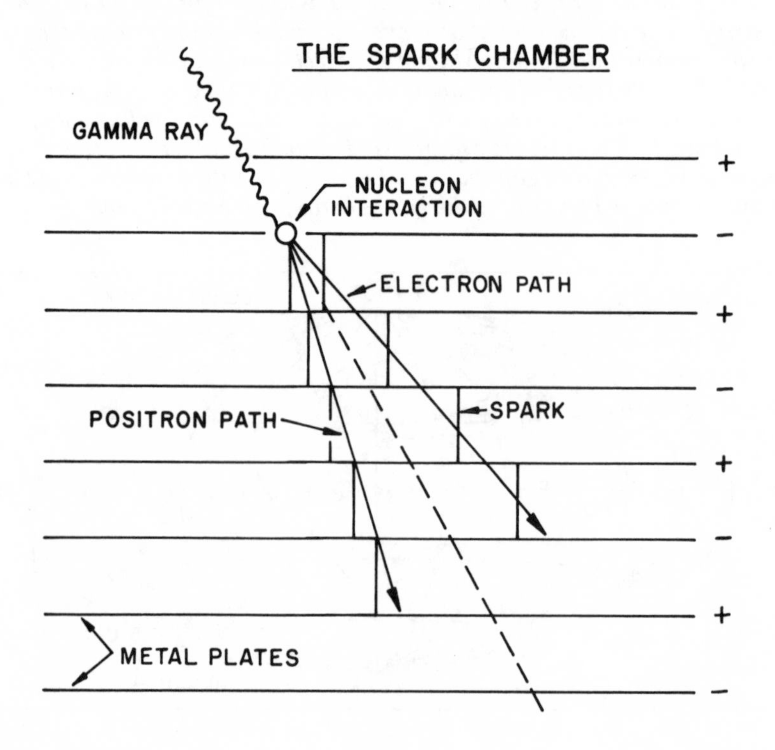

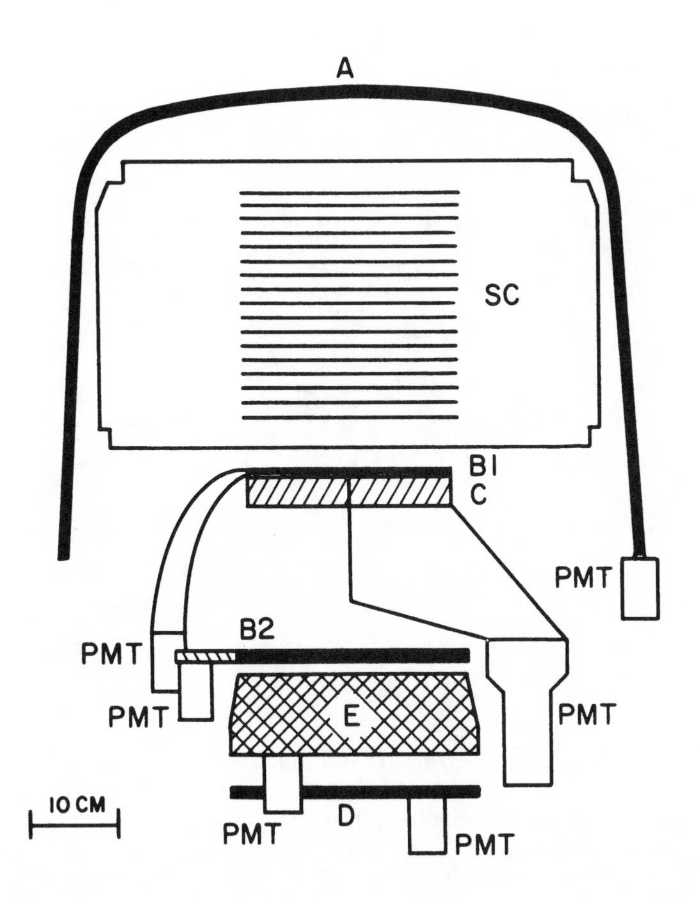

FIGURE *29(b). The* COS–B *satellite spark chamber. Gamma rays were converted to electron–positron pairs in any one of 12 tungsten sheets, interspersed between the upper twelve gaps. The particles travel in nearly the same direction as the incoming gamma rays. Each gap was bounded by two orthogonal grids of parallel wires. A high voltage of several kiloelectron volts was applied across the pairs of grids, causing spark discharges to mark the tracks of the particles. The spark chamber was triggered within 1 microsecond whenever coincident signals were registered from the telescope consisting of the Cerenkov counter (C), the scintillation counter (B2), and the total absorption counter (E).*

A plastic scintillator guard counter (A) vetoed trigger signals produced by background charged particles. The sensitive area was 24 centimeters–by–24 centimeters. (European Space Agency.)

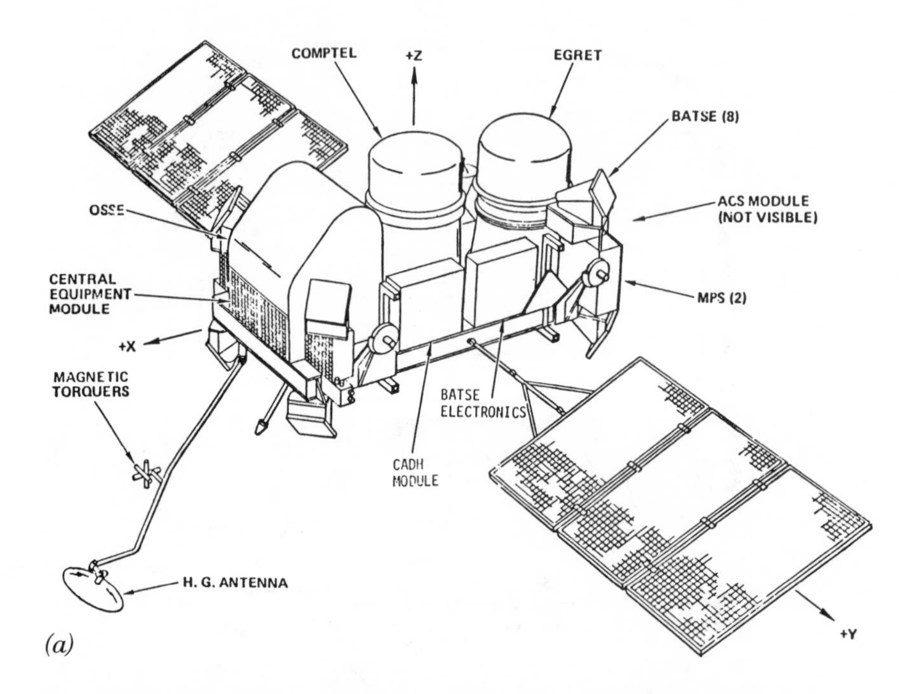

(a)

FIGURE *30. The NASA* Compton Gamma Ray Observatory (CGRO), *launched in 1990, is more than 10 times as sensitive as* COS–B. *(a) schematic diagram of the arrangement of the four principle instruments: (1) EGRET (Goddard Space Flight Center) has a wide field of view, based on a spark chamber, for from 20 to* 3×10^4 *million electron volts. (2) Comptel (Max Planck Institute and University of New Hampshire), a double Compton telescope covering 1 to 30 million electron volts, will provide the highest spatial resolution, better than 1 arc minute. (3) The Oriented Scintillation Counter Experiment (OSSE) (U.S. Naval Research Laboratory) consists of four large phoswich scintillation detectors mounted on gimbals to allow selected pointing. Its energy resolution is about 8% at 0.66 million electron volts. (4) BATSE (Marshal Space Flight Center) is a Burst and Transient Spectrometer Experiment that monitors a large portion of the sky continuously. (NASA.) (b) Completed OSSE instrument. Four identical shielded and collimated scintillator crystals provide a total sensitive area of 2310 square centimeters with a 3.8° by 10° field of view. The gimbal mounts permit simultaneous measurements on and off sources to subtract the background. The energy range is from 0.1 to 10 million electron volts. (U.S. Naval Research Laboratory.)*

short distance further on, each of the fast-moving particles suffers a close encounter with a gas molecule and radiates a new pair of gamma rays. This geometrical progression through the atmosphere is sustained by a steadily growing number of energetic particles and gamma photons, so that the shower builds up to thousands of particles and photons spread over hundreds of square meters at the ground level.

On their way through the atmosphere, shower particles create brief flashes of bluish light that last only a few nanoseconds. The Soviet physicist Pavel Cerenkov discovered this luminescence in 1934. It occurs whenever a high-energy particle passes through air or denser material such as water at a speed greater than the velocity of light in the medium. More specifically, the light is emitted by electrons of the medium disturbed by the passage of the high-energy radiation. This *Cerenkov light* is the electromagnetic analog of the shock wave generated at the nose of a bullet or the bow wave created by a speedboat traveling faster than the speed of wave propagation in water. In air, from 200 to 300 photons are emitted per centimeter of track. The direction of emission is at a small angle, as though each gamma ray carried a microscopic searchlight to illuminate its path with a disk of light.

On moonless nights the Cerenkov flashes are easily detected by an optical telescope that mirrors the light onto a phototube. At the Whipple

Observatory in Arizona, a conglomeration of 248 paraboloidal mirrors combines to make a large light bucket, 10 meters in diameter, that focuses onto a hexagonal array of 37 phototubes (see Fig. 31, p. 95). The intensity can be mapped over the Cerenkov light disk and the direction determined to less than 0.25 degree of arc.

One of the great surprises of recent years was the discovery of very high energy gamma rays from Cygnus X-3. First identified as an x-ray source from a rocket flight in 1966, its soft x-ray spectrum was deficient, a symptom of absorption caused by the rays having traveled a distance of about 40,000 light-years. It was detected in the infrared but not in the visible because it lies in a particularly dusty direction of the Milky Way. When NASA launched the x-ray astronomy satellite in 1970, many x-ray sources, including Cygnus X-3, were found to vary periodically, suggesting pairs of stars locked in binary orbits. The periodicities could best

FIGURE *31. The 10-meter reflector of the Whipple Observatory atop Mount Hopkins in Arizona is the largest camera in the world. It consists of 248 mirrors that focus onto an array of 37 phototubes. Flashes of Cerenkov light set off by air showers can be mapped to identify the directions of arrival of the gamma rays that set off the showers. (Smithsonian Institution, Whipple Observatory.)*

be explained by the eclipsing effects of the paired stars as they circled each other. For Cygnus X-3, the period was 4.8 hours.

Thus far, there was no cause for special attention to Cygnus X-3. Then on 2 September 1972, a remarkable outburst of radio emission occurred. By chance, Philip C. Gregory, at the Algonquin Radio Observatory in Ontario, casually pointed his telescope to Cygnus X-3 while waiting for his main target of the evening to come over the horizon. To his surprise, Cygnus X-3, which was ordinarily a modest radio source, had brightened 1000 times to become one of the strongest radio sources in the sky.

The event set up an impromptu international campaign of observations. Gregory called Robert Hjellming at the National Radio Astronomy Observatory in Green Bank, West Virginia, who confirmed the Canadian observation. The two astronomers called every observatory they could, and within a couple of days nearly all the world's radio astronomers were focusing on Cygnus X-3. The radio surprise was quickly augmented by the detection of x-rays and gamma rays. Cerenkov radiation flashes seen by Soviet observers in the Crimea set very high energy limits on the gamma rays, greater than 10^{12} electron volts, energy higher than has ever been achieved in a manmade particle accelerator. The Soviet results were so surprising that most astronomers viewed them with great skepticism. But in recent years energies as high as 10^{16} electron volts have been confirmed with Cerenkov telescopes at other observatories. These high energies cannot be explained by conventional models of cosmic-ray acceleration. A few point sources such as Cygnus X-3 could supply all the higher-energy cosmic rays in the Galaxy, thus solving the baffling problem of their place of origin.

Neutrinos: Ghostly Space Travelers

The neutrino is one of nature's most elusive particles, literally capable of passing through the mass equivalent of light-years of lead without interaction. Yet ingenious detectors have been devised for neutrino astronomy, and the potential for detection is encouraging. Thus far, a serious effort has been directed toward solar neutrino astronomy. Hydrogen fusion in stars like the Sun should, in theory, produce floods of low-energy (about 1 billion electron volts) neutrinos. We expect about 70 billion neutrinos from the Sun to strike each square centimeter of the Earth every second. But efforts to detect these ghostlike particles have until now been a baffling disappointment.

Neutrinos are a fairly recent discovery in the zoo of atomic particles. Wolfgang Pauli proposed their existence in 1930 to explain observations of radioactivity known as *beta decay*, a process in which a neutron spontaneously converts to a proton and an electron. While the proton remains bound in the atomic nucleus, the electron (a beta particle) is ejected. If

no other particles were emitted, the electron would have a unique energy consistent with the law of conservation of energy. Instead, the electrons have a spectrum of energies from zero up to the maximum predicted by the conservation law. Pauli's answer to the dilemma was to postulate an undetected particle—the neutrino—that carries off the missing energy. Almost in jest, Pauli required that this particle have no electric charge and zero or very small rest mass; properties that would make the neutrino almost undetectable. Indeed, Pauli thought that no one would ever detect his imaginary particle. He told his friend, astronomer Walter Baade: "Today, I've done the worst thing for a theoretical physicist; I have invented something which can never be detected experimentally" (meeting of the American Physical Society and the American Association for the Advancement of Science, Pasadena, 16 June 1931).

Pauli originally called his particles *neutrons* (the neutron had not yet been discovered). The name *neutrino* came from Enrico Fermi at a conference in 1933. He explained that the neutrons discovered in 1932 by Sir James Chadwick were big particles, whereas Pauli's neutrons were very small and should therefore be called neutrinos since the *ino* ending is the Italian for diminutive, as in *bambino.* A full 20 years passed before experimental verification was made. Frederick Reines, Clyde L. Cowan, and their partners finally observed neutrinos in the very high flux stream emerging from the nuclear reactor at Savannah River, Georgia. Beams of neutrinos are now available at several accelerators, and neutrino-induced reactions are widely studied.

Experimental evidence points to the existence of three different "flavors" of neutrinos. Just as the electron has two heavy relatives, the muon

FIGURE *32.* Wolfgang Pauli (1900–1958). *The Pauli exclusion principle in essence postulates that two electrons in an atom can never exist in the same state. The consequences of this simple principle for astrophysics are profound and make it possible to understand the "degeneracy" pressure that supports a white dwarf star when all internal heat generation has ended. For his theory Wolfgang Pauli was awarded the 1945 Nobel Prize in physics.*

In 1931 Pauli addressed the problem of radioactive beta decay in which the emitted electron exhibited a wide range of energies. He concluded that there must be an electrically neutral particle of small or negligible mass that carried away energy and momentum to balance the requirements of energy and momentum conservation. Enrico Fermi further developed Pauli's ideas and named the mysterious, undetected particles neutrinos.

When Pauli hypothesized the existence of the neutrino nobody had any idea how to confirm it experimentally. The neutrino remained hypothetical until 1953, when Fred Reines and Clyde L. Cowan set up their detector at a nulear production reactor that provided neutrinos in such great abundance as to be detectable.

and the tau particle, physicists expect that there are muon and tau neutrinos as well as electron neutrinos. Furthermore, theory suggests that neutrinos may oscillate from one form to another millions of times a second as they travel and that they have a small but finite mass. If so, neutrinos could constitute a major part of all the mass of the universe.

As neutrino astronomers try to capture their elusive particles, they must build their telescopes deep underground in order to avoid the confusion introduced by the general background of cosmic rays. For 10 years Raymond Davis, Jr. of the Brookhaven National Laboratory operated an enormous detector tank 1 mile underground in the Homestake gold mine of South Dakota. Davis's target was the chlorine in 100,000 gallons of dry-cleaning fluid, perchlorethylene. The basic reaction for detection was capture of a neutrino by chlorine 37 to produce argon 37. The argon isotope is radioactive, with a 35-day half-life, and is readily detected by electron counting in a proportional counter. First, however, the isotope must be removed by flushing the tank with helium. The removal was demonstrated to be more than 90% efficient.

Only about one atom of argon 37 is produced every 3 days. This result is only one-third the theoretical prediction. The discrepancy has startled solar physicists and challenged accepted models of stellar interiors and neutrino physics. If the solar temperature were 1 million degrees cooler

FIGURE *33.* Enrico Fermi (1901–1954). *Enrico Fermi had genius for both theoretical and experimental physics. He received the 1938 Nobel Prize in physics for his discovery that a neutron could induce radioactivity when shot into a nucleus and, in particular, for his studies of nuclear reactions brought about by slow neutrons. In 1926 Fermi, recognizing the significance of the Pauli exclusion principle, derived a statistical mechanics theory of an electron "gas" in solids. The Fermi electron gas pressure is sufficient to stabilize a white dwarf against further collapse when the core density is about 10 tons per cubic inch. In a neutron star it is the Fermi pressure of neutrons that stabilizes the star.*

In 1933 he resolved the problem of beta decay, the expulsion by the nucleus of high-energy electrons in the process of radioactivity, by taking into account the role of a hypothetical neutrino. Fermi proposed that just as an atom changes from a higher-energy state to one of lower energy by emission of a photon, so does a neutron (a high-energy state) change to a proton (the low-energy state) by emission of an electron (beta particle) and a neutrino, which is needed to conserve energy and momentum. Fermi's idea was so unconventional that the editors of Nature *rejected his paper when it was submitted for publication.*

Fermi's seminal theory of beta decay introduced a new force, the weak interaction, that completed the quartet of basic physical forces in nature: the long range forces of gravity and electromagnetism and the strong and weak forces that operate inside the nucleus.

than in accepted theoretical models, the theoretical discrepancy would disappear. But such a large error in calculated core temperature is almost impossible to reconcile with the observations of solar photospheric temperature. Since Davis's detector was sensitive only to electron neutrinos and not to muon or tau neutrinos, some theorists suggest that electron neutrinos transform to one of the other two types in the course of 8 minutes of travel from the Sun to the Earth. Almost all high-energy elementary particles have ephemeral qualities, so it is not surprising that theorists find reasons to predict such behavior in neutrinos. But persistent efforts to demonstrate neutrino oscillations in the laboratory have thus far failed.

The chlorine experiment remains baffling, and neutrino astronomers are turning to a gallium-71 detector. Gallium is sensitive to more abundant low-energy neutrinos from the Sun than the higher energies that are captured by chlorine. There is, however, a major practical obstacle: To record one neutrino capture a day would require 50 tons of gallium, essentially all the world production in a year, at a cost of about $25 million. Impatient solar physicists have been speculating about cornering the gallium market for 1 year, running their experiments for a few years, and then returning the gallium to the world market when finished. Other target schemes are also being investigated using lithium 7, bromine 81, technetium 99, and indium 115; each of these isotopes is sensitive to a different energy range. It is not unrealistic to hope that all of these targets eventually will be used for neutrino spectroscopy as well as simple detection.

Whereas solar neutrino astronomy has reached an unsettling impass, galactic neutrino astronomy succeeded in the most impressive fashion when Supernova 1987A exploded in the Large Magellanic Cloud in February 1987. In another remarkable example of serendipity, a great flash of neutrinos was recorded, unpremeditated, by instruments built for an entirely unrelated scientific experiment. The full story of SN1987A will be told in Chapter 4.

For detection of very high energy neutrinos from the Galaxy, an ambitious project now in prototype design is DUMAND, the Deep Underwater Muon and Neutrino Detector, which is designed to operate in an enormous volume of water. It would look like a cube and measure the length of three football fields on a side. Seven hundred fifty-six photomultipliers would be strung on 36 vertical lines, 21 tubes per string, spaced 25 meters apart. Each string would be 50 meters distant from its nearest neighbors. The great array would operate by observing the Cerenkov radiation produced as high-energy muons and neutrinos interact with water. When fully developed, the plan is to locate DUMAND at a depth of about 5 kilometers in the Pacific Ocean off the coast of Hawaii. This huge "all sky camera" would carry out high-energy neutrino astronomy of cosmic-ray interactions as well as of discrete celestial sources such as supernovae.

The mechanical problems of suspending the DUMAND array so deep in the ocean, where it would be subject to the vagaries of ocean currents that could tangle the dangling string lattice, are a great challenge to ocean engineers.

Ripples in the Curvature of Spacetime

Visible light, radio waves, and x-rays are ripples in an electromagnetic field generated by accelerated electric charges. By way of analogy, we might expect accelerated masses to generate gravitational waves. And just as electromagnetic waves jiggle charged particles, gravitational waves should accelerate masses. According to Einstein's theory of general relativity, gravitational waves oscillate perpendicular to the direction of motion and travel with the speed of light, as does electromagnetic radiation. In the language of relativity, a gravitational wave ripples the curvature of spacetime and deforms any mass that it traverses.

Gravitational interaction between two ordinary masses is incredibly weak compared to electrical interactions. Consider, for example, two electrons separated by 1 millimeter. In order for the electrical force to be reduced to the same strength as the gravitational force, the electrons would have to be separated by 100 light-years. Imagine a locomotive placed in a super centrifuge and spun around so fast that it is on the verge of disintegrating. The power radiated in gravitational waves would still be far below any possibility of detection. Gravity becomes a force to be reckoned with only when the masses are as large as the Sun and planets. As the Earth orbits the Sun, it glows with gravitational radiation; but the total power is a mere 200 watts, about the energy radiated by a room light. It would take a mass as large as the Sun, crushed to the size of a neutron star and spun up to almost the speed of light at its surface before the gravitational wave radiation might be detected.

In spite of the difficulties, physicists expect that within 1 or 2 decades gravitational wave astronomy will supplement astronomy in the electromagnetic spectrum. Because the Earth is almost transparent to gravitational radiation, the detectors will have an essentially unimpeded view of the heavens at all times. But gravitational wave detectors are very difficult to design when we begin with little knowledge of what the gravitational wave sky looks like. Until there are better clues to the strongest types of gravitational wave sources, it is difficult to design optimum detectors for them. We can make intelligent guesses now as to which sources are the most likely strong gravitational wave radiators at which frequencies, but the truth may surprise us.

In 1916 Einstein himself attempted to calculate the gravitational radiation from a binary star. He concluded that "the loss of energy from gravitational radiation is so small that in all foreseeable cases it has a negligible practical effect" (Stephen Boughn, "Detecting Gravitational Waves," *American Scientist* 68, 1980, p. 174). At the time that was a reasonable conclusion because compact objects such as white dwarfs and neutron stars had not yet been discovered. The shortest binary periods then known were on the order of days. Since then, binaries with periods measured in minutes have been detected. Because the energy-loss rate via gravitational waves varies with the sixth power of the frequency, it makes an enormous difference if the period is very short.

Scientists believe that observation of gravitational waves would open a new astronomical window and give special kinds of information about exotic sources such as collapsing stellar cores, colliding neutron stars or black holes, decaying binary star systems, and perhaps other objects of a nature still unknown. In contrast to the weak gravitational glow of the Earth, the collapse of a massive star into a black hole would send a whistling burst of gravitational waves across the Galaxy. Collapse theoretically occurs in a fraction of 1 second and releases the enormous energy equivalent of 1 solar mass. At a frequency of about 1000 hertz, the gravitational energy reaching the Earth from a collapsing black hole as far away as 1000 light-years could match the electromagnetic energy delivered from the Sun to the Earth in the course of a day: This is a truly enormous gravitational signal, yet it is still very difficult to measure.

In 1974 Joseph H. Taylor and Russell A. Hulse, observing at the Arecibo radio telescope in Puerto Rico, discovered a pulsar in a binary star system (PSR 1913+16). The radio pulsations are produced by a 1.4–solar mass neutron star spinning at 17 rotations per second (period = 0.059 second). It provides a precise clock in orbit about its unseen companion, believed to be another neutron star.

Of the first 200 radio pulsars discovered, only 3 were members of binary star systems. Taylor's pulsar was particularly special because the orbital period was only 7.75 hours, which meant that the separation of the stars was very small. General relativity predicts that two stars racing around each other in such a tight orbit will radiate prodigious amounts of gravitational radiation. The loss of energy must lead to a slow but steady collapse of the orbit that would be evidenced by a decrease in period; that is, the stars circle each other faster and faster as time goes on. Since the pulsar "clock" drifts by less than one-billionth of 1 second in 4 years, Taylor and his colleagues could time the orbital period by the Doppler variation in the signal with high accuracy. Over more than 2 decades of observation, the speedup has been 72 microseconds per year, in excellent agreement with Einstein's theory. Although an indirect detection, it is the best evidence we have of a source of gravitational waves. As they eventually approach collision, the two neutron stars, locked in an ever-tightening gravitational grip, will produce a chirping burst of gravitational waves. The frequency will rise from a deep subsonic boom of a few cycles per second to a whistle of about 1000 cycles per second. But these death throes won't occur for another 100 million years.

In the 1950s a serious approach to the practical problems of detection of gravitational waves was undertaken by Joseph Weber at the University of Maryland. His first gravitational wave antenna, built in 1967, was a 1300-kilogram aluminum cylinder studded with piezoelectric transducers to convert minute changes in the length of the bar into electrical signals in much the same manner as a ceramic phonograph pickup. Such a bar tends to vibrate like an organ pipe at its resonant frequency, about 1660 hertz.

The energy carried in a gravitational wave is characterized by the strain it induces when it encounters a massive object. *Strain* is defined as the change in length divided by the total length. A supernova explosion in the Milky Way could produce a strain of 10^{-17}. Over a rigid bar 1 meter long, that would be a quiver of only 10^{-15} centimeter, about one-hundredth of the diameter of an atomic nucleus. Even if this fantastic sensitivity could be achieved, it would hardly satisfy a gravitational wave astronomer. If supernovae occur roughly once every 30 years in the Milky Way, few physicists would care to gamble on an experiment to detect an event that might require a wait of tens of years. For the astronomer to have a significant number of events to work with, the detector must be able to pick up supernovae in distant galaxies. In the Virgo cluster, which contains some 2500 galaxies, a supernova might explode as often as every week to a month. Since the distance from Earth to the Virgo cluster is about 1000 times the distance to the center of the Milky Way, to detect a supernova in the Virgo cluster would require a detector about 1 million times as sensitive as that mentioned earlier. At the present time, scientists are optimistic about obtaining a capability of monitoring supernova explosions in the Virgo cluster with a strain sensitivity of 10^{-21}.

Tuning in on Gravitational Waves

An interesting possibility for achieving very high sensitivity is to tune a gravitational antenna to a source of known period and phase, such as a fast pulsar. Until a few years ago, the pulsar in the Crab nebula, with a period of 33 milliseconds, was the fastest known. Then a radio pulsar with a period of only 1.6 milliseconds, PSR 1937+214, was discovered, corresponding to a rotational frequency of 642 hertz. If it were to spin-up just a little faster, toward a period close to 1 millisecond, it would approach relativistic instability as its surface velocity neared the speed of light. In spite of the fantastic tensile strength of its crust, it must succumb to overwhelming forces as it closes in on the speed limit. Quivering and quaking, it would dissipate enormous energy in gravitational waves. One plan now under consideration is to build a proportional counter x-ray detector of very large aperture, about 100 square meters, which could be attached to the space station. With timing capability in the microsecond range, it could survey the sky and search out very fast x-ray pulsars. The discovery of such objects would then provide a frequency key for ground-based gravitational wave detectors to focus their search for accompanying gravitational wave radiation.

Some experimenters claim to have already detected signals that may be gravitational waves, but their claims are suspect. The evidence of marginal signals will remain controversial until a Milky Way supernova flashes in the sky and registers simultaneous strong responses in several gravitational wave detectors around the world. Supernova 1987A in the large Magellanic Cloud was 100 times too far away to give gravitational physicists what they are waiting for. None of the world's gravitational wave detectors gave any clear signals.

Inspired by the advanced concepts of telescopes, sensors, and space technology, astronomers are poised for an unprecedented scientific assault on the mysteries of the universe. Goethe's words express the urgent challenge: "What you can do or dream you can, begin it: Boldness has genius, power, and magic in it" (Prelude, "On the Stage," from *Faust*, 1806).

PART II

ASTRONOMICAL DISCOVERY

CHAPTER 4

The Sun

Thou sun of this great world both eye and soul.

—JOHN MILTON,
"Paradise Lost," Book V, 1667

To observe the Sun from outer space was an astronomer's dream from the time of the earliest speculations about modern rockets. The dream came alive in 1946 when captured World War II German V-2s rose from the New Mexico desert with astronomical instruments in their nose cones. I was lured from a quiet, traditional laboratory mode of research by the spine-tingling adventure of high-altitude experiments with rockets. My first experience led to a lifelong addiction. The Sun was a brilliant opening target for the new field of space astronomy, and even the crude early efforts produced important scientific discoveries. Much of my story of the Sun is built on personal reminiscences of life with rockets early on.

The Sun is the astronomer's Rosetta stone (if we can call a ball of gas a stone) in which we can read clues to the nature of the billions of stars in the Galaxy. At a distance of only 92 million miles, we feel the warmth of its radiation, whereas other stars seem to be only cold, sparkling jewels in distant isolation. The Sun's face is so close that we can discern most of its finer features and the patterns of turbulent weather that roil its surface. When the Sun was very young, gales blew from its surface with hurricane violence. The gentle wind that now blows and occasionally gusts from the

Sun rustles the magnetic field that guards the Earth and lights up the beautiful polar auroras. Younger giant stars in the Milky Way blow winds 100,000 times as violent.

The blackness of the night sky conveys a sense of the emptiness that pervades the enormous reaches of space between the stars. But cosmic space is not entirely empty; in comparison to the best vacuum that can be produced in the laboratory, it is only extremely rarefied. Its principal components are hydrogen, helium, and a sprinkling of dust, which are concentrated in clouds that are the embryos of new suns. These condensations include the debris of ancient stellar explosions, thus providing for a reincarnation of the past in the naissance of new stars. Our Sun, for example, is a second-generation star, and we are recycled from all the elements heavier than hydrogen and helium that were synthesized in our ancestor stars.

About 5 billion years ago the Sun was spawned by a giant cloud of dust and gas about 1 trillion miles in diameter that floated in a void between the stars of the Milky Way. Gravity within the cloud caused it to contract upon itself just as the force of gravity at the Earth pulls all things toward the ground. Once the cloud began to collapse, it continued to fall in on itself, perhaps assisted by the push of a shock wave from a nearby exploding supernova. As gas and dust particles drew closer together, their mutual gravitational attraction increased and the collapse accelerated. Toward the center of the cloud, the crush squeezed harder and harder, and a dense central core accumulated. Whatever small spin characterized the initial cloud accelerated as the cloud contracted. Centrifugal force made the cloud flatten out toward the edges until it took on the shape of a fried egg with a bulge near the center, out of which the Sun would be created. A gradually thinning disk extended outward from the Sun and eventually condensed into the coterie of planets.

Collapse generated heat just as the air in a bicycle pump becomes warmer as it is squeezed into a tire. The temperature grew greater and greater toward the center of condensation until the dust melted and vaporized. In a matter of perhaps 10 million years from the time collapse began, the pull of gravity was balanced by the pressure of hot gas, perhaps 10 billion times the pressure of the atmosphere around us, and collapse slowed down. At that stage the protosun had formed, completely hidden inside a cocoon of surrounding cooler gas that absorbed the visible light radiation and permitted only an infrared heat glow to escape. For millions of years the protostar continued to radiate heat but did not cool off because, in the process of radiating, it dissipated the pressure that was needed to hold off further gravitational collapse, and that kept it warm.

For 10 million years the temperature in the core of the protosun rose steadily. It was then that nuclear reactions were ignited in the core and the mature Sun was born. With this great new source of thermonuclear energy, all the heat necessary to balance the Sun against further collapse became available. From that moment forward, the Sun's nuclear furnace preserved a state of equilibrium in which the Sun produced exactly the amount of heat and pressure needed to support its overlying layers.

As more time elapsed, the dust and gas that surrounded the newly born Sun gradually thinned out and became more transparent to visible light. Slowly, the Sun's light emerged from the cocoon within which the Sun had formed. One hundred billion stars in the Galaxy have evolved through processes of star birth similar to that of the Sun. Far beyond the Milky Way there are perhaps another 100 billion galaxies, each containing a similar quota of stars.

From Galileo to Einstein

Solar astronomy began with Galileo's invention of the telescope, but scientific understanding of the nature of the Sun came very slowly over the next 200 years. As late as the beginning of the nineteenth century, Sir William Herschel, the preeminent astronomer of his day, believed that solar radiation escaped from a hot, thin cloud layer. Occasional gaps in the cloud would enable astronomers to see cooler regions below, and these, he believed, were the dark sunspots that marked portions of the disk of the Sun. He was convinced that the Sun was solid and protected from the intense heat overhead by lower-level clouds that even made it conceivable for the Sun's surface to have a habitable temperature. Herschel's reputation was so great that such bizarre views persisted for a half-century after his death.

Until the coming of age of atomic energy, attempts to understand the source of solar power were mostly wild speculations. As early as 467 B.C., the Greek philosopher Anaxagoras assumed that a large iron meteorite that fell on the Peloponnesus came from the Sun and took this to mean that the Sun was made of molten iron. Archbishop James Ussher calculated in the seventeenth century that the world began precisely in 4004 B.C. His date was widely accepted until about 1 century ago. In the nineteenth century many astronomers still thought of the Sun as a mass of burning coal, even though chemical burning would have sufficed to maintain a flow of energy from the Sun for only a matter of a few thousand years. It was not until 1869 that an American physicist, Jonathan Lane, advanced the concept that the Sun was a ball of gas bound by gravitation and supported by a central source of energy. Since it was also speculated that the Sun had condensed out of an enormous nebula of gas and dust that grew hotter and hotter as it collapsed, it was not surprising that some scientists in the nineteenth century held that the Sun, even after it formed, continued to radiate as a result of steady shrinkage. The German physicist Hermann von Helmholtz and his British colleague, Lord Kelvin, believed that the Sun could extract sufficient energy from its own collapse to shine with a steady brightness. They estimated that a modest decrease in diameter of only 150 feet per year would be sufficient. Such a slow rate of contraction would be barely perceptible over the span of human civilization but would suffice to keep the Sun shining at its present rate for about 22 million years.

The contraction theory, however, contained serious contradictions. If the Sun had been shrinking for some tens of millions of years at the required rate, it originally would have been so large that Earth would have been engulfed. The Earth could not have formed until the Sun was considerably smaller than the Earth's orbit. Thus, it would be impossible for the Earth to be more than some 10 million years old. On the other hand, geologists and biologists of the time already had strong evidence that the Earth, and necessarily the Sun, was more than hundreds of millions of years old. Still another theory suggested that energy derived from the capture of comets could heat the Sun, but that idea also was soon rejected as far too inadequate. Even if planets themselves spiraled into the Sun and were cannibalized one at a time, the energy delivered would keep the Sun shining for no more than 50,000 years.

In 1861 the Swedish physicist Anders Jonas Ångström became the first to identify hydrogen in the Sun from the evidence of the dark spectrum lines. For a long time, however, scientists were puzzled that the spectrum was comparatively weak, implying a low abundance of hydrogen. In 1929 the American astronomer Henry Norris Russell explained that the temperature of the photosphere was not high enough to excite strong visible emission from hydrogen, and he determined that the composition of the Sun had to be about 75% hydrogen and 25% helium. (H and He are the two simplest atoms. All the heavier atoms exist in the Sun, but their total adds up to less than 1% of the Sun's mass.)

If the Sun is composed primarily of hydrogen and helium, then the only possible nuclear reaction that could supply energy would be hydrogen fusion to form helium. Stars that are in the burning phase of their life cycle are said to be on "the main sequence." Nearly 85% of the stars in the Galaxy evolve along the main sequence, deriving their energy much as the Sun does. Giant stars consume their energy so quickly that they can remain on the main sequence as normal stars for only a few million years. Very small stars get along on a much lower rate of hydrogen consumption and can shine for more than 100 billion years.

The foremost architect of the early twentieth century concept of a star was Sir Arthur Eddington. On the basis of the simplest physical principles of the competition between forces of thermal expansion and gravitational contraction, he modeled the internal constitution of the Sun. Because he required a core temperature of some 40 million degrees Kelvin, he proposed in 1920 that some form of nuclear energy must be at work in the solar furnace at that high temperature, but he could not come up with a plausible hypothesis to explain it. He wrote with uncanny prescience: "If indeed, the sub-atomic energy in the stars is being freely used to make their great furnaces, it seems to bring a little nearer to fulfillment our dream of containing this latent power for the well-being of the human race or for its suicide" (*Observatory 43*, 1920, p. 353). But Eddington's contemporaries believed that unless the energies of the nuclei were orders of magnitude greater than Eddington's theoretical 40 million degrees Kelvin the nuclei would still repell each other. They were skeptical of the possibility of fusion. Eddington stubbornly held his ground and snapped back testily, "It has, for example, been objected that the temperature of the stars is not great enough for the transmutation of hydrogen into helium . . . but helium exists, and it is not much use for the critic to urge that the stars are not hot enough for its formation unless he is prepared to show us a hotter place" ("The Source of Stellar Energy," *Nature*, May (Supp.) No. 2948, 1926, p. 30).

Discoveries in the physics of nuclear transmutations were progressing rapidly in the 1920s in the Cavendish Laboratory, Cambridge, under the leadership of Lord Ernest Rutherford. George Gamow, a young Soviet physicist working with Rutherford, developed the concept of *tunneling*, by which protons of the comparatively low energy of Eddington's solar-core temperature could penetrate the classical repulsive barrier of tens of billions of degrees and come close enough to fuse. When Sir James Chadwick discovered the neutron he showed how, at the temperature of the solar core, neutrons can be formed from collisions between electrons and protons, and two protons and two neutrons can combine to make a helium nucleus (alpha particle). In the latter process, the masses don't add up to balance the equation, and 0.7% of the mass of the fusing particles is converted to thermonuclear energy. The flood of energy from the Sun proclaims the truth of Einstein's law of the equivalence of mass and energy:

$$E = Mc^2 \text{ [Energy} = \text{Mass} \times \text{(velocity of light)}^2\text{]}$$

Nuclear fusion produces 10 million times the energy generated by chemical burning of an equal amount of coal. Although it would take 10 trillion atomic fusions to provide enough energy to light a 50-watt bulb for 1 second, the solar furnace burns at such a ferocious rate that about 2 trillion trillion trillion hydrogen atoms are annihilated every second. The resulting flow of energy brings about 200 trillion kilowatts of sunshine to the Earth. Yet since the Sun began to radiate visibly 4.5 billion years ago, only about 4% of its hydrogen has been converted to energy. At this rate, it could continue to shine brightly for more than 100 billion years, but other factors will limit its life to about 10 billion years.

Gamma rays and neutrinos are generated by the energy of nuclear reactions in the core of the Sun. The neutrinos fly out of the Sun almost uninhibited, at close to the speed of light, but the gamma rays weave a tortuous path involving innumerable collisions before they escape the solar interior. As the gamma rays bounce from one atomic encounter to another,

the average wavelength degrades toward the red end of the spectrum; that is, from x-rays to ultraviolet, to visible and infrared. The melee of particle and wave interactions in the solar interior that leads to the slow, chaotic outward diffusion of gamma rays was vividly described by Eddington in his 1927 book, *Stars and Atoms:*

> We can now form some sort of picture of the inside of a star—a hurly burly of atoms, electrons and aether-waves. Disheveled atoms tear along at a hundred miles a second, their normal array of electrons being torn from them in the scrimmage. The lost electrons are speeding a hundred times faster to find new resting places. . . .
>
> And what of all this bustle? Very little. The atoms and electrons for all their hurry never get anywhere: they only change places. The aether-waves are the only part of the population which accomplish anything permanent. Although apparently darting in all directions indiscriminately, they do on the average make a slow progress outwards. There is no outward progress of the atoms and electrons; gravitation sees to that. But slowly the encaged aether waves leak outwards as through a sieve. An aether wave hurries from one atom to another, forwards, backwards, now absorbed, now flung out again in a new direction, losing its identity, but living again in its successor. With any luck it will in no unduly long time (ten thousand to ten million years, according to the mass of the star) find itself near the boundary. It changes at the lower temperature from gamma-rays to light rays, being altered a little at each re-

FIGURE *34.* *For 40 years* Sir Arthur Stanley Eddington *occupied the Plumian chair of astronomy at Cambridge University, where he became the foremost proponent of astrophysics of his time. He developed a fundamental model of the internal constitution of stars in terms of the transformation of mass to radiation well before the development of the details of the proton–proton cycle for the fusion of hydrogen to helium. His theory of the high-temperature regime in the core of the Sun recognized the importance of radiation pressure and the highly ionized state of the gas.*

Eddington was one of the leading exponents of Einstein's theory of general relativity and organized the expedition that led to one of the most dramatic proofs of it: the observation of the bending of light at the 1919 solar eclipse. He also calculated that the density of matter in a white dwarf was so great that a ton could be "put into a match box." Accordingly, he persuaded Walter S. Adams in 1924 to measure the displacement to the red of the spectral lines of Sirius B that confirmed Einstein's prediction of a gravitational redshift.

> birth. At last it is so near the boundary that it can dart outside and travel forward in peace for a few hundred years. (p. 27)

The specific details of the nuclear reactions that occur in the core of the Sun were worked out by Hans Bethe and Charles Critchfield in the United States and by Karl-Friedrich von Weizsächer in Germany in 1938. At the 15-million-degree temperature of the solar core, the so-called proton–proton cycle dominates. The first step is a fusion of two protons to form a deuteron (a nucleus of heavy hydrogen consisting of a proton and a neutron), accompanied by the release of two subatomic particles, a positron (antielectron) and a neutrino. In the next step, the deuteron fuses with another proton to form light helium (helium 3, two protons plus one neutron). Finally, two light helium nuclei fuse to form a single normal helium nucleus (helium 4, alpha particle), and at the same time, two excess protons are released to repeat the process. When the initial and final masses are compared, a small amount of mass is missing, transformed to thermonuclear energy.

Raymond Davis, Junior's quest to catch the will-o'-the-wisp neutrinos was described in Chapter 2. To the dismay of all the believers, the Davis experiment did not give the answer they expected. The neutrino flux is only one-third that predicted by theory. The dilemma would be resolved if the solar-core temperature were only 14 million degrees kelvin instead of 15 million degrees kelvin, but the temperature deduced from the flow of solar radiation and from the Sun's volume and weight doesn't allow much uncertainty in the theoretical solar model. Various speculative hypotheses have been put forward. Perhaps the Sun's core temperature oscillates and today's solar luminosity was generated millions of years ago at a higher temperature, whereas the neutrinos fly out promptly and reveal the present core temperature. Most bizarre is the suggestion of a black hole that creates energy in the core of the Sun by a nonnuclear process. Of course, there is no direct evidence for a black hole in the Sun.

Some Vital Statistics

Because the Sun is so near, it appears to be the largest and brightest object in the sky. Actually, its rank is only yellow dwarf, Class G-2, one of the smaller and fainter dwarf stars in the Galaxy. Our star is approximately 365,000 miles in diameter, about 109 times the diameter of the Earth. Giant stars sometimes have diameters hundreds of times wider, and dwarf stars may be no larger than planet Earth. The Sun's mass is 2×10^{33} grams, or 2×10^{27} tons, 330 thousand times the mass of the Earth; just a middle-class star. Although the Sun is a gas ball throughout its volume, the average density is 1.4 times that of water. In its outer layers, the density of the solar atmosphere decreases to much less than that of terrestrial air, but in the core its density is 160 times that of water, or roughly 60 times that of ordinary rock. The innermost 3% of the Sun contains two-thirds of its total mass, and it is there that the temperature and density combine to provide the conditions for nuclear reactions to occur.

Inside the Sun, three regimes can be identified: The thermonuclear core reaches to about one-fourth the solar radius; the radiation zone, in which energy is carried out from the core by x-rays and ultraviolet light, extends to 80% of the radius; and the convection zone, which transports heat by turbulent motion, comprises the outermost 20%. The energy that bubbles to the photosphere is converted again, into radiation that floods away from the Sun. Convection occurs only in stars that, like the Sun, have surface temperatures of less than 8000°K. At higher temperatures hydrogen becomes fully ionized and so transparent that convection cannot compete with radiative transfer. In small main-sequence stars with surface temperatures of less than 3000°K, the convection zone must occupy almost the full interior. It is the convective churning of electrically conductive plasma in stars such as the Sun that generates intense magnetic fields.

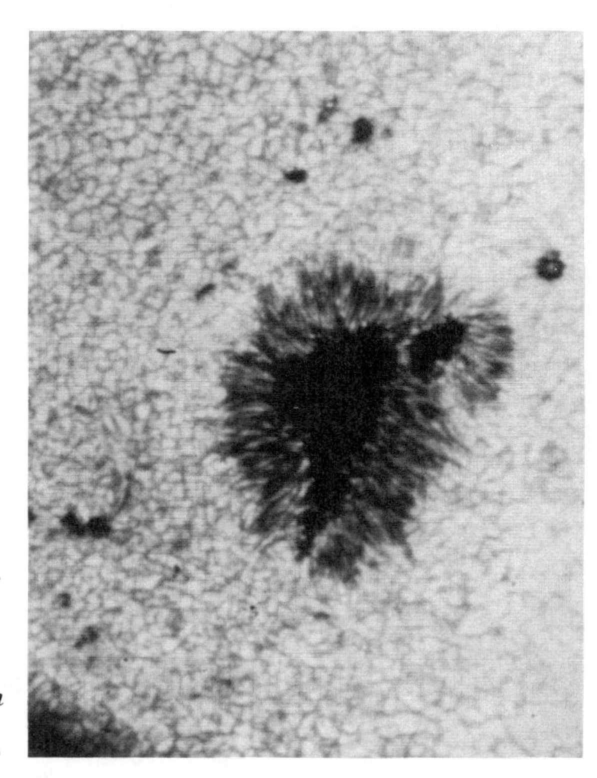

FIGURE 35. *A sunspot and the surrounding granular appearance of the solar surface are revealed in this photograph obtained from a balloon-borne telescope on 17 August 1959 at an altitude of 80,000 feet. (Courtesy Martin Schwarzschild, Princeton University.)*

The Sun's photosphere, or visible surface, is a zone of granulated texture that resembles the pattern of rice grains in a bowl of soup. It is only a few hundred kilometers thick and has a temperature of about 6000°K, which is too hot for any solid or liquid matter to exist. One may be puzzled that a ball of gas can exhibit a "surface." In ordinary air gas has virtually no opacity to visible light, but if the air temperature were 6000°K the visibility would be only a few feet. Even though the hot gas at the Sun's surface is very thin, its opacity is so great that the edge of the Sun appears sharp.

In a pot of boiling water, bubbles come zigzagging from below, bumping and merging into larger bubbles, to burst at the surface. Similarly, the Sun's coarsely mottled surface is believed to overlie a convection zone in which hot gas bubbles, seething at temperatures of thousands of degrees, rise in columnar fashion to deliver heat to the surface, after which cooled gas sinks, to be heated again. A solar bubble may be 1300 kilometers across. At any moment, the surface is covered with about 1 million bubbles that rise at speeds of several thousand kilometers per hour, then subside in a matter of minutes.

Spectroscopic observations made with telescopes show that the surface of the hot Sun "shivers" in a variety of modes. These shivers are subtle and complex. Most obvious is a motion with a periodicity of about 5 minutes, in which the surface heaves as much as from 1000 to 2000 kilometers. But the motion does not involve the entire photosphere simultaneously. Like the surface of a choppy sea, small local regions bubble independently of each other, and the 5-minute period is only a rough average. Recent studies also indicate long-period oscillations of small amplitude. Theoretical models predict such movements and relate them to the Sun's interior structure, much as earthquakes generate seismic waves that tell us about the interior of the Earth. When all the vibrational modes of the Sun's "heart throbs" are resolved, it may become possible to draw fundamental inferences about the amounts of heavy elements in the interior and about how density and temperature vary with depth.

Some solar physicists are beginning to suspect that conventional static models of the solar interior may be too simplified and that more-dynamic models might explain the missing neutrino problem. Present models do not include any mixing of the outer layers with the core region of nuclear burning. If such mixing occurs, the solar luminosity might vary on time scales ranging from 1 to 10 million years.

The Sun in Old Age

The Sun is now sedately middle-aged. Some 5 billion years hence, the aged Sun will undergo two dramatic evolutionary transformations. In comparatively rapid succession, it will become a red giant and collapse into a white dwarf.

The surface of the Sun appears to be white hot at 6000°K. A red star typically will have a temperature closer to 3000°K. If the star is a red dwarf the size of the Sun, it must necessarily appear to be very much dimmer. To explain the brightness of many red stars therefore requires that their size be giant. To compensate for their low surface brightness—that is, the comparatively small amount of light emitted per unit surface area—they must be very much larger than the sun. In this class are the familiar red giants Betelgeuse and Antares.

How will the Sun evolve to the red giant stage? Over billions of years it will steadily deplete the hydrogen in its core, and the helium formed by fusion will concentrate toward the center. Although much unburned hydrogen remains in the outer regions, there are no downdraft currents to circulate that fuel into the core fusion zone. Starved of fuel, the nuclear furnace banks down and the core begins to collapse under the weight of the overlying layers. Helium created by hydrogen fusion sinks to the center, but the temperature does not immediately rise enough to ignite helium fusion.

As more and more helium accumulates in the core, it becomes hotter and hotter until the temperature and pressure reach the threshold of helium fusion into the heavier nuclei of carbon, nitrogen, and oxygen. With accelerating fusion of helium, more and more heat is delivered to the outer regions of the star. Expansion takes off at a far greater rate as the Sun leaves the main sequence and balloons into a red giant.

When the outer portions of the Sun expand and cool they take on a reddish hue. Even though the emission per unit area decreases, the surface grows so quickly that it more than compensates for the coolness. When the Sun's diameter has increased 100 times, its surface will be 10,000 times as great; the total power radiated will be far higher than it was during the Sun's normal stage of burning, despite the fact that the surface is cooler.

Fusion of helium produces far less energy than the earlier hydrogen-burning stage, and the supply of helium runs out much faster than the supply of hydrogen did earlier on. All in all, the energy available from helium fusion is only 5% of that from hydrogen. But the red giant radiates energy at such a ferocious rate that it cannot survive very long, perhaps 1 or 2 million years. Because their lifetime is so short, we see few red giants compared to all the other stars; they comprise 1% of the total.

When nuclear fuel runs out and there is no source of heat to hold the star in balance against its own gravity, it must collapse. The red giant does so in a very short time. As it collapses, it heats up. Any hydrogen that remains on the outer fringes may then get hot enough and compressed enough to ignite nuclear fusion. When that happens, a violent explosion blows the outer portions of the star into space, and an expanding shell of gas known as a *planetary nebula* surrounds the collapsing star. The shell appears brightest as we look through the edges where the thickness is greatest in the line of sight, and it takes on the appearance of a large smoke ring.

The core of the Sun will shrink and grow hotter until a brilliant blue-white dwarf is formed. Its radiation pressure will continue to drive the planetary nebula far into space until it eventually thins out and disappears into the thin debris of gas and dust of interstellar space. After about 100,000 years, all that is left to see is the highly compressed core remnant, a white dwarf. A piece of white dwarf material the size of a sugar cube can weigh hundreds of tons. Over tens of billions of years the white dwarf will cool until its light is extinguished. The fossil star will eventually become a black rock, lost in the cold void of space.

I had a dream, which was not all a dream.
The bright sun was extinguished, and the stars
Did wander darkling in the eternal space,
Rayless, and pathless, and the icy Earth
Swung blind and blackening in the moonless air.

—LORD BYRON,
"Darkness," 1816

Solar Activity

Sunspots

Aristotle described the Sun as a disk of pure fire without blemish. A fleeting glimpse of the Sun with the naked eye does indeed give the impression of a uniformly bright disk, but if one looks carefully through a smoked glass, it is possible to detect small dark blotches called *sunspots* on its sublime face. When Galileo first pointed his telescope toward the Sun in 1610 and clearly resolved sunspots, he aroused the anger of theologians of his time, for whom sunspots meant blemishes in the handiwork of God. They refused to look through his telescope for fear that they might be convinced the spots were there. Galileo himself quickly concluded that the spots were attached to the Sun, although he first thought they might be clouds above the Sun's surface. It was soon recognized that spots move across the face of the Sun from left ro right in about 14 days, consistent with a rotation period of about 27 days; but the rotation has the peculiar property that spots at the equator rotate faster than do those at the poles. All manifestations of intense solar activity are connected with sunspots.

Sunspots are cooler regions (about 4000°K) embedded in very strong magnetic fields and surrounded by the much brighter 6000°K gas of the photosphere. Although they look black by contrast, sunspots are actually as bright as the full Moon. Some of the largest are about 40,000 kilometers across, 3 times the diameter of the Earth; smaller spots range all the way down to pores only 1000–2000 kilometers in diameter. The number of sunspots varies with a periodicity of roughly 11 years, although some cycles have been as short as 7 or 8 years and as long as 18 years from minimum to maximum and back to minimum. At the minimum of the sunspot cycle, it is not unusual for the disk of the Sun to be almost entirely free of spots. At solar maximum, 100 spots may cover the disk on any day. Small individual spots may last for 1 day or less; large groups and very large spots often persist for several months.

Most schoolchildren have experimented with bar or horseshoe magnets and iron filings, repeating a demonstration first performed in the nineteenth century by Michael Faraday. The filings are sprinkled on a sheet of paper placed over the magnet. As the paper is tapped, the filings move into alignment in a pattern that resembles lines of magnetic force emerging from the North Pole and arching around to the South Pole. The magnetic field of the Earth can be modeled as though a bar magnet were located in the core just slightly offset from the axis of rotation. A small toy magnet made of steel may have a field strength of some tens of gauss close to the pole; the Earth's general north-south field is considerably less, about 0.2 gauss. The Sun also exhibits an average dipole field, and its strength is only about 1 gauss.

As a magnetic star the Sun is a weakling compared to magnetic white dwarfs and neutron stars that exhibit fields millions to trillions of times as strong. Sunspots themselves have far stronger magnetic fields than the average Sun field, often in excess of 1000 gauss. The powerful magnetic fields are generated deep inside the Sun. Because of the solar rotation, the lines of magnetic force are dragged into ascending and descending streams of gas that become twisted into whorls and seem to rise like smoke rings toward the surface. They finally break through the photosphere and loop back again in tight arches to form pairs of spots at the feet of the loops.

After the initial excitement of the discovery of sunspots, no systematic study of their variability ensued. It is recorded that some Italian astronomers remarked on possible connections between sunspots and weather; they noted that a period of unusual drought accompanied an absence of spots in 1623.

The English astronomer William Herschel suggested in 1801 that sunspots were linked to weather. He found that extreme spottedness of the Sun correlated with the price of a bushel of wheat on the London market, which in turn was connected by the law of supply and demand to the weather. His idea predated by a half-century the discovery of the sunspot cycle. However, Jonathan Swift made a similar conjecture as early as 1726.

The discovery of sunspot cycles is attributed to Heinrich Schwabe, a pharmacist in Dessau, who started to keep careful records in 1826. His objective was to distinguish the transit of a planet across the face of the Sun, but to his great surprise his records of spots revealed a definite periodicity.

Since Galileo's first telescopic observations, there have been two stretches of time totaling more than 150 years when very few spots were noted. From 1645 to 1715, a period known as the Maunder Minimum and spanning the reign of Louis XIV, the "Sun King," there is little record of sunspot observations in Europe, and coronal streamers were almost totally missing when eclipses were observed. It was not that seventeenth-century astronomers were uninterested in sunspots. The Astronomer Royal, John Flamsteed, noticed a spot in 1684 and wrote, "These appearances, however frequent in the days of Scheiner and Galileo, have been so rare of late that this is the only one I have seen in his face since December 1676" (*In* Simon Mitton, *Daytime Star*, 1981, p. 132). Walter Maunder himself, in a retrospective study, reported to the Royal Astronomical Society in 1890 that "for a period of about 70 years, ending in 1716, there seems to have been a remarkable interruption of the ordinary course of the sunspot cycle. In several years no spots were seen at all and in 1705, it was recorded as a most remarkable event that two spots were seen on the sun at the same time, for a similar circumstance had scarcely ever been seen during the 60 years previous" (*Monthly Notices of the Royal Astronomical Society 50*, 1890, p. 252).

As persuasive as these observations seem, there is currently much effort to look into European diary sources, some of which cite sunspots during the Maunder Minimum. In 1975 Chinese astronomers began a general survey of ancient astronomical records. During a search of 8000 collections of private and public records, they discovered 13 reports of sunspots during the Maunder period. It seems that the Maunder Minimum must have been a period of few sunspots, although certainly not spot-less.

Some scientists have been strongly tempted to connect the Maunder Minimum with the simultaneous occurrence of the Little Ice Age in Europe. However, any evidence of a relationship between sunspots and weather is very shaky and is regarded with skeptical reserve by environmental scientists.

Above the Limb of the Sun

The finely bubbling surface of the Sun tells little about the dramatic panorama of activity that unfolds at greater heights. In 1889, at the youthful age of 21, George Ellery Hale made a profound mark on solar astronomy with his invention of the spectroheliograph, an instrument that could isolate a narrow range of color—a single spectrum line—in which to image

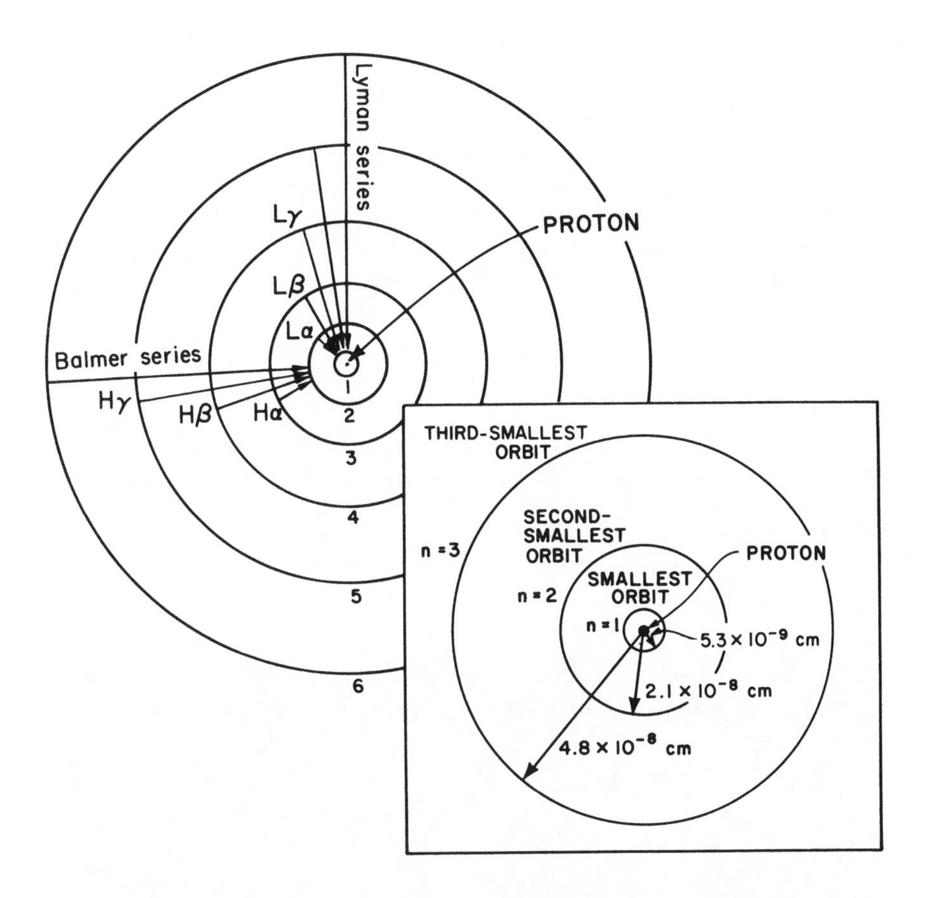

FIGURE 36. *Hydrogen is the simplest atom, consisting of one electron revolving about a proton nucleus. In 1913 the Danish physicist Niels Bohr developed the modern quantum theory of the hydrogen atom, in which the electron can circulate the nucleus only in distinct orbits that are denoted by integer quantum numbers, each corresponding to a specific energy state. A jump of the electron from a higher to a lower orbit is accompanied by the emission of a photon, a quantum packet of radiant energy. Conversely, absorption of radiation can occur in discrete energy jumps from a lower to a higher orbit. Transitions from orbit 2 to orbit 1 produce the Lyman alpha spectrum line at 1216 angstroms in the far ultraviolet. Transitions from 3 to 2 produce the red Balmer alpha line at 6563 angstroms.*

Atoms of higher atomic number have many orbital electrons and their spectra are correspondingly more complex. The spectrum emission and absorption lines are unique "fingerprints" of the elements.

the Sun. In the red light of hydrogen or the violet of calcium, the images were strikingly different. Fine details were resolved that were otherwise blurred in the complexity of colors of all the elements combined in white light.

By 1936 Hale's instrument was refined to a high degree of perfection by Robert H. McMath at the University of Michigan. McMath made time-lapse movies of the Sun in hydrogen light that showed a startling transformation from an almost smooth surface to one of rocketing ejections of prominences, whirling and surging formations twisting in great arches to heights of 150,000 kilometers. Just above the rim of the Sun, the granules of the photosphere transform into a fountainlike structure. Moving at speeds between 15 and 25 kilometers per second, jets of gas called *spicules* reach heights as great as from 10,000 to 15,000 kilometers. These spicules have been likened to a spray of the photosphere. The altitude regime in which

FIGURE 37. *Faint, far-reaching streamers of the solar corona are revealed at times of total solar eclipse. The wispy coronal forms shine by scattering the light of the photosphere. As shown here, at solar sunspot maximum the streamers surround the disk like the petals of a dahlia. When the Sun is quiet, the streamers concentrate toward the equatorial plane. (Photo by Serge Koutchmy.)*

they occur is the chromosphere, a thin shell dominated by the red light of hydrogen (Balmer alpha line, 6563 angstroms) in which the temperature rises abruptly from the 6000°K surface to as high as 200,000°K.

Still higher up, the chromosphere blends into the corona, where the solar gas suddenly becomes 1000 times thinner. So transparent is the corona that the stars shine clearly through it when the dazzling light of the disk is blocked out at times of eclipse; comets traverse it without noticeable effect. At sunspot maximum, eclipse photographs show a corona shaped like a pearly white dahlia, with the black moon at its center. When the sunspot cycle reaches its minimum, great equatorial streamers distort the symmetry, reaching millions of miles into interplanetary space. With modern instruments space scientists can sense these long fingers of solar gas brushing the near-space environment of the Earth. The corona is highly structured, not just a ballooning bag of gas.

Because it is so thin, the white light of the corona is not brighter than the full Moon. But the image is altogether reversed in x-rays and far-ultraviolet light. The corona and chromosphere then shine out brightly, and the photosphere is almost black. The message of x-ray and ultraviolet astronomy brought back from rockets has given us a new Sun.

The Leap into Space

From the ground we can observe only the visible light of the Sun, a small range of its radio noise, and some of its infrared heat rays. Radio scientists in the late 1920s were aware of an electrified region at altitudes above 100 kilometers. This is the ionosphere, which behaves as a mirror for the reflection of broadcast radio waves. Short-wave radio signals skipped around the Earth, bouncing between the ground and the ionosphere over distances far beyond the line of sight. The source of this ionization was a mystery. No solar radiation in the visible or near-ultraviolet spectrum could ionize the known constituents of air, yet the ionosphere waxed and waned over the course of a day, maximizing when the Sun was at its zenith. Using observations of the height of the ionosphere and the calculated atmospheric density, Edward Olson Hulburt, head of the Physical Optics Division of the United States Naval Research Laboratory in Washington, D.C., was impressed with the good fit of x-ray absorption to the height of the ionosphere.

In 1929 John A. Fleming, then director of the Department of Terrestrial Magnetism of the Carnegie Institution, wrote to Hulburt. At that time Robert H. Goddard's experimental rockets offered promise of eventually reaching heights perhaps as high as 100 kilometers. Hulburt responded to Fleming's request for ideas about how to make scientific use of such rockets. Among many suggestions including studies of atmospheric composition, pressure, temperature, and ionization, Hulburt mentioned that, "In the daytime, a photographic plate covered with black paper, etc., sent up would record the presence of x-rays" (letter to Fleming).

But Goddard's development of rockets was suspended during World War II. The most notable inventions during the war years in the United States were the bazooka, a small antitank rocket launched by infantrymen from a tube resting on the shoulder, and the JATO, Jet Assisted Takeoff rocket, which was mounted on aircraft to increase its lift on takeoff. Nothing remotely comparable to the V-2 was available in the allied countries. Hulburt's intuition about solar x-rays eventually was proven correct when direct measurements by means of high-altitude rockets became possible after World War II.

It was my good fortune to join the ranks of astronomers at this exciting juncture in the early rocket years just after World War II. When the war ended in Europe, V-2 rockets were being produced in Nordhausen, Germany, in the largest underground factory in the world. It was manned by thousands of prisoners from France, Poland, the Netherlands, and the Soviet Union working under the most brutal conditions. The site was within the Soviet zone, and it was nip-and-tuck for the United States army to get there first and abscond with the last of the V-2s from under the noses of the advancing Soviet forces. Many parts had suffered damage from Nazi demolition efforts, looting of storage areas, and exposure to weather. Of the total United States Army haul, only two V-2s could be reassembled from originally matched parts. Enough components were brought to the United States, however, to put together about 100 flightworthy rockets.

After crossing the ocean aboard Liberty ships, about 300 freight carloads of V-2 parts arrived in New Mexico in August 1945. A fleet of flatbed trucks hauled the cargo across the mountains from Las Cruces to the White Sands Missile Range. While the Army set to work with the rockets, scientific users were invited to fill the space of the 2000-pound warhead compartment with instruments in place of the lead ballast or concrete needed to balance the rocket. At the United States Naval Research Laboratory (NRL), the Rocketsonde Branch was established January 1946 to begin preparing scientific payloads for atmospheric, ionospheric, and cosmic-ray research. Enthusiasm for use of the rockets quickly spread to the Optics Division, which was led by Edward Hulburt. At that time I was head of the Electron Optics Branch and enjoyed an enticing diet of space-science discussions at lunches with Hulburt. His wide interests included solar control of the ionosphere, airglow, zodiacal light, gegenschein, and the escape of particles from the exosphere.

Early in 1946 Hulburt casually gathered a few of us to spread the news that NRL had secured a promise to fly scientific instruments in some of the captured V-2 rockets. He had dusted off a small quartz spectrograph that had seen service in auroral research during the second International Polar Year, 1932–1933. "Wouldn't it be a good idea," he suggested, "to

FIGURE *38. A V-2 rocket before launch at the White Sands Proving Ground, New Mexico. About 45-feet tall and 5 feet in diameter, the rocket was fueled by 10 tons of alcohol and liquid oxygen. A successful flight could reach 170 kilometers and last about 450 seconds. (U.S. Naval Research Laboratory.)*

fly that little old Hilger spectrograph that has been to so many places around the world?" Carried above the ozone layer that absorbs all solar ultraviolet below 3000 angstroms, it could observe the solar spectrum to wavelengths as short as the transmission limit of quartz.

Richard Tousey and other colleagues at NRL accepted Hulburt's challenge but soon realized that his simple approach wouldn't work. Within 3 months an innovative design for a spectrograph was developed in cooperation with the Baird Instrument Company in Cambridge, Massachussetts. It used lithium floride optics that transmitted shorter wavelengths than quartz. On 10 October 1946, on its second try, the 1½-foot-long spectrograph brought back the first solar ultraviolet spectrograms extending down to 2200 angstroms. By comparison with modern instruments, it was flea-weight and toylike, but it worked and opened a new era of rocket astronomy.

Other early participants in the V-2 upper-air research program at the White Sands Missile Range included the Air Force Cambridge Research Laboratory, the Army Signal Corps, The Johns Hopkins University Applied Physics Laboratory (APL), Princeton, Harvard, and the University of Michigan. Each organization tackled its own set of experiments and met with varying degrees of success and frustration. The novelty of conducting upper-atmospheric research with rockets presented unprecedented technical problems. Some of the early experimenters quickly gave up, but others persisted and succeeded in taming the maverick rockets.

The V-2 was a great gift to solar astronomy, but it came with a variety of handicaps. The rocket was 14 meters tall and its motor was so large

FIGURE 39. *A succession of ultraviolet spectra taken aboard a V-2 rocket 10 October 1946, as the rocket penetrated the ozone layer. The spectrum at 55 kilometers was the first ever obtained of the extension of solar ultraviolet short of the ground level ozone cut-off at 3000 angstroms. (U.S. Naval Research Laboratory.)*

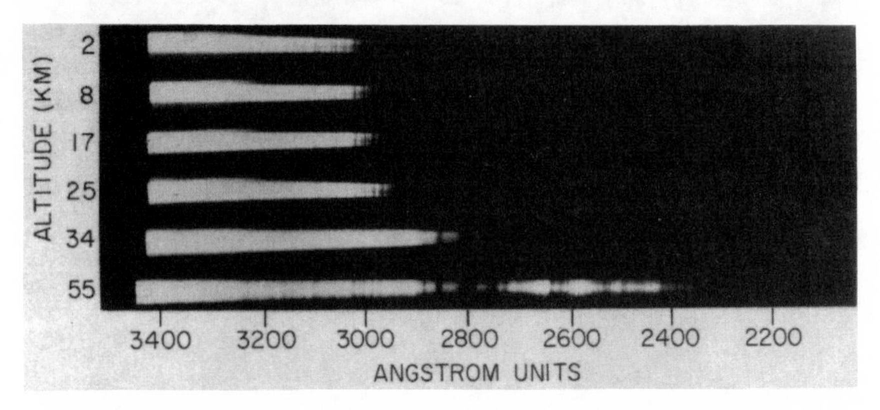

that a person could crawl through the nozzle. It had been designed as a weapon, not as a tool for high-altitude research. In the very early days at White Sands, one had to be something of an acrobat to put one's instruments aboard the V-2. The top of the rocket was reached by a long extension fire ladder. A little later a gantry was supplied, and from then on access to the rocket was simpler and safer.

After witnessing the short-lived explosive burning and rapid acceleration that characterized the firing of small military rockets, the slow majestic rise of the V-2 and the sudden vanishing of the rumbling roar of the rocket in the eerie aftermath of burnout was a breathtaking experience. Silent, snaking vapor trails marked the passage of the rocket through the stratosphere, and sound returned only near landing, when shock waves reverberated from the mountains. But rarely did the V-2 fly like an arrow straight into the sky, even though it was equipped with gyro-controlled jet vanes for guidance. As often as not, the rockets tumbled and faltered. Some burned up furiously on ignition; one took off in horizontal flight to land on the edge of Juarez over the Mexican border; another fell not very far from tourists wandering over the pristine sands of the White Sands National Park at the edge of the rocket range. The unauthorized Juarez landing had international repercussions that resulted in a temporary closing of the border. On the day after it landed, street urchins were hawking souvenir parts in the markets of Juarez.

The first five of the huge rockets all returned nose-down in streamlined flight and buried their completely pulverized remains in craters about 10 meters deep. Not a trace could be found of the first spectrograph flown by NRL in June 1946. The scientific crew members dug and dug inside the hole until they found—water. In the few buckets of debris that were sifted from tons of overlying sand, there was no trace of the armored steel film cassettes. To lessen the impact damage, later rockets were severed at the tail and nose sections by pyrotechnics as they reached 50-kilometers altitude on descent. The separated parts then tumbled down at the comparatively gentle speed of about 100 meters per second, and the cassettes were recovered successfully.

I was one of the first American scientists to have the opportunity to use the V-2 rocket. My colleagues were all physicists; none of us had been exposed to formal education in astronomy. Astronomers at the time were busy pursuing studies with ground-based telescopes and shunned the electronic gadgetry, radio telemetry, and especially the risks associated with rockets. It took an adventurous breed of scientist to spend months to a year preparing a payload that would be perched atop a huge rocket motor, suffer a violent shock upon takeoff, and disappear into the blue yonder, eventually to return to Earth with devastating impact. The tension of countdown, and apprehensions about malfunctions after the launch, when there was no recourse, were enough to discourage faint-hearted souls.

My comment about physicists taking the lead should be tempered by the recognition that only a handful of physicists moved into space physics and astronomy in the early years. After an early success in measuring solar x-rays, I reported the results in a paper submitted to the *Physical Review*. Sam Goudsmit, then editor of the *Physical Review*, wrote me in return that he would publish the paper this time, but that few physicists were interested in such work; he urged me to find another journal in which to report future research. I subsequently sent papers to the *Astrophysical Journal*, where they were accepted for publication without quibbling; that is how I became an astronomer. Today, the American Physical Society has its Division of Astrophysics and is a major forum for the discipline.

In spite of the forbidding odds, high-altitude research with rockets provided dramatic mixtures of exaltation and trauma for early practitioners, a sense of high adventure, of bold gambling with big stakes and great risks. We would flirt with disaster while anticipating unexpected discoveries from the exploration of unknown regimes. Scientists who fly their telescopes on rockets are still youngsters at heart, exhilarated by the suspense of countdown and the blast of liftoff, much as they had once thrilled to exploding a giant firecracker.

NRL was not alone in its early attempts to measure the solar ultraviolet spectrum. John J. Hopfield and Harold E. Clearman at The Johns Hopkins University Applied Physics Laboratory obtained excellent results only 6 months after NRL. But it was immediately apparent that extension of the solar spectrum to shorter wavelengths would require a means of pointing the spectrograph at the Sun from a stabilized platform in order to prolong the exposures. The change from the NRL method of simply counterrotating the payload about the rocket spin axis to two-axis stabilization that compensated for both spin and pitch was developed by the University of Colorado for the Air Force. It was a major technical development, but it took 6 years. The biaxial pointing control eventually became the most important single contribution to instrumentation in solar astronomy until the orbiting solar observatories of NASA were developed 1 decade later.

Prelude to Solar X-Ray Astronomy

Around 1940 scientists inferred from simple physical considerations that the solar corona must be very hot. It was recognized, for example, that to sustain an extended atmosphere hundreds of thousands of kilometers above the photosphere against a force of gravity 28 times as strong as at the surface of the Earth required that the coronal gas temperature exceed 1 million degrees kelvin.

More specific evidence of high temperature came from examination of the coronal spectrum at times of eclipse. The light of the corona exhibits

FRITZ ZWICKY

Although California Institute of Technology astronomer Fritz Zwicky's name appears frequently throughout this book, my first contact with him was not in astronomy but in the course of rocket development at White Sands. He wanted to be the first to put an artificial body in orbit, and he proposed using the V-2 as the first stage from which to launch small metal slugs weighing no more than 1 ounce. The slugs were to be shot out by shaped charges that could impart escape velocities of 7 miles per second. At suborbital velocities, the slugs would be heated to vaporization and would create luminous trails mimicking meteorites. It is estimated that about 1 billion meteorites enter the atmosphere each day. At the time that Zwicky's experiment was performed, meteorite trails were studied for what they revealed of the physics of the upper atmosphere and the possibility of radio communication via reflections from their ionization trails. There is no record of any conclusive result from Zwicky's experiment, although the launch, carried out in December 1946 with the assistance of NRL and APL scientists, was successful. No repeat was attempted.

Some of Zwicky's proposals were so outlandish for his time that many of his fellow astronomers refused to take him seriously. For example, he talked in the 1930s of sending a rocket to the Moon and retrieving a sample of lunar soil. He was only 30 years premature. More important for astronomy was his concept of a neutron star, which he proposed shortly after the discovery of the neutron in the 1930s. With Walter Baade, he suggested that the end product of the supernova process would be a tiny, incredibly dense star composed entirely of neutrons crushed into contact with each other. At the time, his idea was brushed aside by most astronomers as just another wildly speculative figment of the Zwicky imagination. It remained for J. Robert Oppenheimer to treat the idea of neutron stars seriously in the late 1930s, and then another 3 decades passed before neutron star pulsars were actually observed.

a continuous spectrum that closely resembles that of the underlying Sun, just as would be expected if the corona shone by scattered light like the halo around a street lamp on a foggy night. Because the discrete spectral lines of the photosphere are almost completely washed out in the corona, it was deduced that the light of the corona was being scattered by free electrons moving at very high speeds. In the process of scattering light, the fast electrons Doppler-shift any sharp spectral lines into a smeared-out blur. Their high kinetic energies implied a temperature of about 1 million degrees kelvin.

Careful spectroscopic measurements of atoms in the corona itself showed broad emission lines that mystified solar astronomers because the wavelengths did not correspond to the spectra of any elements known on Earth. They named one of the brighter lines at a green wavelength of 5303 angstroms *coronium* and attributed it to some peculiar substance unique to the Sun. But such a hypothesis violated the evidence of the periodic table of Mendeleev, which accounted for all the known elements and left no place for coronium.

During the Sixth General Assembly of the International Astronomical Union in 1938, spectroscopist Bengt Edlén showed photos made in his laboratory at Uppsala, Sweden, of the heliumlike and hydrogenlike spectra of oxygen and nitrogen atoms stripped of all but their last one or two electrons. These spectrum lines fall in the soft x-ray range from 10 to 30 angstroms and were not thought to be of much astrophysical interest. His results were not even mentioned in the draft report of the meeting.

Edlén finally solved the puzzle of coronium in 1940, when he identified the strange spectral lines as atoms as common as iron, nickel, and calcium that had been stripped of as many as one-half of their normal complement of electrons by atomic collisions at the high temperature of the corona. He subjected iron electrodes to powerful high-voltage discharges to produce iron vapor ions that had lost as many as 13 of the normal array of 26 orbital electrons. The mystery green line of the corona then appeared.

The spectral lines of highly ionized atoms that radiate from the hot corona in the visible region are known as "forbidden" lines because they derive from atomic transitions that are relatively rare and correspondingly weak. By contrast, the fundamental transitions of iron stripped of 13, 12, 10, and 9 electrons, and of highly stripped nickel and calcium, lie in the soft x-ray spectrum. Together with the continuum radiation, strong-line emission accounts for a substantial input of x-rays capable of ionizing the upper atmosphere. Edlén's spectral diagnostics confirmed the high temperature of the corona and prepared the stage for an early triumph of x-ray astronomy: the direct measurement of the Sun's x-rays. After initial successes with broad-band photometry of solar x-rays in 1949, followed by a full decade of monitoring the solar cycle variability of x-rays, my colleagues and I flew a Bragg crystal spectrometer in 1963 and recorded the x-ray spectrum lines of oxygen stripped of all but 1 or 2 of its normal compliment of 8 electrons, nitrogen left with only 2 electrons, and iron that had lost 16 electrons. Edlén's predictions were fully confirmed after 25 years. When I sent him a preliminary copy of our coronal spectrum he acknowledged with great pleasure.

The Search for Solar X-Rays

During the first few years of V-2 activities our NRL team endured failure after failure to achieve simultaneously accurate spectrograph pointing, high altitude, and good recovery. I wondered if some simpler approach to obtain a rough map of the solar spectrum to the shortest wavelengths might achieve earlier success. But I was then almost totally committed to work associated with the atomic weapons program, and rocket experiments had to take second priority.

An opportunity to test my approach finally came in 1949 when I prepared a collection of photon counters sensitive to narrow bands of the ultraviolet and x-ray spectrum and coupled them to radio telemetry so that data were received in real time as the rocket flew. The V-2 flew to 151 kilometers. As the radio signals came in, we first read solar Lyman—

FIGURE 40. *The first measurement of the penetration of solar x-rays into the upper atmosphere made by a V-2 rocket in 1949. The 8-angstrom x-ray signal was modulated by the spin of the rocket as the Sun came into view once each roll period. X-rays were first detected at about 90 kilometers and reached peak intensity at about 130 kilometers. (U.S. Naval Research Laboratory.)*

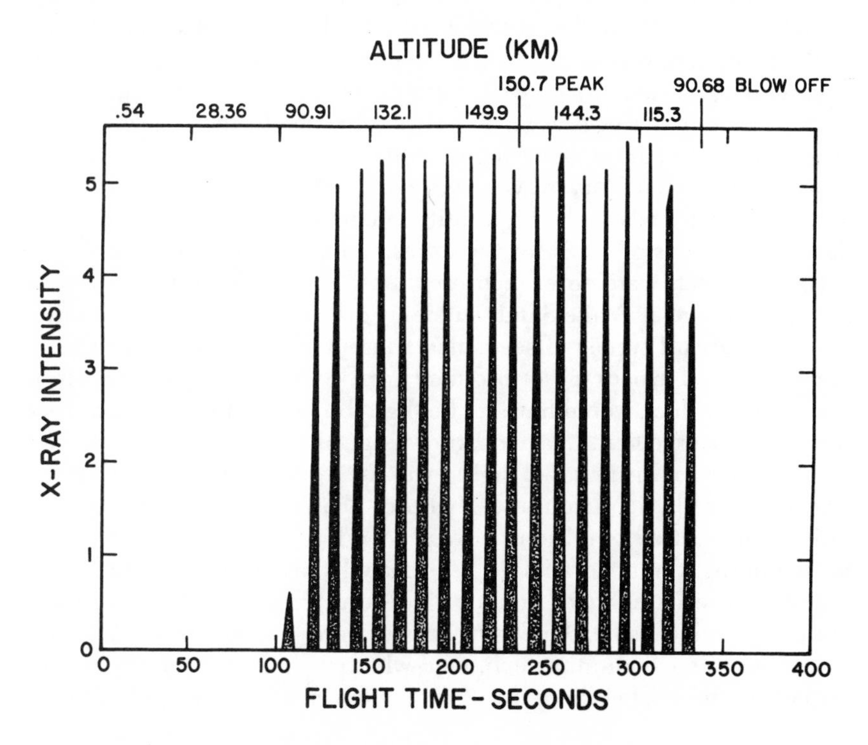

Alpha radiation in the extreme ultraviolet at a height of about 75 kilometers, followed quickly by soft x-rays above 85 kilometers. In this first simple set of measurements, Lyman–Alpha was identified as the source of the lowest-level ionization, the D-region, and soft x-rays as a principal factor in production of E-region ionization from 85 to 150 kilometers. Ultraviolet absorption near 1450 angstroms revealed the persistence of molecular oxygen to altitudes far above the level predicted by the theory of its dissociation into atomic components, with important theoretical consequences for the high-altitude structure of the ionosphere. These simple measurements solved much of the mystery of the Sun's control of the ionosphere that had baffled radio scientists for a quarter-century.

Our success established a high priority for further flights of photon counters, and I immediately set out to build a much more elaborate payload for a 1950 V-2 flight. On that occasion beginner's luck deserted me; the rocket barely lifted off the pad; it "walked" toward the blockhouse, where it fell over and burned so furiously that smoke and flames engulfed the entire structure, a frightening experience for those of us inside.

In the micromanaged environment of today's big rocketry, it is hard to recall the casual coordination between shared experiments in the early V-2s and the reliance on individual initiative. In the first few years of operations experiment preparations were usually rushed, and glitches in execution were more common than not. There was great dismay and embarrassment, for example, when a sophisticated cloud chamber instrument for cosmic-ray studies was flown with the lens cap still on the camera.

Because there was so much space aboard the V-2, it was parcelled out to several experimenters on each flight with little regard for compatability. Packets of frog's eggs or seedlings were frequently thrust on the launch crews late in preparations for flight, with pleas to tape them into any free space. Biologists at Harvard who wished to determine the effects, if any, of exposure of seeds to the high-altitude environment provided specially developed strains of seeds for the experiment. After the first two flights, the seed packets were never recovered. At the next opportunity, the technician in charge at White Sands suddenly discovered to his consternation that the Harvard supply of seeds had run out. He rushed to Las Cruces and purchased a package of ordinary corn seeds and installed them in the rocket. As luck would have it, the flight broke all altitude records and recovery was excellent. I have no knowledge of the Harvard analysis.

At the time of my first V-2 flight I was not aware that Herman Yagoda, a cosmic-ray physicist from the Air Force Cambridge Research Laboratories, had sent an emulsion package to White Sands with instructions for mounting it in any available V-2 space. It was our practice to test the detectors by exciting them from outside with gamma rays from a 5-milligram radium capsule. Unwittingly, we were also exposing Yagoda's emulsions; after flight, they exhibited what would have been an incredibly large number of electron tracks for the few minutes exposure to cosmic

(a)

Figure 41. *Sketches of (a) the Aerobee launch tower and (b) the Viking rocket hangar at the White Sands Missile Range. (H. Friedman, 1950.)*

(b)

rays at high altitude. Yagoda was completely baffled, and we were innocently unaware of the damage we had wrought. This comedy was repeated again before the connection was made and we realized what had gone wrong with Yagoda's experiment.

Jim Van Allen, while still with APL, conceived the development of an inexpensive two-stage research rocket, the Aerobee, that could be assigned to a single investigator for each flight. It used an acid-analine sustainer motor atop a solid propellant booster, and the combination was fired out of a 140-foot tiltable tower. The name was coined from *Aero*, for the Aeroject Company that built it, and *bee* for the Bumblebee family of APL rockets. But the Aerobee started with a jolting 15-gravities acceleration and barely reached between 80 and 90 kilometers. At best, it could penetrate the D-region, the lowest portion of the ionosphere. When the first production run became available, there was little demand. I asked for one and was offered three. Our mode of operation became extravagant. I planned on using three each time we went to the field, and counted on success by statistics. After a couple of rounds of barely scratching the height of penetration of x-rays, we developed a stretched version of the Aerobee by simply lengthening the tank section. Immediately, it exceeded V-2 altitude and the Aerobee took on a new glamour for research. The rocket evolved through several versions of Aerobee-Hi and became the workhorse of research rockets. During the 1950s a series of Aerobee rocket flights showed that the intensity of x-ray emission followed the sunspot cycle.

Solar Flares

Of all the forms of solar activity, flares are the most spectacular and create the strongest impact on the terrestrial environment. In 1859 two English astronomers, Richard C. Carrington and R. Hodgson, described a remarkable burst of light within a large group of sunspots. The flash lasted about 5 minutes and spread rapidly over some 50,000 kilometers. For the next few nights, brilliant auroras stretched to middle latitudes, and compass needles wiggled erratically. Carrington thought he might have witnessed the plunge of a large meteor into the Sun; it was a popular notion around the turn of the century that the capture of meteors by the Sun was a major source of its energy.

Superficially, the flash of a flare suggests a lightning stroke discharging electricity accumulated in the flaring region, but a flare releases the power of billions of hydrogen bombs. Such energy can be drawn only from the intense magnetic fields in the solar atmosphere. Geophysicists had been puzzled by the association of flares with sudden ionospheric disturbances that often led to prompt blackouts of short-wave radio communications that could persist for hours. In the visible red light of hydrogen, a flare could cover a few square arc minutes of the solar disk and brighten tenfold as compared to the surrounding photosphere. It seemed reasonable to suppose that an accompanying brightening in the far-ultraviolet Lyman lines of hydrogen could produce the ionization of the upper atmosphere that was responsible for the radio blackout. But a simple calculation indicated that the energetic requirement was far from satisfied. Instead, we could show theoretically that a flash of x-rays was a far more reasonable explanation.

My personal interest in studying solar flares was whetted by a chance observation made during the course of entirely unrelated work in 1949. Following the Bikini Island atom bomb test, the NRL became involved in developing a technical surveillance scheme to monitor Soviet nuclear weapons development. My colleagues and I established a network of stations designed to detect radioactive fallout carried by winds from the first high-altitude explosion over the Soviet Union. One component of the sys-

FIGURE 42. *The great flare of 7 August 1972, photographed in the red light of hydrogen at the Big Bear Solar Observatory. The flare covered several one-thousandths of the area of the disk. Its x-ray emission created violent ionospheric and magnetic storms. (Photo by H. Zirin, Big Bear Observatory.)*

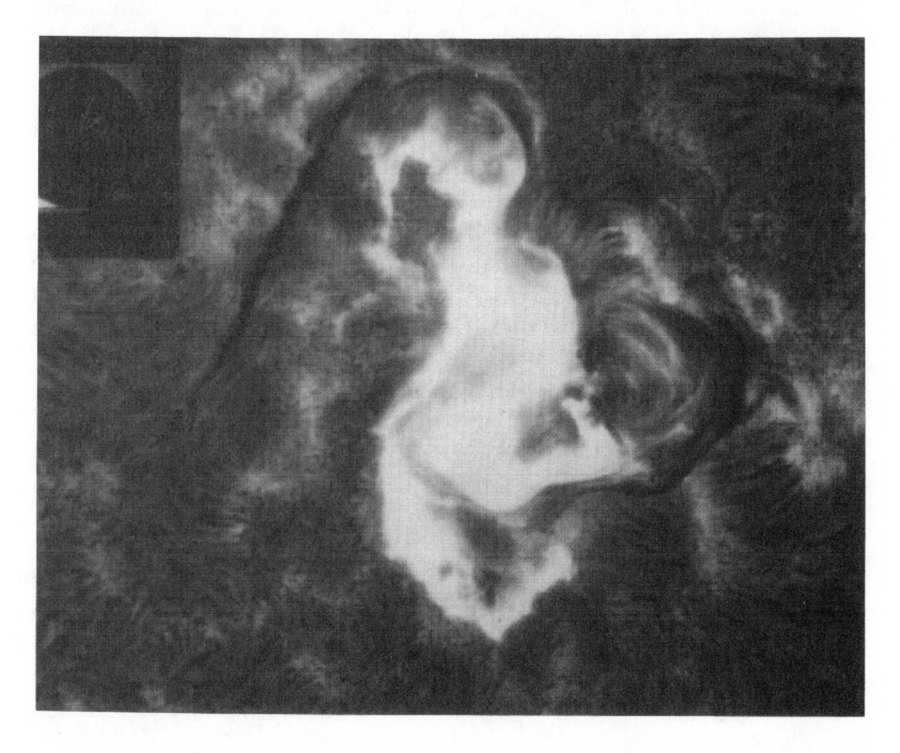

tem was a large bank of Geiger counters operated automatically at a set of air weather stations. When the solar flare of November 1949 occurred, the greatest on record at that time, we obtained excellent measurements of the penetration of high-energy solar cosmic rays from Arctic to Antarctic latitudes. Unfortunately, on that particular project I was bound by secrecy. I resolved then to pursue studies of the flare phenomenon in the unclassified rocket program.

The Aerobee was ill-suited to the study of solar flares. Before firing, its fuel system had to be pressurized with helium. Preparation went slowly, and once the rocket was ready it had to be fired within 1 hour. It offered little chance of catching unpredictable flares.

In the early 1950s Lieutenant Lee Lewis, attached to the Office of Naval Research, and Jim Van Allen proposed a novel approach to cheap rocketry: the Rockoon. A solid propellant Deacon rocket, 14 feet long and 6 inches in diameter, was suspended on a Skyhook balloon and floated to 25 kilometers, where it could be fired by radio command. Here was an inexpensive system that could be kept in the air all day and fired at the moment that the outbreak of a flare was detected.

Without recourse to an expensive orientation system on the balloon, the rocket's impact point could not be predicted within a circular area of about 100 miles radius, and winds contributed further uncertainty. Launching from shipboard far out to sea offered range safety and facilitated the handling of the balloon. While the Skyhook balloon was being inflated, surface winds had to be less than 10 knots. By cruising downwind, the ship could achieve nearly zero relative wind conditions for inflation, and the balloon would stand straight up with its suspended payload.

Our NRL group worked up plans for a flare-study program to be conducted with Rockoons from the deck of a naval vessel in the Pacific and gained support for an expedition in 1956, the pre–IGY (International Geophysical Year) year. We sailed on the U.S.S. *Colonial*, a Landing Ship Dock (LSD), from San Diego, and began operations about 400 nautical miles off the coast of California. Each morning we launched a balloon trailing a string of retro-reflectors, below which dangled the rocket, which we tracked by radar. Then began the chase. We could not permit the Rockoon to get out of range of our radio cutdown command, but upper-air winds were faster than the lumbering LSD. It helped that the winds reversed during the course of the day so that we could let the Rockoon travel away from the California coast in the morning, then catch it coming back in the afternoon.

It was characteristic of those days that one never had a fully tested payload until the night before firing the rocket, and often not then. One of the most bothersome technical problems was the igniter of the Deacon rocket. In its military version it was fired at normal atmospheric temperature and pressure from the ground. But that ignitor refused to function in the frigid thin air at 25 kilometers. Jim Kupperian, who recalled some juvenile experience mixing saltpeter and charcoal to make gunpowder with a junior chemistry set, volunteered to serve as our pyrotechnics expert. It must have appalled the ship's gunner's mate to watch such amateurish fiddling with explosives, but after a couple of disappointments Kupperian

FIGURE 43. *A Deacon rocket suspended from a Skyhook balloon. (U.S. Naval Research Laboratory.)*

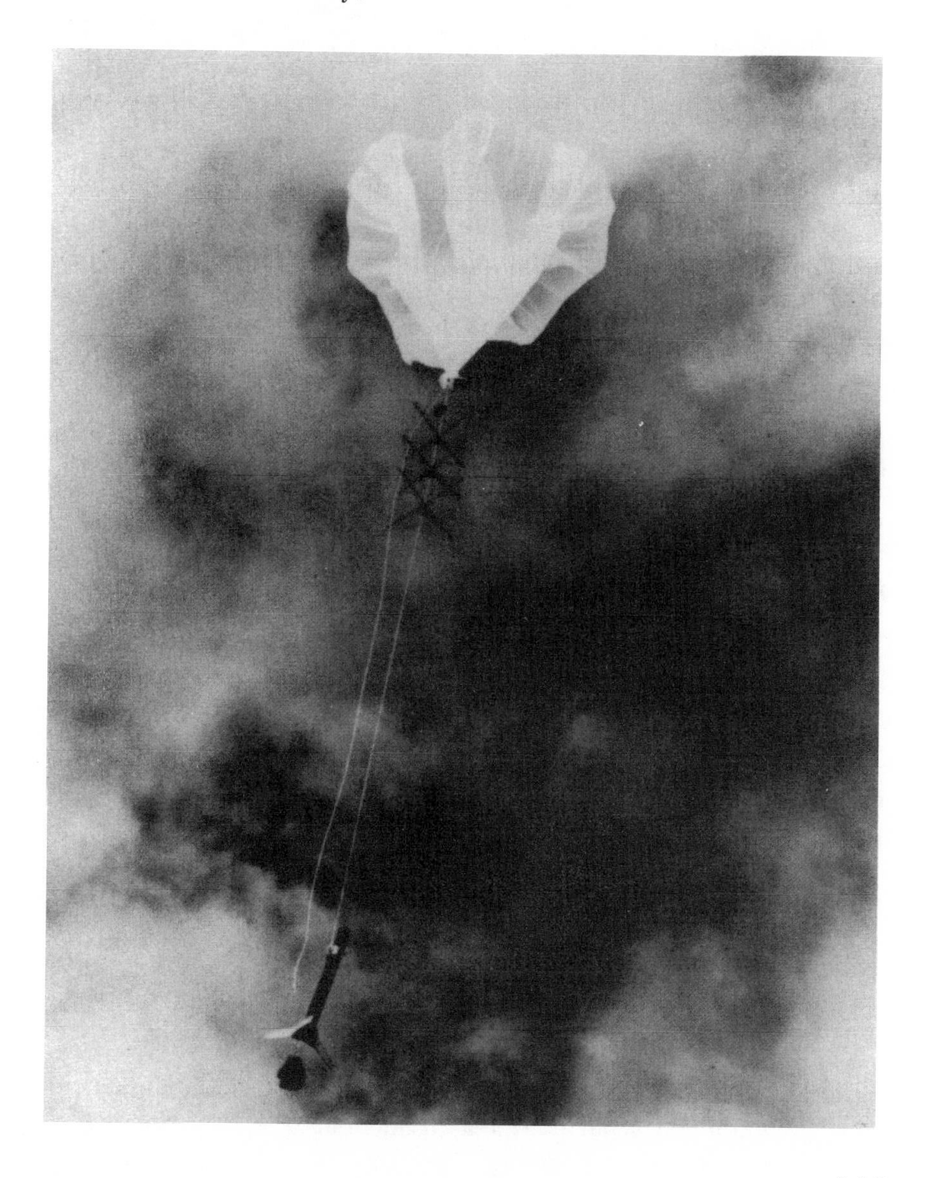

found the right formula and his igniters fired successfully. Another nice feature of these experiments was that the telemetry data appeared immediately in analogue form on a roll of pen-recorder paper. We would sit at the stateroom table with a cup of coffee and the entire record and, in a few minutes, the results of the experiment would be apparent.

We went to sea with 10 Rockoons and planned to launch 1 a day. The first 3 days produced no flares. Sunday was a day of rest, and with no Rockoon aloft to observe, the Sun produced two very good flares. Murphy's law was in complete control. During the fourth try we were all at lunch, except for Bill Nichols and Bob Kreplin, when a flare erupted. Bob reacted quickly and fired the rocket. It caught a weak Class 1 flare, but the observation of a strong x-ray flux above 70 kilometers was clearcut. We had proven that solar-flare x-rays are the key to the production of radio fadeout.

One year later Rockoons were out of style. In place of the balloon, a Nike solid-rocket booster lifted the Deacon to the stratosphere. The two-stage combination was fired from a simple rail launcher. My NRL team set up on San Nicholas Island, overlooking a beach on which hundreds of seals basked in the sunlight. No environmental impact statement was required, and we simply prepared to shoot our rockets over Seal Beach.

We needn't have worried about harm to the seals, but we should have had more concern for members of the press. The first test shot was scheduled for the opening day of the IGY. As usual, the Nike fired perfectly, but the second-stage Deacon did not. After separation, the Nike caught up with the Deacon and tore it up. Pieces of metal fell around the launch area, but the press was unaware that anything had gone wrong. With appropriate fanfare, the newspapers announced the first "successful" rocket launch of the IGY. Subsequently, we did have a number of very successful flare x-ray measurements.

When I studied the records of the Rockoon expedition of 1956 in more detail, I was intrigued with evidence of nonsolar hard x-ray radiation detected by a scintillation counter as the rocket rose above the maximum of cosmic-ray ionization near 25 kilometers. The residual radiation increased with altitude. Kupperian and I suspected that the origin might be cosmic. That observation spurred us to undertake a search for galactic x-rays. I will return to that story later on.

For those of us at NRL who had set our sights on launching the *Vanguard* satellites in connection with the IGY, the anticipation of getting the first astronomical observations from a space observatory was quickly dispelled by the surprise of *Sputnik* and the *Vanguard* test-launch disasters that followed. *Sputnik* was orbited on 4 October 1957. Public reaction was initially mild, reflecting President Eisenhower's comment that "It does not raise my apprehensions one iota about national security." The numbness wore off quickly, and public figures began to decry the shameful situation with the usual litany of accusations of Administration penny-pinching, short-sightedness, and general stupidity. Harry Truman blamed the McCarthy Era for having deprived America of its best brains. By implication, the *Vanguard* team was a bunch of second-raters. The Senate Preparedness Committee, chaired by Lyndon B. Johnson, immediately began an inquiry into the status of United States rocket and satellite programs. White House pressure forced *Vanguard* Project Director John Hagen to advance the date of the first launch of a 6-inch-diameter test sphere with a radio transmitter from February 1958 to December 1957.

By Wednesday, December 4, thousands of spectators, a new breed of birdwatchers, converged upon Cape Canaveral. No longer was the forthcoming launch just the first test in a complex development program. The press played it up as a full-fledged satellite launch, America's answer to *Sputnik*. Further delays added to the tension. When the rocket flame was ignited on December 6, the vehicle rose only a little over 1 meter and faltered. As it fell back, the fuel tanks exploded and the rocket crumpled to the ground, enveloped in hellish flames and billowing smoke. From the top of the three-stage rocket, the silvery 6-inch test satellite plummeted 25 meters through the flames and bounced on the concrete deck. There the wounded bird, its antenna badly bent, radiated a futile signal at 180 megahertz.

A wave of outrage struck the country. Senator Johnson labeled the failure most humiliating. Public chagrin poured out in scores of letters. Many suggested that we scientists ought to find work more appropriate to our talents. The Soviet delegation to the United Nations laughingly offered to include us in the U.S.S.R. program of technical assistance to developing nations. Authorization of a parallel satellite-launching effort by the Army Ballistic Missile Agency, using the Jupiter rocket, followed quickly, and that success story is well known. My solar radiation payload had been assigned first flight in the *Vanguard* plan, and James Van Allen's cosmic-ray experiment had second position. His payload was removed from *Vanguard* and assigned to the *Explorer I*, launched on 30 January 1958, which was to discover the Van Allen radiation belts.

The *Vanguard* crew persisted and finally succeeded in orbiting the grapefruit-sized test sphere in March 1958. In retrospect, it was a remarkable achievement, having started from scratch only 2½ years earlier. My long-awaited scientific opportunity came on 28 April 1958, but our first instrumented *Vanguard* satellite turned out to be submersible instead of orbital. In September, *Vanguard III* was orbited with my x-ray and Lyman–Alpha detectors aboard, but it was swamped by Van Allen Belt radiation. What was good for Jim was poison for me. My colleagues and I had to wait for the first Navy *SOLRAD (Solar Radiation)* satellite, launched in June 1960, before we succeeded in monitoring x-ray and Lyman–Alpha emission from the Sun. The *SOLRAD* program was conducted with 10 launches over a period of 1 decade and provided a complete history of the behavior of solar x-rays and ultraviolet over a solar sunspot cycle. In the same time frame, NASA conducted the much more advanced *Orbiting Solar Observatory (OSO)* series. By the beginning of the 1970s, with the foundation of manned experience in the *Apollo* program, solar physicists were ready for the first long-duration missions involving men with an astronomical observatory: the *Skylab* program.

Solar Eclipse: Darkness at Noon

Cold of the sun's eclipse
When cocks crow for the first time
hapless, and dogs in kennel howl
Abandoning the richly-stinking bone,
And the star at the edge of the shamed
and altered sun shivers alone,
And over the pond the bat but not the
swallow dips
And out comes the owl.

—Edna St. Vincent Millay,
"Huntsman What Quarry?", 1939

Serious studies of the solar corona became possible with the introduction of photography late in the nineteenth century, but the corona could be observed only when a total eclipse masked the blinding light of the solar disk. Only then did the faint, wispy structure of the crowning solar atmosphere stand revealed. Some 50 years after the introduction of photography for eclipse observations in 1882, Bernard Lyot, a French astronomer, succeeded in designing a telescope that artificially blocked the view of the disk so that the corona could be photographed well beyond the limb of the Sun at any time. His first observations were made from atop the Observatoire de Pic-du-Midi, 9415 feet above the snow-covered landscape of the Pyrenées. Lyot's type of "coronograph" has been installed on several mountain tops around the world where observations can be made on a clear day out to about 20 arc minutes from the edge of the disk.

Still, a total solar eclipse provides the most ideal conditions for scientific observation from the ground as well as the most awe-inspiring natural spectacle in the sky. Small wonder that eclipses have attracted scientific adventurers to remote parts of the world. Some were shipwrecked. In his determination to reach the coast of Africa, French astronomer Pierre Janssen risked German rifle fire as he flew in a balloon over the German army as it besieged Paris during the Franco–Prussian War. In many instances the end of a difficult journey was marked by poor weather that spoiled the view. Janssen's heroic effect was foiled by rain. One Scotsman held the record of traveling 75,000 miles to reach the observing sites of six eclipses but saw only one.

At NRL, Hulburt had established a tradition for solar-eclipse studies.

He had successfully led an expedition to Brazil in 1947 and organized another to Khartoum in the Sudan in 1952. For the latter, I had looked forward to joining Hulburt, but my visa was not forthcoming from the Egyptian government. In any event, the expedition suffered from numerous logistics problems. Lyot, in particular, was frustrated by mishandling of his equipment. He suffered a fatal heart attack on his way back to France. My chance to become an eclipse astronomer in the style of the new space astronomy finally came in 1958 during the IGY.

For high adventure, nothing in my experience surpassed the solar-eclipse expedition that I led to the South Pacific in 1958. IGY funds were mostly spent or fully committed by early 1957, and proposals for new ventures had little chance of support. The total eclipse expected on 12 October 1958 offered an intriguing target for solar astronomy with rockets. I pleaded for the opportunity to carry out an expedition to no avail until *Sputnik* started to play its tune overhead. Suddenly, research money began to flow again. Instead of following the prescribed route through the National Academy of Sciences Panel on Rockets and Satellites for the IGY, I approached Admiral Rawson Bennett, Chief of Naval Research, and obtained his promise of $70,000 and the use of a Navy vessel, the U.S.S. *Point Defiance*, to carry out a joint expedition of ground-based astronomers and our rocket team. The plan was to rendezvous with the eclipse on a small coral atoll with the lyrical name of Puka Puka, a few hundred nautical miles southwest of Samoa in the Danger Islands. That expedition was the tonic that I needed to erase the disappointment of *Vanguard*.

We arrived at the Danger Islands about 50 days before the event. The island's sole article of commerce was coconuts, which were collected when ripe to produce a cargo of copra (dried coconut meat) for the annual visit of the tramp steamer from New Zealand. Green coconut meat was the staple sustenance for the natives, supplemented by chickens and pigs. Coconut milk was used for drinking and cooking. To fish, the natives had to paddle their outriggers beyond the coral reef. Nothing more mechanically sophisticated than a bicycle existed on the island. No commercial aircraft flew overhead to bring them distant signs of the modern age.

When our 175-meter landing ship dock anchored outside the coral reef and a helicopter took off to land on the beach, the natives fled to the shelter of the coconut groves. In a remarkably short time, however, they adjusted to the sights of amphibious crawlers bearing equipment ashore and of a Marine demolition team blasting a channel through the reef. The Polynesians became devoted attendees of the nightly movie entertainment provided on shore; they relished the lovemaking and cheered the action, especially the scenes of natives in outriggers paddling out to the submarine in the film of Jules Verne's *Nautilus*. Within weeks they absorbed much of the basic culture of twentieth-century Western civilization. A few days before the eclipse, the natives threw us a good luck party featuring exotic delicacies that I had difficulty swallowing. Jack Evans, leader of the ground-based astronomy experiments, and I responded with an impromptu, highly simplified tactical briefing about eclipses, rockets, and solar physics.

On the island, Jack Evans and his astronomers set up the traditional apparatus for studies of an eclipse while, aboard ship, my NRL team mounted six Nike–Asp rockets on rails affixed to simple tripods. A Nike is not very different from a 1000-pound bomb, and our spacing was only 6 meters from rocket to rocket. We gambled that firing one rocket would not damage its neighbor and that no catastrophic explosion would sink our proud ship. I could not decide whether the skipper was extremely permissive because he had never witnessed a Nike rocket firing close up or because he had naive trust in the wisdom of scientists. Our operations plan was to launch the rockets in timed sequence so as to observe the progressive masking of active regions on the Sun by the passage of the Moon.

As the morning of eclipse day dawned, clouds obscured the sky and cast a pall on the preparations of Jack Evans' group of astronomers on Puka Puka. Jim Ring of Manchester University in England painted a dripping cloud on the plywood wall of his instrument shack and per-

FIGURE 44. *Herbert Friedman briefing Puka Puka children in advance of an eclipse in 1958. (U.S. Naval Research Laboratory.)*

formed a medicine dance exhorting the clouds to disperse, all to no avail. It rained throughout the eclipse on Puka Puka. Just 20 nautical miles away, the NRL rocket team had a clear sky and a perfect view of the eclipse from the deck of the *Point Defiance* as our rockets were ready to fly high into the eclipse shadow.

At 7:30 A.M. the Moon bit into the edge of the Sun at first contact and we prepared to initiate the sequence of firings. Sailors crowded forward to get a better view of the rockets on the helicopter deck. The shattering blast and burst of flame that accompanied the first firing shook up the spectators, who quickly retreated to safer distances. After the smoke cleared, rocket number 2 took off reassuringly, with no evidence of damage from the first launch. I felt a surge of relief. As the darkness of eclipse deepened, we fired rockets 3 and 4. Everything seemed to be going beautifully. The fifth rocket had been christened *Miss Fire*. True to her name, she refused to respond when the firing button was pressed. Don Brousseau, our electronics expert, guessed that the pull-away umbilical connector had been shaken loose. John Lindsey stepped forward so that he could be seen holding the firing key aloft, and Brousseau scrambled up the tripod to the umbilical connector and pressed it tight. The rest of us watched with butterflies in our stomachs. Brousseau then dropped to the

FIGURE 45. *Six Nike–Asp rockets mounted on the helicopter deck of the U.S.S.* Point Defiance *and ready for firing during the course of 1958 eclipse off the Danger Islands. (U.S. Naval Research Laboratory.)*

FIGURE 46. *A Nike–Asp rocket fired from the deck of the U.S.S.* Point Defiance *off the Danger Islands in the South Pacific at the onset of the total eclipse of 10 October 1958. (U.S. Naval Research Laboratory.)*

deck and raced to the shelter of the nearest 5-inch gun tub, about 15 meters away. The firing key was reinserted and the button was pushed, but again, nothing happened. Time was passing, and I decided to try rocket 6. But it, too, failed to ignite. Once more we went through the agony of Brousseau repeating his daring act, and this time he succeeded. Number 6 roared off to the cheers of all the crew.

Our task was completed and we could enjoy the spectacle of the Sun returning from behind the Moon. By that night we had the full message of the eclipse. X-ray emission was concentrated over active regions, and 13% of the x-ray corona was exposed at totality.

Stubborn rocket 5 was rechecked that afternoon and rescheduled for another firing attempt the next morning in order to provide background data. Under no strain or pressure, we went through a leisurely countdown and had a perfect launch. To our great surprise the telemetry showed enormous x-ray intensity; by an incredible coincidence, we had unknowingly launched at the peak of a great solar flare and were observing the best x-ray flare data yet obtained.

Within the next few years, x-ray studies of the Sun made rapid progress. In 1960 we obtained a crude x-ray photograph of the Sun with a simple pinhole camera the size of a cigarette carried on an Aerobee rocket. Most of us can recall the childhood experience of making a pinhole camera out of a shoe box. Not surprisingly, the Paris edition of the *Herald Tribune* carried a story under the headline "Sun Photo Taken by Shoebox." The two-axis pointing control could not correct for rotation of the camera about

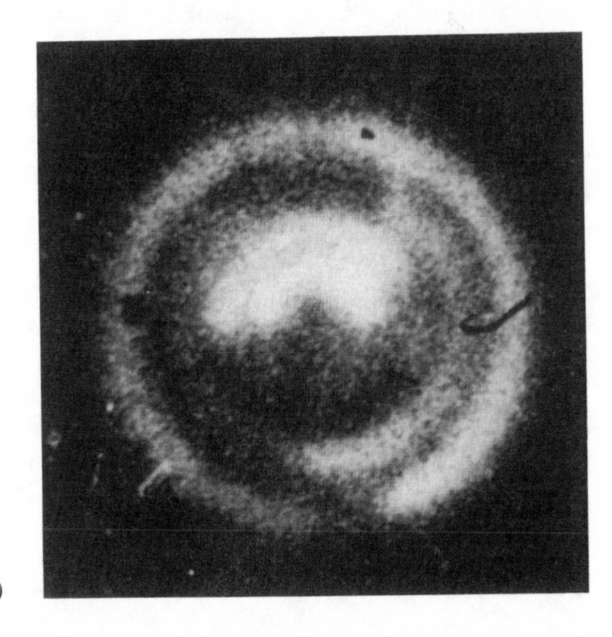

FIGURE 47. *First x-ray picture of the sun taken with a pinhole camera aboard an Aerobee rocket in 1960. (U.S. Naval Research Laboratory.)*

the direction to the Sun; the image was correspondingly blurred, but its significance was clear. At least 80% of the x-ray emission came from localized condensations in the corona, covering no more than 5% of the area of the disk. Soon afterward, scientists of the NASA and American Science and Engineering Corporation were able to construct reflecting x-ray telescopes of the type described in Chapter 3, to produce high-resolution photographs of x-ray–emitting regions in the solar corona.

Perhaps the most bizarre adventure of the new breed of eclipse observers and the best demonstration of their tenacity took place 7 March 1970, when an eclipse shadow fortuitously crossed Wallops Island, Maryland, an established rocket range. Plans were approved for a fleet of rockets carrying a wide variety of experiments to be launched in conjunction with that eclipse, and I sat in the viewing stands to explain things to congressional guests. A sophisticated set of instruments was designed by Guenter E. Brueckner of the NRL to photograph the ultraviolet flash spectrum of the eclipse over a very wide wavelength range (400–1600 angstroms) that had never been observed before. The flight trajectory and timing were perfect after launch, and telemetry showed that all the instruments were working. But as the rocket descended, the radio message reported that the parachute attached to the rocket payload and flotation device had not opened; the partially separated payload plunged in free-fall into the ocean. Ships at sea near the impact zone attempted to locate the impact position with hydrophones, but an area search for the NRL payload was unsuccessful.

On his way back to Washington, D.C., feeling greatly depressed, Brueckner suddenly realized that his film had a chance to survive even at the bottom of the ocean because of the cold water temperature. The water depth at impact point was 6000 feet, deeper than any previous record of ocean salvage operations. The Navy had developed a device known as CURV III (Commandable Underwater Recovery Vehicle), which was designed to dive to such a depth and pick up larger objects than the NRL payload, but it had never been demonstrated before. CURV III was flown from California to Norfolk and carried out to sea on an old submarine tender. The leader of the CURV III team estimated the NRL chances of recovery as about 1 in 10,000 because the probable location was too uncertain. The search radius of CURV III was only 100 meters, and it took several hours to cover such an area. One could expect that several payload-sized objects on the ocean bottom would be picked up by sonar and inspected by CURV III's television system. For each of these inspections, CURV III would need to approach within 3 meters of the object.

CURV III was lowered to the ocean floor and started its search, but after 8 hours only water pipes, empty shells, and drums were found within the perimeter of the first search circle. The second search was extremely lucky. CURV III located and picked up the payload within a few hours. Back at Norfolk, the film cameras were removed from the badly bent rocket cylinder. Hacksaws were used, and all the work was done in ice water to protect the film until it could be developed at NRL the following day. Fortunately, one set of films survived in good shape because the internal parts of that camera were coated with Teflon, which slowed the electrolysis between the film and the camera's metal housing. Although there was considerable film fogging, the spectroscopic details were clear. The evidence proved that the transition zone between the chromosphere and corona covered a large range of altitudes rather than the thin layer that had been assumed in previous solar models. This evidence of the structure of the transition zone was used to develop a model that 15 years later was confirmed by high-resolution observations from *Spacelab II*.

The era of individual initiative that characterized this experience persisted through the decade of the 1960s. The first *Orbiting Solar Observatory (OSO-1)* began a new epoch in solar research in 1962. The NRL groups were fortunate to have instruments aboard *OSO-2*. There were x-ray detectors from my group and an ultraviolet spectroheliograph and white-light coronagraph from Richard Tousey's group. We reminisce about the kind of proposal to NASA that satisfied the management requirements in those days. The proposal consisted of two drawings and 12½ double-spaced pages, the half-page being the cost section. In the text I predicted that it would be possible to observe "the expulsions through the corona of the clouds of plasma that are known to reach the Earth following solar flares. . . ." That was all the scientific justification we needed. What a contrast to the stacks upon stacks of documentation that would be required to accompany a proposal just 1 decade later!

Skylab

Skylab and its astronautical crews were a carryover of *Apollo* experience and hardware. The 7 years following the last mission to the Moon were devoted to planning, designing, and testing *Skylab*. A great battery of solar telescopes was placed on the Apollo Telescope Mount (ATM) attached to the body of the *Skylab* workshop. These instruments were the most advanced ever put into operation in space and had resolution capable of revealing finer details of the solar surface than ever before. They watched the solar atmosphere continuously for nearly 9 months, in contrast to the fleeting moments, perhaps 100 minutes all told, of previous ground-based eclipse observations. Seated at their consoles, the astronauts could bring up television displays of the corona in ultraviolet x-ray, hydrogen-alpha, and white-light wavelengths as they chose. Decisions were made in real time about where to point the instruments and when to activate exposures on film that was later returned to Earth.

While orbiting the Earth, the *Skylab* team was in constant communication with other solar astronomers on the ground throughout the world.

A total of 150 scientists in 17 countries were informed daily of the planned observations so that they could coordinate their own ground-based observations to the best advantage. All told, three crews of astronauts manned the *Skylab* during the course of the mission. The first launch was on 14 May 1973 and the crew spent 171 days aboard. The second mission lasted 59 days, and the third, 84 days. In all, 160,000 solar images were recorded on film, and the total catch of transient phenomena, such as flares, surges, and eruptive prominences, eclipsed all previous experience.

Early in this chapter I described how it used to be thought that the corona was a huge bag of gas tied to the Sun by gravitation. In order to explain its large extent, it had to be assumed that the coronal temperature was in the million-degree range. This concept was dispelled quickly during the *Skylab* mission. The x-ray and ultraviolet images showed that although the coronal plasma was truly in the million-degree range, it was tied to the Sun by closely knit magnetic loops that covered much of the solar disk. Where the loops were missing, large dark holes appeared in

FIGURE 48. *An x-ray photograph of the Sun taken with the telescope aboard* Skylab *in 1973. A dark coronal hole snakes its way from the Sun's north pole across the solar equator toward the south pole. Coronal holes delineate the areas in the solar corona from which solar wind escapes. (American Science and Engineering Corp.)*

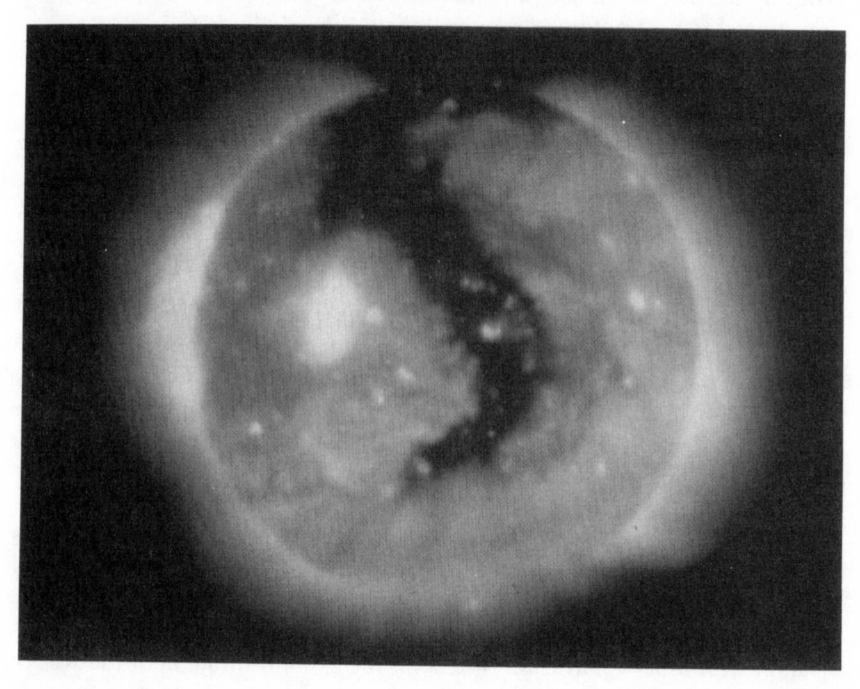

the corona. These are the routes by which the solar wind blows out of the Sun into interplanetary space.

When the magnetic loops first emerge from within the solar convection zone, they range in size from as large as 30,000 kilometers down to the optical resolution limit of about 1000 kilometers. The loops eventually disperse over time and open into space. High above the Sun, bottle-shaped gas streams following the open field lines neck down into tight jets; some project almost as far as the distance to the Earth. As the Sun rotates, the pattern of streaming gas brushes by the Earth every 27 days.

Among the many new discoveries made by *Skylab* instruments was the phenomenon of *bright points*, small point like sources of x-ray and ultraviolet emission. At any time thousands of bright points dot the solar disk, even covering the poles and appearing in coronal holes. They sparkle on a time scale of minutes, sometimes flaring to 10 times normal intensity. Most significant is the possibility that magnetic flux carried in bright points may be at least as great and possibly greater than that of sunspots. As

FIGURE 49. *An image of the Sun in the extreme ultraviolet light of the resonance line of singly ionized helium at 304 angstroms, obtained aboard the NASA* Skylab *space station in 1973. A huge prominence constrained by twisted ropes of magnetic field lines stands above the limb of the Sun. (U.S. Naval Research Laboratory.)*

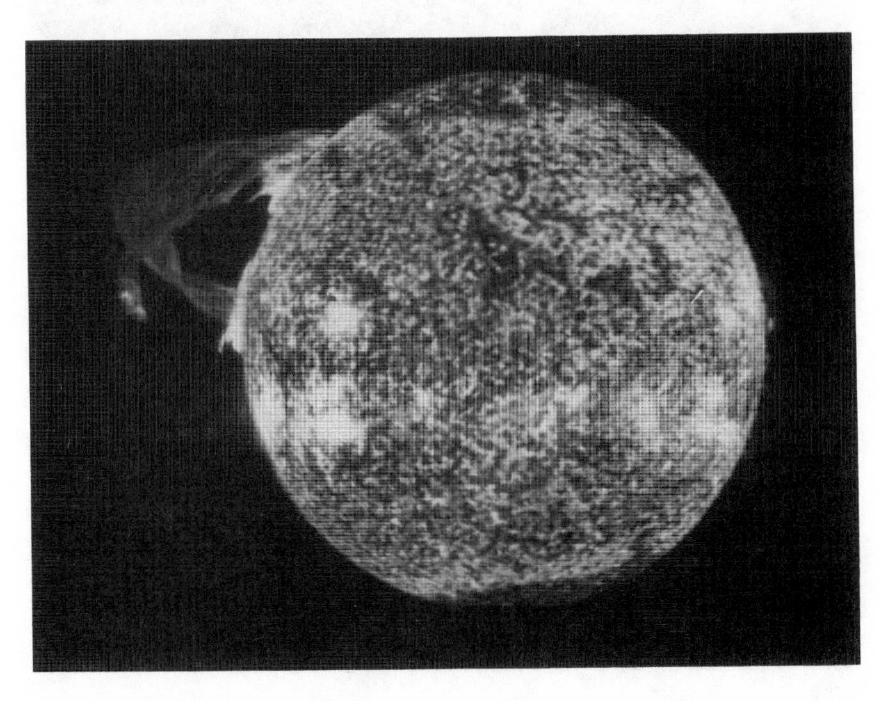

sunspots wane, the magnetic flux from bright points increases enough to balance the overall activity of the Sun.

In 1951 German astronomer Ludwig Biermann proposed that comet tails were driven away from the Sun by interaction with streams of particles in a solar wind. American physicist Eugene N. Parker then calculated in 1958 that the hot solar corona cannot be static, but must expand in the form of a wind into space. By 1962 several scientific spacecraft confirmed that the concept of a solar wind was correct. The wind blows from the lower corona at speeds that increase to about 1 million miles per hour at a distance of 20 solar radii and beyond. Fluctuating in time and in space, the wind is supersonic throughout the interplanetary medium. Although this description of solar wind may suggest hurricane force, the wind is very thin and can only be sensed by delicate instruments in space and by the ruffling effect it has on the outer atmosphere of the Earth. The

FIGURE 50. *A coronal mass ejection observed by the coronagraph aboard the* Solar Maximum Mission *spacecraft on 14 April 1980. In sequence there emerges a bright outer loop of material followed by a dark cavity, under which there appears a bright loop of erupting prominence material. (R. M. MacQueen, High Altitude Observatory.)*

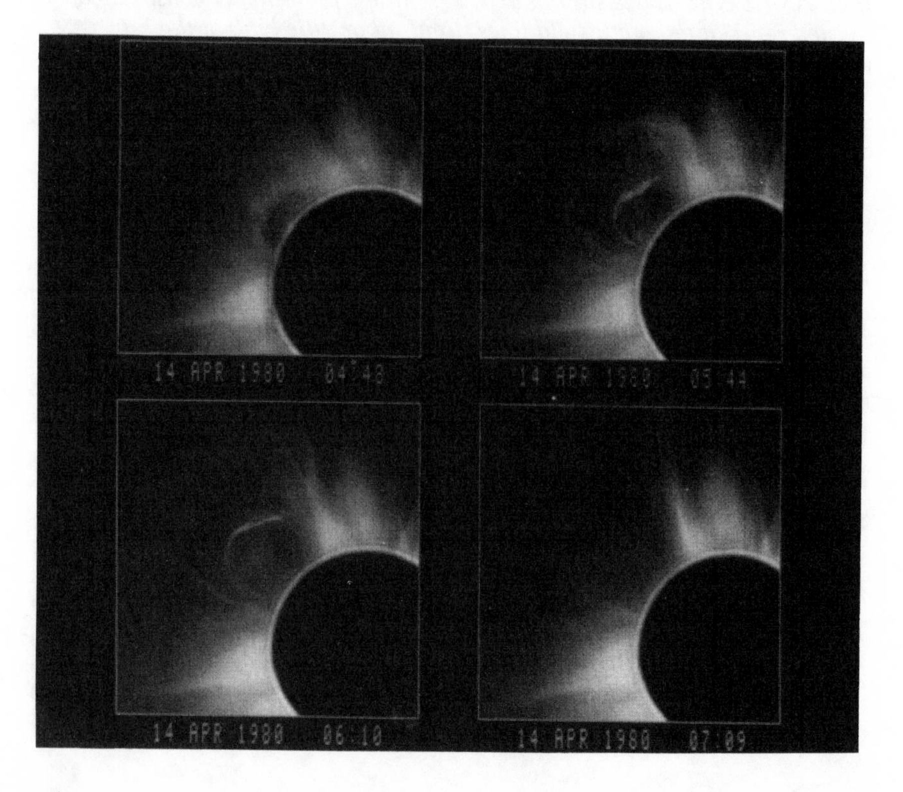

solar wind carries away about 10 trillion tons of gas per year. Still, this flow is only a trickle when measured against the total mass of the Sun. In contrast, the winds that blow from young giant stars can be 1 million times as strong.

Orbiting coronograph instruments flown since 1973 on *Skylab* and later missions have revealed a large amount of information about enormous coronal transients. The High Altitude Observatory instrument aboard *Skylab* and a NRL experiment called SOLWIND on the *STP 72-1* Air Force satellite showed transient expulsions of as much as 10 billion tons of gas, ballooning along at speeds ranging from 150 kilometers per second to 900 kilometers per second. About one-half of these exploding bubbles of gas leaving the Sun are accompanied by prolonged x-ray signatures that often last 9 or 10 hours. Shocks travel with these globs of plasma through the interplanetary medium, where they have been observed by solar satellites orbiting at distances of from 60 to 200 solar radii from the Sun's limb.

One of the most remarkable of all of the planetary explorers, *Pioneer 10,* has traced the solar wind beyond the orbits of Uranus and Neptune. Since it was launched on 2 March 1972, *Pioneer 10* has survived far beyond all expectations. In 1981 it passed the 25–astronomical unit mark and has since been heading in the direction of the star Aldebaron at a speed of close to 3 astronomical units (1 astronomical unit = 92 million miles) per year. Scientists await a signal that it has reached the stagnation boundary, where the solar wind encounters the interstellar gas. Their best guess now is that it is now at a distance of about 100 astronomical units, far beyond the orbit of Pluto. The power supply of the 500-pound craft will soon die, causing it to become deaf and mute before the year 2000. Although it can never return home, it may outlive the Earth.

The Solar–Stellar Connection

No longer is the Sun studied in its own isolated glory, set apart from the faint, pointlike stars that fill the Milky Way. Astronomers can now relate many phenomena of solar physics—the chromosphere and corona, sunspots, flares, photospheric oscillations—to counterpart behavior in other stars. In spite of the low levels of stellar brightness (less than one-trillionth that of the Sun even for the nearby stars), modern observations of other stars can detect their faint x-ray, ultraviolet, and radio emissions, and map in surprising detail the movement of dark cool spots, magnetic disturbances, and outbursts of plasma.

Olin Wilson of the Mount Wilson and Las Campanas Observatories has studied long-period cycles in stellar activities since the 1930s. He discovered early on that stars have activity cycles that closely resemble the 11-year cycle of the Sun. From 1966 to 1977, he used the 100-inch Mount

Wilson telescope to monitor the behavior of 91 stars. On some nights he slewed the large telescope to as many as 47 stars. The body of information he collected is a great resource to solar–stellar astronomers. His goal was to develop a better understanding about how heat is transferred from the convective region beneath the photosphere and corona.

To measure the rotation rate of stars, astronomers observe the Doppler effect, a shortening of the wave lengths on one edge of the star rotating toward the Earth and a lengthening on the opposite side. In one group of five main-sequence stars smaller and cooler than the Sun, Wilson found a whirlwind rate of 6.9 days in one and a lazy 43-day spin in another. In characteristic chromospheric emission lines of calcium, he confirmed a clear link between chromospheric activity and rotation rate. Older stars appear to rotate more slowly and to have lower chromospheric activity.

By extrapolating from the spin rates of stars that are younger and older than the Sun, astronomers conclude that the Sun may have been spinning 3 times faster when it was 500 million years old. If so, violent flares were very frequent, great outpourings of solar wind flowed from its surface, and spectacular auroral displays lit up the polar night.

> The Sun and Moon are the windows of his house;
> the whole universe is his courtyard.
>
> —Liu Ling (Third century)

CHAPTER 5

Red Giants, White Dwarfs, and Supernovae

Furthest, fairest thing, stars,
free of our humbug,
each his own, the longer known, the
more alone,
wrapped in emphatic fire roaring out
to a black flue . . .
Then is Now. The star you steer by
is gone.
. . . .
Sirius is too young to remember.

—Basil Bunting,
"Briggflats," 1978

As the Earth revolves around our homey Sun, astronomers point their telescopes to the billions of stars in the Milky Way and in the galaxies far beyond. In that spectacular panorama we study the drama of the birth and death of stars. The history of the starry universe is written over all the sky!

About three-fourths of the family of stars bear a close resemblance to the Sun, but the younger and older relatives can be very much brighter or dimmer, heavier or lighter. In spite of the vast range of size and luminosity, most stars in the prime of life fit on a simple evolutionary track that connects the brilliant hot blue stars of large mass at one extreme to the dim cool red stars of small mass at the other end. This main sequence of hydrogen-burning stars exhibits a simple relationship between luminosity and temperature. Surprisingly, although the luminosity varies one hundred thousandfold, the range of mass is comparatively small.

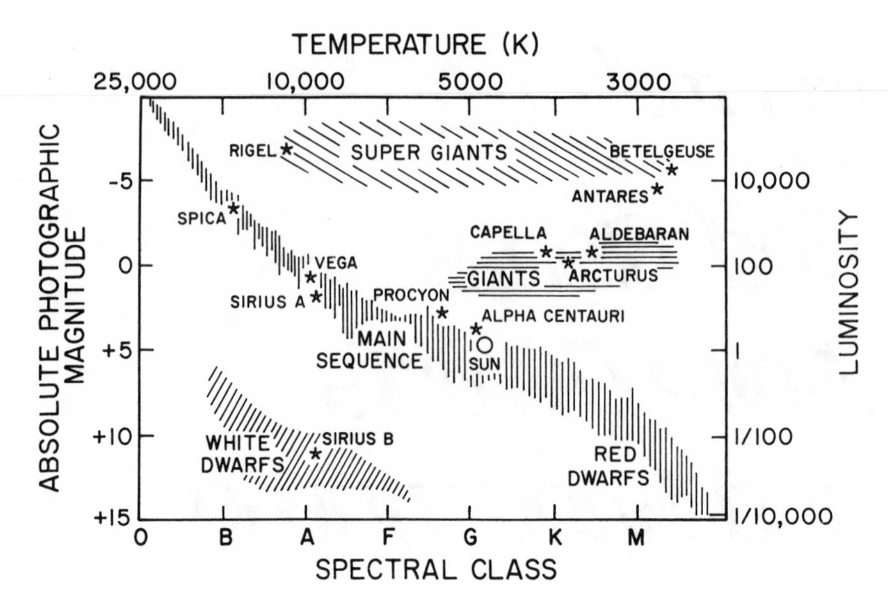

FIGURE 51. *The Hertzsrung–Russell color luminosity diagram. The absolute magnitudes or luminosities of the stars are plotted against the spectral class (color) and the temperature. Most stars by far lie on the main sequence and move along it as they burn up hydrogen. Red giants evolve off the main sequence, to the upper right, whereas white dwarfs descend into the lower left. Our Sun is near the middle of the diagram, a yellow dwarf of type G-2.*

Above and below the main sequence are the red giants and white dwarfs, stars that evolved off the main track as they entered the terminal phases of stellar life. Supergiants are so enormously bloated that they average only one-thousandth of the density of the air we breathe. White dwarfs are no bigger than planets, but they are harder than diamond and their densities are measured in tons per cubic inch.

Not only do different classes of stars exhibit a wide diversity of mass and luminosity, but each type has its own distinctive properties of evolution and variability. Some head rapidly toward explosive catastrophe; others, like our Sun, are content to age slowly and sedately. Many puff and shrink in oscillatory cycles while expelling mass spasmodically in strong stellar winds. Cosmologist George Gamow once remarked that classifying stars is much more difficult than specifying human characteristics, "Whereas all humans have approximately the same life expectancy," he said, "the life expectancy of stars varies as much as from that of a butterfly to that of an elephant" (*A Star Called the Sun*, 1964, p. 145).

The most abundant stars in the universe are the dim, low mass red dwarfs, destined to be the eldest survivors of the stellar community. With masses only a few tenths that of the Sun, they burn their smaller stores of hydrogen fuel in an unhurried fashion and can continue that way for trillions of years, long after the brilliant hot blue stars have disappeared. Eventually the red dwarfs must cool to brown, and finally to dead black cinders.

Space may contain an abundance of brown dwarfs; these are not quite massive enough to ignite nuclear fusion. The threshold mass needed to sustain fusion is about 8% of the mass of the Sun. A brown dwarf might be roughly the size of Jupiter. In known stellar populations, low-mass stars far outnumber stars such as the Sun, which suggests that they might account for most of the mass of the galaxy. These brown dwarfs could conceivably make up much of the dark matter of the universe that is inferred from its gravitational influence on the visible stars (see Chapter 6).

For more than 100 years astronomers have been searching for a faint brown dwarf companion of the Sun capable of perturbing the orbits of the outer planets Uranus and Neptune. In recent years there has been speculation that such a partner in an extended elliptical orbit could penetrate a great cloud of comets stored far on the outskirts of our planetary system, well beyond the orbit of Pluto. As the small star passed them by, some comets would shake loose and fall into the inner Solar System, perhaps to crash into the Earth. Proponents of that theory are motivated to account for evidence of the periodic extinctions of varieties of living species, including the dinosaurs about 60 million years ago, by postulating catastrophic impacts that would have cloaked the Earth in clouds of dust and blotted out the Sun.

The largest stars in the sky are the red supergiants. Red supergiants evolve from stars that have completed their phases of hydrogen and helium burning. When the Sun reaches the supergiant stage, it will be large enough to engulf the Earth. One of the most familiar of the red supergiants is Antares, in the constellation Scorpius. Even though it is more than 300 light-years from us, its diameter has been precisely measured by means of a geometrical coincidence. From time to time it passes behind the Moon and is eclipsed. Ordinarily, when a small star passes the edge of the Moon its light is extinguished abruptly. For Antares the occultation takes as long as 0.5 second, which means that its diameter must be about 400 million kilometers, or 250 times the size of the Sun. Unlike the very steady Sun, most supergiants are highly variable. Antares waxes and wanes in brightness by a factor of 2 every 5 years. Supergiants of a class known as *Cepheid variables* oscillate in brightness much more dramatically and rapidly, with a clocklike regularity. In Chapter 7 we shall describe how they are used as standard candles to measure the scale of the universe.

In the heart of every red giant lurks a white dwarf. Over the relatively short time of 100 million years, a red giant sheds most of the mass outside its nuclear-burning core. Some 1500 planetary nebulae have been observed that appear to be the luminous shells of gas expelled by red giants. As the nebula expands outward at high speed, the white dwarf at its center becomes exposed. The star is so hot that its light is mainly ultraviolet and excites the colorful fluorescence of the nebula. Like a giant neon sign in space, it marks the end of the red giant and the emergence of the white dwarf.

Rigel, in the winter constellation Orion, is a familiar example of a massive blue giant. The bluish-white star shines as brightly as 64,000 suns. Its mass is 40 times and its diameter 25 times that of the Sun, and its surface temperature of 13,000°K is more than twice as great. A simple calculation, dividing its radiant power by its mass, shows that Rigel is using up its store of hydrogen fuel 1600 times as fast as the Sun is. At that spendthrift pace it will exhaust its energy reserves in a matter of mere millions of years, as compared to the multibillion-year life expectancy of the Sun. From time to time a giant star such as Rigel will explode with the awe-inspiring brilliance of an entire galaxy of 100 billion stars, becoming a supernova.

Astronomers have wondered for years how massive a star could be. There is much evidence for stars 50 or 60 times as heavy as the Sun, and possibly for a few as massive as 100 suns. Whereas the pressure to balance gravity in the Sun comes from the heat generated by nuclear fusion in its core, the support of the outer regions of a supermassive star comes from the radiation pressure of ultraviolet photons. Eddington's theoretical models showed that the balance between radiation pressure and gravity in such stars is unstable; any substantial perturbation would set the star pulsating violently until it came apart.

More recent theoretical studies, however, have indicated that vibrational instabilities could be limited by the production, inside the star, of shock waves that dissipate energy and damp down the pulsations. If true, a supermassive star could survive on the edge of instability for perhaps as long as 1 million years while steadily shedding its mass. But the theoretical picture is still unclear, suggesting that stars up to 30 solar masses should be stable but that any star as heavy as 1000 solar masses would be violently unstable.

Some nebulae in the Milky Way, such as 30 Doradus (the Tarantula) and the enormous nebula that contains Eta Carina, are so luminous that astronomers have inferred the existence of supermassive ultraviolet stars to light them. If all the nebular luminosity were excited by a single star, that star would need to be as massive as from 2000 to 3000 suns. Eta Carina, in particular, is suspected of being the most massive star in the Galaxy. Existing near the limit of instability, Eta Carina must have shed a great deal of mass in violent outbursts over its recent history. In 1840, when it erupted briefly to become the second-brightest star in the sky, it must have expelled as much as 1 solar mass of material in a great cloud of gas and dust. It now radiates about 5 million times the power of the Sun, and its original mass is estimated to have been about 200 solar masses.

Some 15 years ago several stars in the same nebula as Eta Carina were discovered to be very hot, luminous, and massive. Recent observations with highly sophisticated new instruments now suggest that Eta Carina itself is multiple, consisting of at least two and perhaps as many as four very closely spaced stars only 0.006 light-years apart. These observations imply that other giant nebulae such as the Tarantula are excited to glow by clusters of stars that include several stars as heavy as 200 solar masses. Still, it is not understood how such tight clusters of supermassive stars are born together in active star-forming regions. But nowhere is there now any evidence for a star as heavy as 1000 solar masses.

Galactic and Extragalactic Superstars

Over the past millennium three supernovae produced such brilliant displays in our galaxy that they were well documented by naked-eye descriptions. These explosions took place in A.D. 1054, 1572, and 1604. Earlier historical references to new stars seen in the southern skies by Middle Eastern observers in A.D. 185 and 1006 exist, but the descriptions are very scant, even though the supernova of 1006 in Lupus may have been one-tenth as bright as the full Moon, the brightest in all recorded history.

Much was written in Chinese chronicles about the new star in Taurus when it blazed forth in A.D. 1054 to create the Crab nebula. Much more brilliant than Venus at its brightest, it could be seen in daylight for 3 weeks. At night its light cast a faint shadow, as is sometimes seen with Venus. As it faded away it remained visible to the naked eye for almost 2 years. Although the Orientals reported the spectacular appearance of the new star, it is puzzling that no reports are known to exist in European or Arabian annals. It might have been confused with Venus by casual observers, or perhaps astronomical reporting was not encouraged in Europe in the Middle Ages, or European skies may have been heavily overcast at the most opportune time. American Indians in Chaco Canyon, New Mexico, saw the new star glowing brightly near the crescent Moon. Filled with awe, they painted their recollection of this apparition on the face of a rock. The image still survives.

Over 500 years later, on the night of 11 November 1572, Tycho Brahe stepped out of his laboratory and was astonished to see a new star that outshone all its neighbors in the constellation Cassiopeia. Remember, in Tycho's time the pattern of stars in the sky was believed to be permanent. Filled with excitement, Tycho accosted passersby who readily confirmed the bright star but couldn't be sure it hadn't been there all the time.

Without benefit of a telescope, Tycho made naked-eye comparisons of his new star until it faded away. He wrote that when it first appeared it was brighter than any of the brightest fixed stars, including Vega and

Sirius, and exceeded the planet Jupiter, then rising in twilight. According to his description, the nova stayed close to maximum brightness throughout November, so brilliant that it could be seen even at noon, and at night shone through clouds that blotted out all other stars. Then, as it gradually faded away, Tycho compared it to progressively fainter stars. By February 1574 it matched the dimmest stars that can be seen with the naked eye, about sixth magnitude. His careful observations were published in a Latin text, *De Nova Stella* ("About the New Star"), in 1573 and received widespread attention. Tycho's records are so good that it is possible to reconstruct precisely the light curve of his supernova in modern magnitudes. The Aristotelian concept of the permanence of the heavens was effectively rejected, and Tycho became renowned throughout Europe. His use of the name *Nova* to describe the transient star is still retained, although we now distinguish between the classes of novae and supernovae.

Only 32 years later, Tycho's former assistant, Johannes Kepler, observed another stellar outburst in the constellation of Ophiuchus. Although not a match for Tycho's star, it was at least as bright as Jupiter. Following Tycho's lead, Kepler compared the maximum brightness of the new star in October 1604 with Mars, Jupiter, and Saturn, and subsequently traced its dimming to a faintness of about fifth magnitude 1 year later, as well as its eventual disappearance in the spring of 1606. The similarities in the light curves of Tycho's and Kepler's supernovae are striking. We now classify them as Type I, the most abundant category, based on numerous examples of supernovae in distant galaxies. Typically the light intensity falls to one-half in about 60 days.

It is unfortunate that the nova of 1604 occurred just a few years before the telescope was invented. No supernovae have been seen in the Milky Way since then. Supernova 1987A recently burst into naked-eye view in the nearby Magellanic Cloud, becoming the closest supernova to be detected in the past 100 years in an external galaxy; S Andromedae (1884), 2 million light-years away, was seen by telescope, but no spectrum was obtained. The reason we see so few supernovae in the Milky Way is that the disk of the Galaxy is laden with enough dust to obscure all but comparatively short-distance events from view. The concentration of dust is so pronounced toward the hub of our spiral galaxy and the plane of its disk that a supernova can be seen with the naked eye only if it is closer than about 20,000 light-years. If it occurs near the galactic center or on the far side it is beyond the range of visibility. When one is looking above or below the plane of the Galaxy, the dust is sufficiently thin that supernovae can be observed in other galaxies.

Following the appearance of S Andromedae, astronomers began to detect substantial numbers of supernovae in distant galaxies; they were, of course, too dim to be seen with the naked eye, even though each explodes with the power of all the tens of billions of stars in an average galaxy. About 50 years ago, in 1934, Fritz Zwicky began an earnest search for supernovae (he coined the name) among the faraway galaxies. In a few years, his study of the large cluster of galaxies in Virgo yielded 12 supernovae.

To search the broad sky for supernovae requires a telescope with a wide field of view. A Schmidt telescope is designed especially for broad sky surveys. Whereas the Hale telescope pierces small portions of the night sky, a Schmidt telescope provides a panoramic view. When Walter Baade arrived at Mount Wilson Observatory in 1931, he brought the design for a Schmidt. Zwicky realized it would be an ideal instrument for his supernova search. Construction of the 200-inch Big Eye had barely begun when Zwicky badgered the optical engineers into building a Schmidt. It included a 26-inch mirror behind an 18-inch corrector lens. In a single exposure it could photograph a field as large as the bowl of the Big Dipper. This "Little Eye" was capable of surveying the entire northern sky at least once each year. First light was achieved in 1936, and Zwicky almost immediately began to find supernovae flashing out in distant galaxies. After 11 years as the only telescope at the Palomar observatory, the Little Eye was superseded by a magnificent new 48-inch Schmidt.

Zwicky's quest was shared by others, and by now some 400 supernovae have been detected. Within the range of our larger telescopes there are about 100 million galaxies in which it is possible to observe supernovae. If the average time between supernova explosions in an individual galaxy is about 50 years, somewhere in the observable universe a supernova must occur about once every 10 or 20 seconds. The greater the power of our telescopes the more interesting does the hunt for these dramatic events become.

Once astronomers are alerted to the outburst of a supernova they measure its light curve—a plot of its brightness variation—and the spectrum of its component colors. Much of the success of the Palomar Supernova Search that Zwicky set up can be attributed to his ability to co-opt the talents of Walter Baade to obtain the light curves and of Rudolph Minkowski and Milton Humason to obtain the spectra. From George Ellery Hale, director of the observatory, he wangled priority for the use of the 60-inch Mount Wilson telescope. This privilege meant that the work of other astronomers could be interrupted for Zwicky's team to observe a bright supernova any time that one burst forth. Observing time is so precious on powerful telescopes that the competition for scheduling is fierce, and there was considerable resentment among other astronomers over the favored treatment of the supernova watchers. The privilege did not remain automatic for long.

Zwicky and Baade were an "odd couple" who carried on a strained collaboration during all the years they studied supernovae with the Schmidt telescopes. Their partnership was very fragile. Zwicky was something of a wild genius, full of brilliant insights and a rough disregard for conventional procedures. He could be rough with anyone who questioned his ideas. Baade was a delicate man who suffered a limp from childhood polio and was so easily excited that he trembled from nervousness and stuttered. At a telescope Baade could exhibit exquisite control, and he achieved photographs of the highest quality. During most of their professional relationship Baade lived in fear of violence from Zwicky, whom he suspected was on the verge of going mad.

Minkowski studied the spectra of supernovae discovered in Zwicky's campaign and found two generic classes, Type I and Type II. About three-fourths belonged to Type I, which are from 2 to 3 times brighter than Type II supernovae and are characterized by an almost total absence of hydrogen lines in their spectra. The Type II supernovae, although not a strictly homogeneous class, all show rich hydrogen-like spectra. A Type I supernova at the distance of the nearest star, 4.5 light-years, would flash to about one-tenth the brightness of the Sun. For the two types of supernovae, the events leading up to the explosion are thought to be quite different.

Clues to the origins of the different types of supernovae come from their locations in galaxies. Type II supernovae are almost always found in spiral galaxies, particularly in their arms. Type I supernovae occur in all types of galaxies, especially elliptical galaxies, but they also appear in the central regions as well as the arms of spirals. These preferred locations tell us something significant. Ellipticals and the central regions of spirals are largely free of dust. Their stars are old and closely resemble the class of the Sun. In contrast, the arms of spiral galaxies are dense with dust and gas and sparkle with the diamond brilliance of young blue stars. The indications are that Type I supernovae are associated with small stars of about the mass of the Sun, and that Type II supernovae are the product of the evolution of large stars many times the mass of the Sun.

White-Hot Dwarfs

You offer cheer to tiny Man
'Mid galaxies Gargantuan
A little pill in endless night,
An antidote to cosmic fright.

—John Updike,
"White Dwarf," 1985

Discovery of the White Dwarf, Sirius B

Although astronomers already knew of the great disparity in the sizes and masses of different types of stars, they were totally unprepared for the true nature of white dwarf stars. Nothing in their intuition suggested the existence of stellar matter 1 million times as dense and hard as steel. As Eddington read the message of Sirius B, it told him "A ton of my material would be a little nugget that you could put in a matchbox" (*Stars and Atoms*, 1927, p. 50). What reply does one make to such a message? The reply that most of us made in 1914 was, "Shut up. Don't talk nonsense." How astronomers solved the mystery of Sirius B is a wonderful tale of astronomical sleuthing.

In 1844 Friedrich Wilhelm Bessel, the German mathematician and astronomer who was busy studying the distances of stars, came to the conclusion that Sirius had a dark companion. He thought that the invisible partner had exhausted its source of visible energy, leaving only the evidence of its gravitational field. In observing the position of Sirius very carefully, he discovered that it moved in a wavy line, suggesting the proximity of a nearby orbiting object that tugged Sirius back and forth across the line of sight. The combination of the straight-line trajectory and orbital motion produced the wavy track. Sirius was a double star, and its dark companion was the first "invisible" star whose existence was confirmed. The orbital period is about 50 years.

When Alvan Graham Clark tested his new telescope on Sirius in 1862 to determine how sharp an image it gave, he observed what he thought was a spurious speck of light produced by some defect in the telescope. He soon rightly concluded, however, that his tiny speck of light was precisely what Bessel's dark companion implied to account for the wiggly track of Sirius. Clark's faint companion of Sirius is now called Sirius B.

After astronomers learned to deduce the temperature of a star from its spectrum, American astronomer Walter Sydney Adams found that the surface temperature of Sirius B was unusually high. That observation posed an obvious puzzle. If Sirius B were as hot as 10,000°K, why wasn't it brighter than Sirius A? The only possible answer was that it must be extremely small and, even though intrinsically very brilliant, too tiny to produce a great deal of light. In fact, Sirius B must be smaller in diameter than the Earth.

Next came the realization that even though it was so small, it had to be extremely massive to produce the gravitational effect that Bessel had detected. Its gravitational pull must be somewhat greater than that of a mass equal to the Sun, yet its mass is concentrated into a volume less than that of the Earth. The density of Sirius B turns out to be about 20 million times as great as that of the Earth, about 1000 tons per cubic inch. Eddington concluded in 1924 that in Sirius B the atoms were totally smashed into nuclei and free electrons and that the nuclei were crushed almost into contact, a condition described as *degenerate matter*.

About 15% of the stars in the Milky Way are believed to be white dwarfs. Because they are so dim, only those that are comparatively near to the Sun can be seen even with powerful telescopes. These highly compacted stars are often very magnetic, with fields that reach tens of millions of gauss. One white dwarf, PG 1031+234, was recently found to exhibit a magnetic field of 700 million gauss. It spins at the rate of 1 revolution every 3 hours, 24 minutes.

The Life Cycle of a White Dwarf

When a comparatively small star such as the Sun has fused most of the hydrogen in its core to helium and the rate of energy generation has slowed down, it must begin to shrink again. Like the energy generated by water falling over a dam, gravitational collapse produces heat and the temperature rises until, at about 100 million degrees, helium nuclei begin to fuse to form carbon nuclei and the nuclear furnace springs to new life.

We might expect the sequence of collapse and nuclear ignition to repeat through fusion of successively heavier nuclei. The exhaustion of helium would be followed by carbon burning and so on. Most stars the size of the Sun and smaller, however, cannot manage the step beyond helium fusion. Their cores become degenerate when the helium has converted to carbon, and a new pressure takes over that does not require high temperature.

Degeneracy pressure is a quantum-mechanical phenomenon that arises from the "uncertainty principle" formulated by German physicist Werner Heisenberg in the 1930s. He postulated that it is impossible to specify exactly where an atomic particle is and its precise velocity at the same time. Classical physics taught that at a temperature of absolute zero every particle in a gas would halt its motion and no pressure would be exerted. But according to Heisenberg a collection of atomic particles can never be brought completely to rest even at absolute zero temperature. The residual movement derives from "degeneracy pressure" and there is no way to make it disappear.

The particular expression of the uncertainty principle that applies to the degenerate gas-pressure of electrons in a dense star is known as the Pauli Exclusion Principle. In essence, Wolfgang Pauli showed that if a collection of electrons is very densely packed, the electrons will resist the squeeze far more vigorously than would be expected from just the mutual electromagnetic repulsion of like electrical charges. In the evolution of stars the exclusion principle plays no significant role until the density of stellar matter reaches millions of times the density of water. It acts only for particles such as electrons, protons, and neutrons, but not for photons, alpha particles (helium 4), nor any other nuclei made up of even numbers of protons and neutrons. Through the exclusion principle, nature arranges for electrons to take over the support of a star just when the fusion of atomic nuclei can no longer produce heat. Even though electrons cannot fuse with each other or any other particles to release kinetic energy, the exclusion principle provides electron degeneracy pressure sufficient to support a white dwarf against gravitational collapse.

White dwarfs are fundamentally different from main-sequence hydrogen-burning stars because they are no longer generating internal kinetic energy. They live on reserves of stored energy left over from billions of years of previous nuclear fusion. As residual kinetic energy diffuses from the core to the surface, there is no source of energy to replace it, and the temperature must drop steadily but very slowly because of the enormous heat capacity of its superdense body and the small surface through which the heat must leak out. Even when the surface temperature is still high, white dwarfs radiate at only about one-thousandth of the Sun's luminosity because they are typically no larger than planets. Most white dwarfs will lead peaceful, uneventful existences for tens of billions of years.

Eventually a white dwarf's light fades away, leaving a degenerate cinder, a black dwarf. In the words of T. S. Eliot, a dwarf dies "not with a bang but a whimper." However, depending on where it lives, for instance as a member of a close binary, the quiet life style of a white dwarf can be interrupted by abrupt catastrophe and star death in a supernova explosion.

Chandra

Theoretical calculations by the Indian-born American astrophysicist Subrahmanyan Chandrasekhar, in 1931, showed that stars weighing less than 1.4 times the mass of the Sun must evolve to white dwarfs and that more massive stars could never become white dwarfs. No exception to this rule has yet been found; the most massive white dwarf on record just barely exceeds the mass of the Sun.

In 1930, when Chandrasekhar graduated from college in India, there was already a substantial body of theoretical work on stellar structure. Sir Arthur Eddington was the preeminent authority, and his book, *The Internal Constitution of Stars* (1926), was the leading treatise on the subject. Chandrasekhar won a scholarship to Cambridge University and set out on his journey to England by ship. To pass time during the slow voyage, he set himself the task of modeling the structure of a white dwarf, taking into account the new quantum mechanics and, in particular, the phenomenon of degeneracy pressure.

By the time he reached the end of his journey he had found a solution to the problem, but the answer was exceedingly strange. The result was simple enough for low-mass stars: gravity was balanced by degeneracy pressure up to a critical limit of 1.4 solar masses. For larger masses there was no solution; it appeared that the star must implode catastrophically. At Chandrasekhar's mass limit the individual particles in the star edge up close to the speed of light and general relativity becomes important; a condition known as relativistic degeneracy. All the new physics of the twentieth century—Eddington's equations of stellar structure, Einstein's theory of relativity, and the uncertainty principle of quantum mechanics —had come into play to explain the white dwarf.

After years of refining his theory at Cambridge, Chandrasekhar concluded that white dwarfs of large mass as compared to the Sun could not exist, but there were prominent members of the scientific establishment, most notably Eddington, who disagreed. Eddington's remarks were highly

acerbic. A white dwarf of more than 1.4 solar masses "apparently has to go on radiating and radiating and contracting and contracting until, I suppose, it gets down to a few kilometers radius where gravity becomes strong enough to hold in the radiation and the star can at last find peace. . . . I felt driven to the conclusion that this was almost a reductio ad absurdum of the relativistic degeneracy formula. Various accidents may intervene to save the star, but I want more protection than that. I think there should be law of nature to prevent the star from behaving in this absurd way" ("Relativistic Degeneracy," *Observatory 58*, 1935, p. 37).

Today we find nothing absurd about total collapse to a neutron star or a black hole in the evolution of stars more massive than the Sun, but Eddington's prestige on the occasion of the 1935 meeting of the Royal Society at which these remarks were made was enormous, and Chandrasekhar was only a young Ph.D. Chandrasekhar's friends offered condolences, but hardly anybody was prepared to dispute Eddington, who persisted in rebuffing Chandrasekhar for the remainder of the decade.

The argument with Eddington cost Chandra (as his colleagues address him) any chance of tenure at Cambridge. He had little stomach for this kind of public fight and decided to put his results into a book and move on to other fields. History proved him right, and Eddington wrong. As Chandra subsequently remarked, "Once one realizes that a white dwarf cannot be greater than a certain mass, the concept of a black hole is immediate. A black hole is a perfectly natural consequence of the principle of relativity."

It is easier to forgive Eddington's stubborn opposition when one learns that a brilliant Russian Nobel Laureate in physics, Lev Landau, had a

FIGURE 52. Subrahmanyan Chandrasekhar (1910–). *After the discovery of the white dwarf Sirus B in 1915, many astronomers tended to believe that white dwarfs were the inevitable end to the evolution of all stars. Subrahmanyan Chandrasekhar, a Ph.D. candidate at Cambridge University newly arrived from India in the early 1930s, developed a theory that contradicted this view. Applying Einstein's theory of special relativity to the degenerate gas interior of a white dwarf, he found an upper limit of 1.4 solar masses for a stable white dwarf. Additional mass would force catastrophic collapse to a state of incredibly high density. This prediction seemed utterly incongruous to Eddington, the leading astrophysicist of the time, who ridiculed Chandrasekhar for his "absurd theory." But Chandrasekhar was proven right. No white dwarf has ever been found with a mass greater than 1.4 suns, and more-massive stars collapse all the way to neutron stars or black holes.*

Chandrasekhar's contributions to astrophysics range widely beyond the subject of white dwarfs, most notably to radiative transfer of energy in stellar atmospheres and to the theory of black holes. In 1984 he shared the Nobel Prize in physics with William A. Fowler.

similar negative intuition. Two years after Chandra discovered his white dwarf mass limit, Landau wrote of making the same discovery; he was unaware of Chandra's earlier work. Landau instinctively rejected his own theoretical result and decided that basically new physics was needed. He concluded that "there exists in the whole quantum theory no cause preventing the system from collapsing to a point. As in reality such masses exist quietly as stars and do not show any such ridiculous tendencies, we must conclude that all stars heavier than [the limiting mass] certainly possess regions in which the laws of quantum mechanics are violated" ("On the Theory of Stars," *Physikalische Zeitschrift der Sowjetunion 1*, 1932, pp. 285–288).

A White Dwarf's Sudden Death

The normally sedate expiration of a white dwarf over billions of years can be interrupted by a catastrophic explosion if it is coupled to a main-sequence star in a binary pair. As mass overflows onto the white dwarf from its large companion, its weight will approach the Chandrasekhar limit of 1.4 solar masses. When it comes to within a few percentages of the critical mass, the temperature reaches more than 4 billion degrees and the star resumes evolution by production of nuclear energy, fusing carbon to silicon and heavier nuclei.

The ensuing nuclear-reaction rates are extremely sensitive to temperature. In one theoretical model, as carbon burning takes off, the core temperature rises, the reaction rate increases, and carbon fusion becomes a runaway process. The star cannot expand to cool and damp down the runaway because the core is supported by degeneracy pressure, which depends only on the core's density. Soon the core becomes so hot that degeneracy disappears. The star then expands desperately in an attempt to stabilize, while carbon fusion sweeps like wildfire throughout the core. In this round of explosive carbon detonation, the burning may last less than 1 second rather than millions of years, and a solar mass of matter is incinerated, not to iron, but to a radioactive isotope of the same atomic weight, nickel 56. The outer layers are ejected at one-tenth the speed of light in a spectacular Type I supernova that reaches maximum light in about 2 weeks.

Because white dwarfs are so dense, their opacity to the transmission of light is very high. For light to escape, expansion would need to thin the absorbing gas one hundred thousandfold. Such an enormous expansion, however, would cool the gas to very low temperature so that it could no longer radiate at all. Then why is the supernova so brilliant? Some source of internal energy is necessary to heat the expanding gas to high luminescence.

The key to the riddle lies in the role of radioactive nickel 56. It decays to radioactive cobalt with a half-life of 6.1 days. The cobalt, in turn, decays with a half-life of 77 days to iron, and in the process releases a flood of gamma rays and positrons. The latter are annihilated when they collide with electrons, producing still more gamma rays. It is the conversion of the enormous amount of gamma-ray energy within the cloud to heat that keeps the supernova bright hot. After several weeks, most of the nickel had decayed to cobalt and the exploded debris has expanded to the size of our solar system. Cobalt disintegrates more slowly, and the shape of the supernova light curve flattens out. Although the exact details of this scenario may be questioned, it provides a gratifying fit to even the crudely drawn light curves of Tycho Brahe and Johannes Kepler for the supernovae of 1572 and 1604.

When a Massive Star Comes Crashing Down

Whereas Type I supernovae require the special environment of a companion star that dumps mass onto a white dwarf, stars more massive than about 8 suns are doomed from birth and die as Type II supernovae in comparatively short order. The enormous burden of mass creates such high pressure that the nuclear furnace must burn at a fierce rate to prevent collapse. In a flaming youth that may last no more than about 10 million years, the star exhausts its nuclear energy reserves. As its hydrogen fuel is consumed, the stellar core becomes hotter and at the same time more compact.

When the temperature reaches about 100 million degrees kelvin, the helium ash that has accumulated from hydrogen fusion can convert to carbon, but fusion doesn't end with a carbon core. Instead, the unbearable weight inexorably forces further shrinkage that heats the carbon above 600 million degrees kelvin. Carbon is then fused to form oxygen, neon, and magnesium. When the carbon is used up, the core again shrinks, and the temperature soars until fusion moves up to silicon and sulphur. Each successive stage is reached at a higher temperature and burns for a shorter time. The final stages of the sequence of collapse can last from 10,000 years to as little as 100 years. Finally, the progression of nuclear fusion up the atomic scale comes to a halt when the core has been converted to elements of the iron group—iron, cobalt, and nickel—each of which possesses a total of 56 protons and neutrons. Star death is then imminent.

Silicon-fused iron (silicon 28 + silicon 28 → iron 56) is the ultimate ash of nuclear burning. Its nucleus is so tightly bound that it is unable to yield energy by either fission or fusion reactions. For example, if iron 56 combined with a helium nucleus to make nickel 60, the product would have more mass than the nuclei of which it is formed. The extra mass must come from the kinetic energy of the collision, and thus the interaction region cools.

In an iron core, therefore, no fusion to heavier nuclei is possible, although lighter elements in the surrounding layers continue to convert to iron, which settles in the core. The diameter of the core is about 1000 kilometers, density is about 1 billion grams per cubic centimeter, and the temperature is several billion degrees kelvin. Outside the highly compacted core, the stellar envelope bloats to the size of a red supergiant.

When the game of nuclear power is up, the core instantly becomes unstable; it disintegrates and collapses in less than 1 second. If the core weighs less than 3 solar masses, its electrons and protons are squeezed together to form neutrons and the entire melange becomes one supermassive nucleus, a neutron star about 20 kilometers in diameter at a density of 1 million billion grams per cubic centimeter. Now it is neutron

FIGURE 53. *The theoretical evolution of a 25–solar mass star by successive stages of nuclear fusion. Hydrogen burns to helium in 7 million years. After the hydrogen is exhausted, contraction heats the helium ash until the temperature exceeds the threshold for carbon fusion. Each stage of fusion is followed by collapse, until ignition of the next-higher temperature stage of fusion. The successive stages become shorter and shorter. After silicon fuses to iron, no further fusion energy can be generated, and gravitational collapse brings on the supernova explosion. (From "How a Supernova Explodes," by Hans A. Bethe and Gerald Brown. Copyright © 1985 by Scientific American, Inc. All rights reserved.)*

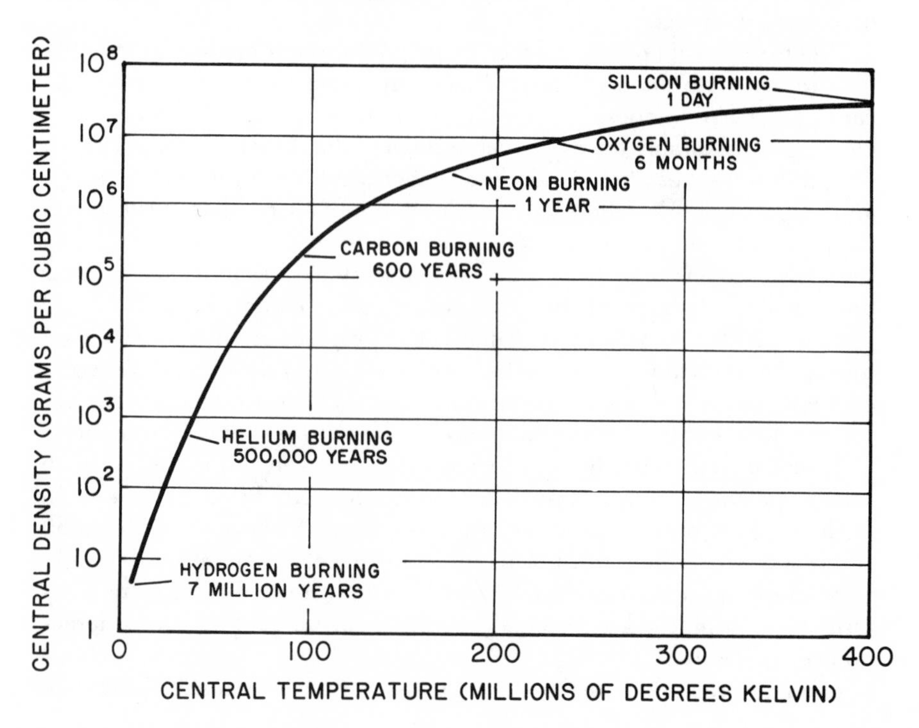

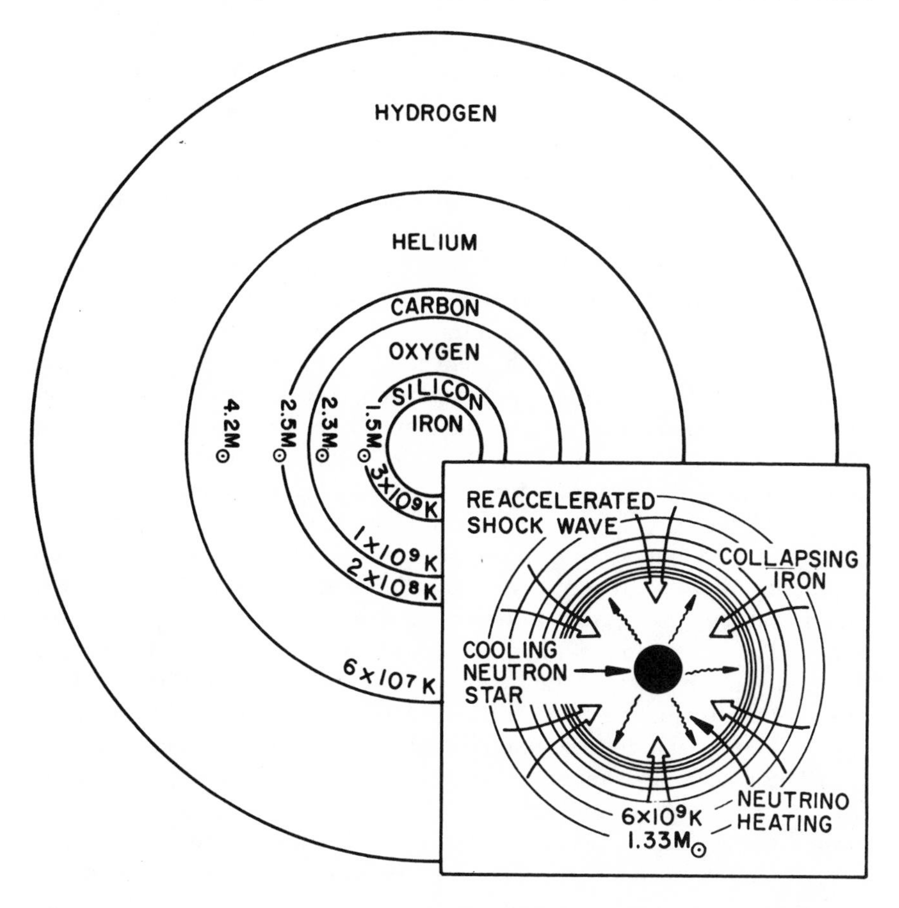

FIGURE 54. *The theoretical structure of a 14–solar mass pre-supernova just on the verge of collapse. The iron core is embedded in concentric shells of silicon and sulphur, oxygen and neon, and carbon that form a tight mantle surrounded by a bloated envelope of hydrogen. Temperature and mass fall off as indicated, relatively slowly in the mantle and rapidly in the envelope.* Insert: *An enlargement of the iron core in collapse to a neutron star. As the out going shock wave starts to stagnate, neutrinos reaccelerate it to the point of explosion. If reacceleration fails, the outer portions of the star will fall onto and crush the newly formed neutron star into total collapse into a black hole. (After J. Craig Wheeler and Ken'Ichi Nomoto.)*

degeneracy pressure rather than electron pressure that staves off total collapse. If, however, the mass exceeds 3 solar masses, nothing can prevent the entire star from crushing itself into a black hole (more in Chapter 5).

With no remaining means of support, the outer regions of the star must follow the collapsing core. Matter crashes down onto the rigid, hard surface of the neutron star, becomes very hot, expands violently, and generates an outward-moving shock wave that drives all before it. The sudden halt of collapse followed by the rebounding shock is called *core bounce.*

As the shock front passes, it leaves behind a mass of strongly heated gas that presses forcibly outward. The temperature shoots up high enough to accelerate the nuclear burning of hydrogen and of the lighter elements in the outer shell. Within 1 second, at 1 billion degrees, oxygen near the surface ignites in a colossal explosion and the supernova flashes forth with the brightness of 100 billion suns for a few seconds. Its light then falls off to about 200 million suns in 2 or 3 weeks.

This core-bounce scenario is attractive but not necessarily correct in detail, and it cannot be said that anybody understands fully how rapid collapse can be turned into rapid expansion. For one thing, no mention has been made here of the role of neutrinos. Even though neutrinos have a very small interaction with matter under ordinary circumstances, conditions in the core of a collapsing star are most extraordinary.

In 1941 George Gamow and his student Mario Schoenberg tried to explain how the emission of neutrinos could account for a sudden drop in pressure that would accelerate the collapse of the stellar core. When a proton captures an electron to become a neutron, a neutrino is also produced. The inverse reaction may also occur, and again a neutrino is released. Gamow named the neutrino-producing reaction the *Urca Process* after the Casino da Urca in Rio de Janeiro where he and his student met to theorize. He felt the name suggested a resemblance between the "rapid disappearance of energy in the interior of an old star with an equally fast disappearance of money from the pockets of the players crowded around the roulette tables in the casino." When he presented the paper for publication in the *Physical Review* (1941) he explained the acronym URCA as "unrecordable cooling agent."

Any tendency on the part of degenerate electrons to resist collapse would be counteracted by the URCA process that uses up the electrons, thus accelerating the rate of collapse. It was proposed that a flood of neutrinos would leave the core to interact with the infalling envelope, and perhaps depositing enough energy to blow away the star's outer regions. Although a great deal of theoretical work has gone into modeling the neutrino role, it remains very uncertain how a blast wave can be generated.

Kip Thorne of the California Institute of Technology suggests that perhaps only a small fraction of stellar collapses lead to supernova explosions. There may be routes to a more quiet death of a massive star that produce somewhat muted explosions and do not light up the sky with a spectacular flash. For example, Casseopeia A is the strongest radio source in the galaxy, an almost spherical nebula that can be attributed to a supernova that exploded about 300 years ago; yet nobody reported a visible flash. It is also puzzling that many radio supernova remnants reveal no central neutron-star pulsar. Perhaps the central compact object formed but was propelled away from the remnant nebula by the force of the explosion.

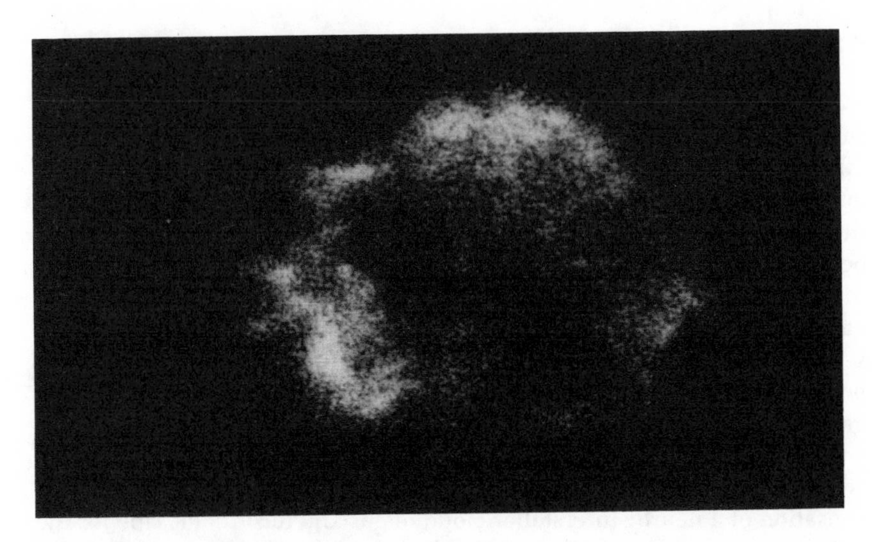

FIGURE 55 (a). Einstein Observatory *X-ray image of supernova remnant Casseopeia A. (Courtesy Steve Murray.)*

FIGURE 55 (b). *A radio image made with the VLA at 6-centimeters wavelength of supernova remnant Casseopeia A about 300 years after the explosion, which went unnoticed. The expanding shell had reached about 11 light-years across. No compact object is detectable at the center of the nebula. After the Sun, Casseopeia is the brightest radio source in the sky. (Courtesy NRAO/AUI; P. E. Angerhofer, R. Braun, S. F. Gull, R. A. Perley, R. J. Tuffs.)*

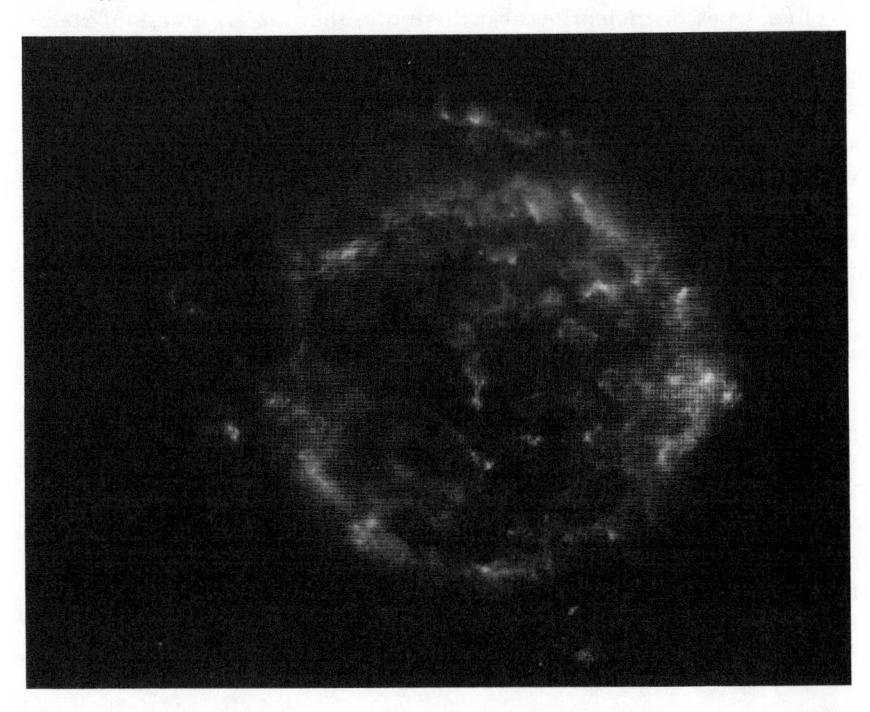

In a young galaxy where most of the mass is still in the form of interstellar gas, the rate of star formation must proceed very rapidly, perhaps 100 times as fast as in the Milky Way. These young stars evolve very quickly into blue giants, and in a matter of only a few million years reach the supernova stage. Under such conditions the rate of explosions could be 100 times that in the Milky Way. Their shock waves must reverberate throughout their galaxies, clashing with each other to heat the interstellar gas to a fantastic 1 billion degrees. This extremely hot gas would radiate x-rays as it blows out of the galaxy in a wind of thousands of kilometers per second. Future x-ray telescopes in space could search out these young galaxies.

The explosion of a supernova brings the course of stellar evolution full cycle. As the shock wave emerges it can accelerate the gravitational condensation of a nearby interstellar cloud of gas and dust that is ripe for the formation of a new star. Robert N. Clayton of the University of Chicago and Gerald Wasserburg of Caltech suggest that the formation of the Sun and its coterie of planets may have been triggered by the compression wave from a nearby supernova. Their evidence comes from examination of microscopic inclusions in a class of meteorites called *carbonacious chondrites* that are believed to have formed when the solar system was born.

Clayton and Wasserburg argue that there is an abnormally large amount of oxygen 16 compared to the heavier isotopes oxygen 17 and 18 in the inclusions of these meteorites than is found in Earthly rocks, lunar soil, or other types of meteorites. Furthermore, there is an excess of stable magnesium 26, formed from the radioactive disintegration of aluminum 26, which has a half-life of only 750,000 years. It appears that these and other anomalous isotope ratios can be explained if they formed in the supernova process and were injected into the interstellar medium by a shock wave that triggered the collapse of the proto-cloud to form the Solar System.

Supernova 1987A: After Four Centuries of Anticipation

All of our theories and speculations about supernovae came to a "moment of truth" in the early hours of 24 February 1987. On that date Ian Shelton, a 27-year-old Canadian assistant at the University of Toronto's Las Campanas Observatory in Chile, made the discovery of a lifetime. Three hundred and eighty-three years after Johannes Kepler sighted the supernova of 1604, a star brighter than any that had appeared before in the Large Magellanic Cloud flashed into view. Although it was not in the Milky Way, it was only 163,000 light-years away.

Shelton's task, the photographic mapping of the Large Magellanic Cloud with the smallest telescope on the mountain, a 10-inch mirror, was almost routine drudgery. His night's work began inauspiciously. The roof of the telescope housing was jammed, and he had to push it back by hand. The winds rose and shoved the roof back against the telescope, knocking it over, but he finally got it all under control and carried out his exposures.

After he developed the plate he was immediately struck by the presence of an intruder star of remarkable brightness. He pulled out the plate of the previous night; only a faint star image appeared at the same position. A plate taken 2 nights earlier was no different. Shelton estimated that the faint star on the first two plates had increased in brightness by more than 1000 times in 1 day. The star appeared to be about fifth magnitude, bright enough to be seen with the naked eye. Sure enough, when he stepped outside, there it was, unmistakably clear, where before nothing could be seen without a telescope. Reacting like Tycho Brahe 4 cen-

FIGURE 56. *Supernova 1987A, which exploded in the Large Magellanic Cloud on 23 February 1987. Its ultraviolet and x-ray light curves are drawn together with a composite curve of the classical supernovas of 1054, 1572, and 1604. Because it began as an atypical blue, rather than a red, giant, its light curve is unusual. At first underluminous, it grew to peak brightness about 20 May when it matched a "standard" Type 2.*

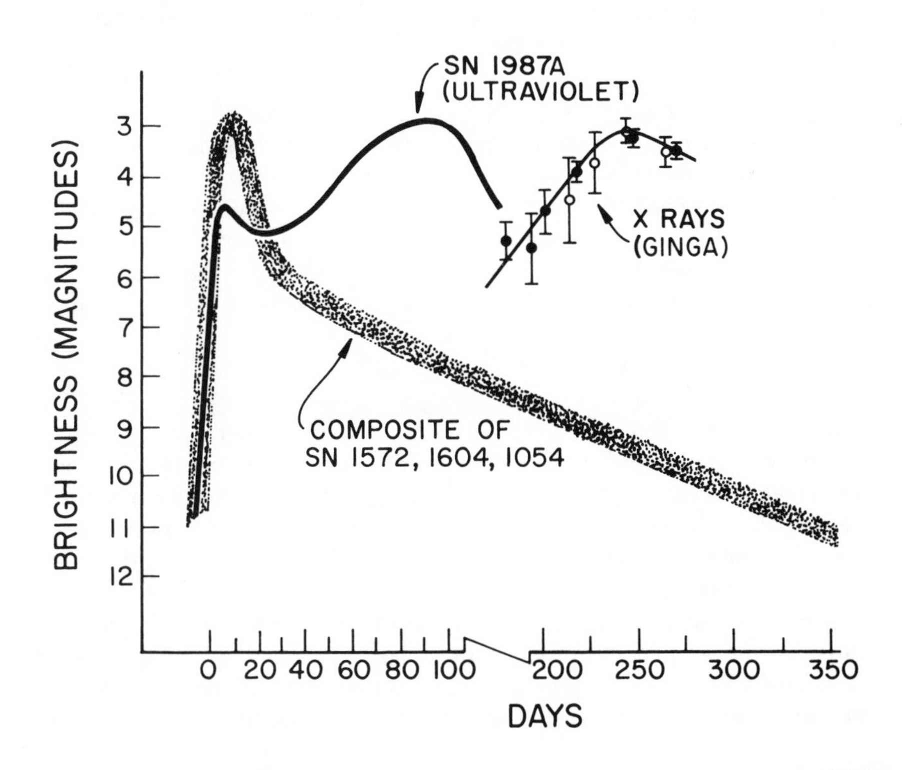

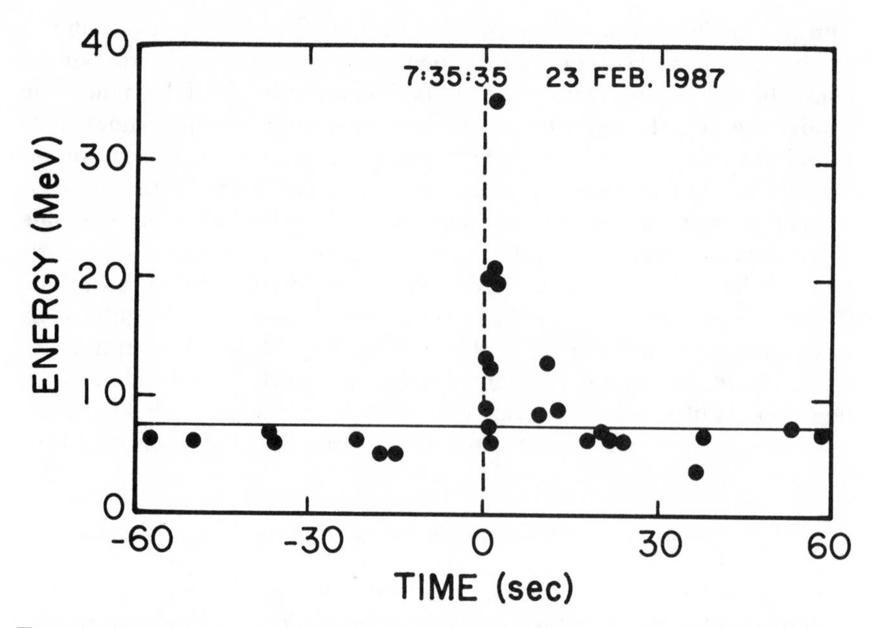

FIGURE 57. *Eleven neutrino events were observed in a 13-second interval at the Kamiokande II detector when Supernova 1987A exploded. Events below the horizontal line are part of the normal background. The total energy in neutrinos was estimated at 4×10^{52} ergs, very close to the theoretical prediction for a massive core collapse.*

turies earlier, Shelton walked rapidly toward the neighboring Carnegie Observatory to share the news with other viewers, who moved outdoors, looked up, and confirmed his discovery. As other observers sighted the supernova, reports of the event spread like wildfire in the astronomical communities around the world. Northern Hemisphere observatories such as Palomar, Kitt Peak, La Palma, the VLA, and radio telescopes at Jodrell Bank have been powerless to observe the supernova because it lies so far to the south. Telescopes in Australia, Chile, and South Africa have the view to themselves.

In the months that followed the initiation of the event, the scientific returns were remarkable. All of the major predictions of theory have been confirmed, but plenty of surprises have also emerged. By the time the mass of observational information has been digested, new light will have been shed on most aspects of the life and death of stars and on the birth of pulsars and black holes.

Since the most critical phases in the observable life of a supernova transpire very rapidly, astronomers rushed to observe with every instrument available. Actually, the first signal to reach the Earth, even before the outburst of light, was a flash of neutrinos, innocently recorded by the massive detectors implemented in recent years to search for evidence of the decay of the proton (predicted by modern elementary particle theory). The number of neutrinos released by the collapse of the core of the supernova was about 10^{58} (1 followed by 58 zeros), as much as 10% of a solar mass converted to pure neutrino radiation. The energy budget of the event was contained 99% in neutrinos, 1% in kinetic energy of the exploded debris, and only 0.01% in visible light. In effect, the great visible-light outburst was only a minor relic of the power of the event. Calculations based on these neutrino fluxes indicated that the temperature in the collapsed core reached 50 billion degrees kelvin.

Altogether, 19 neutrinos were registered at two dectectors: 11 at the Kamiokande II detector near Kamioka in Japan, and 8 at the IMB detector in the United States (a collaboration of the University of California at Irvine, the University of Michigan, and the Brookhaven National Laboratory). These detectors are enormous in volume, as they must be to catch even a few of the thousands of trillions of neutrinos that hit them. At Kamiokande the detector is a cylindrical tank of water 4 stories high that holds 3000 tons of water and is buried in a zinc mine far underground; the 7000-cubic-meter IMB detector holds 9000 tons of water in a cavity excavated deep in a salt mine near Cleveland, Ohio.

Neutrino interactions in the water produce particles that travel at velocities faster than light in water. Like the shock wave of sound that accompanies an aircraft flying faster than the speed of sound in air, these particles generate bluish light known as Cerenkov radiation at a small angle to their path (see Chapter 3, p. 91). To detect the light flashes, the walls of the tanks are lined with the blue-sensitive phototubes.

The 19 neutrinos detected were virtually a torrent compared to the 2 or 3 neutrinos per year expected from proton decay. Furthermore, their arrival times were precisely measured and contained rich clues to their energy distribution. At Kamiokande the first 8 neutrinos came within 2 seconds of each other, followed by the last 3 neutrinos 10 seconds later. The neutrinos at the IMB detector arrived within 8 seconds of each other. Both the IMB and Kamiokande detectors were triggered at almost the same time. The sensitivities of the two detectors centered on different energies of neutrinos, so differences in arrival times provide evidence of the neutrino's mass. Neutrinos of zero mass would travel at the speed of light regardless of their energy, just as photons of visible light and x-rays travel at the same speed. If they were produced at the same time they would arrive at Earth simultaneously. However, if they have mass they must travel at less than the speed of light and the more energetic neutrinos would arrive first.

John Bahcall of The Institute for Advanced Study at Princeton and Sheldon Glashow of Harvard University quickly calculated that the supernova neutrinos had masses of less than 11 electron volts, and probably no mass at all. But the issue of neutrino mass is not that easily resolved because there are three kinds of neutrinos: electron, tau, and muon. The two detectors were sensitive only to electron neutrinos.

As we shall see in Chapter 8, the possibility of neutrino mass has profound significance for the total mass of the universe. There appears to be far too little mass to slow down the expansion of the visible universe and eventually bring about reversal to collapse. The number of neutrinos in the universe is theoretically so large than even a small neutrino mass could determine the eventual fate of the universe. The earliest estimates of an upper limit to the mass of the neutrino from the evidence of arrival times at Kamiokande and at IMB appear, at least, to rule out speculation that electron neutrinos dominate the mass of the universe.

For years to come, astronomers will monitor the shape of the intensity curves of the various radiations from SN 1987A. The remarkable *International Ultraviolet Explorer*, in its ninth consecutive year of successful operation, followed the evolution of the spectrum of the supernova. When the supernova exploded, its elemental structure was layered with hydrogen and helium on the outside and with heavier nuclei in shells down to an iron core. As the gaseous shells expand and thin, the opacity decreases and the various elements are revealed in the spectra like an onion that is peeled away layer by layer. About 3 weeks after the explosion, helium spectra began to emerge, and as time passes, deeper-lying constituents will appear.

If most of the heat that powered the early luminosity had come from the radioactive decay of nickel 56 to cobalt 56 (half-life = 6.1 days), it all would have been gone by June. What remains is the energy derived from the decay of cobalt 56 to iron, which has a half-life of 78 days. For almost 1 year, the light curve of the supernova faithfully hewed to that rate of dimming.

Scientists are greatly interested in following the emission of x-rays and gamma rays. Early on, the explosion nebula is so dense that x-rays cannot escape. My American colleagues have been deeply disappointed that we have had no capability for x-ray measurements since the *Challenger* disaster, but fortunately, two sets of spacecraft instruments were placed in orbit by scientists of other nations soon after the explosion and were trained on the supernova. The Japanese satellite *GINGA* observed x-rays 180 days after the explosion, and almost simultaneously x-rays of somewhat higher energy were detected by instruments installed on the Soviet space station *MIR* by scientists from the Max Planck Institute in Munich and the University of Birmingham, in collaboration with Soviet partners. Sophisticated gamma-ray instruments will be flown on balloons from Australia to search for the expected emissions from radioactive iron, nickel, and cobalt. Particularly strong spectrum lines radiated by cobalt 56 at 847 and 1238 (kiloelectron volts) have been detected by the NASA *Solar Maximum Mission* satellite. Preliminary estimates indicated that the supernova produced a mass of cobalt 56 about 70 times the mass of Jupiter. Will the x-rays soon reveal a fast-spinning neutron star? Or will the evidence suggest a black hole?

It came as a surprise that the preexplosion progenitor star, Sanduleak (69° 202), was a blue rather than a red supergiant. Born about 20 million years ago, it weighed as much as 20 suns. Some astronomers believe that when Sanduleak reached the phase of burning helium to carbon, the star's outer layers swelled to red supergiant size, several hundred times as large as the Sun. But its gravity could not prevent a strong stellar wind from blowing away as much as one-fourth of its gas envelope, and it shrank down to blue giant size, about 50 times the diameter of the Sun. It has appeared on all photographic plates taken since the end of the nineteenth century and was classified as a blue giant of spectral type B3 with a surface temperature of 20,000°K. Throughout the past 100 years, it outshone the Sun 100,000 times while collapse and explosion were imminent.

When Shelton first photographed the supernova, its brightness was fifth magnitude, only a few percentages of the brightness of a normal Type II supernova. Within 1 week it dimmed rapidly to about the temperature of the Sun's surface. Early in March the supernova stopped fading and started to grow brighter. By May it had climbed to third magnitude, and casual observers in the Southern Hemisphere could see it clearly even in moonlight, when the Magellanic Cloud itself could not be discerned. It reached a peak luminosity of 250 million suns and then went into a steady decline.

Theorists have ready explanations for the light-curve behavior in terms of the progenitor being a blue star much smaller than a red giant. After the early burst of ultraviolet radiation, radioactive isotopes provided the main source of energy to preserve the luminosity. Perhaps energy is even now being pumped into the supernova nebula by a fast-spinning pulsar. As time goes on, astronomers will acquire a full history of this supernova in unprecedented detail, and all theories of nucleosynthesis, eventual collapse, and explosion should be greatly refined.

Lesser Lights in the Explosive Firmament

The transient outbursts, known as *novae,* that occur in binary-star systems are not as spectacular as supernovae but are much more frequent. They surge in brightness for days to weeks and then fade to normal again. Although for centuries Oriental astronomers diligently recorded the outbursts of novae, Western astronomers failed to take note of them. Today the study of novae ranks high on the list of astronomical priorities, encouraged to a large extent by the discovery of x-ray novae and the recognition that x-ray transients are associated with double-star systems and the transfer of mass between one star and its companion via an extremely hot accretion disk.

Astronomers classify optical novae as *classical, recurrent,* and *dwarf.*

The classical novae have been observed only once in recorded history and are believed to recur in cycles of thousands of years. Recurrent novae have more modest outbursts in which the luminosity may increase perhaps every 50 years between hundreds and thousands of times. Dwarf novae produce still-smaller outbursts, but more frequently, every month or so. Nova Sagittae has a period of only 81 minutes. Imagine two stars, each roughly the mass of the Sun, circling each other in little more than 1 hour and so close that their surfaces rub.

Robert P. Kraft began a systematic study of close binary systems in the early 1960s and confirmed that most novae occurred in pairs, with typical orbital periods of about 4 hours. Each pair consisted of a normal and red main-sequence star like the Sun and a compact white dwarf. The nova eruption appeared to be associated with the small star. Popular theoretical speculation at the time held that hydrogen spilled directly from the red star onto the white dwarf and accumulated on its surface. As the hydrogen diffused into the core region an explosive situation developed that led to thermonuclear runaway, a condition similar to what we described for the Type I supernovae except that the fuel is hydrogen rather than evolved carbon.

As studies of close binaries continued with more-sensitive detectors, it was noted that the stars flickered over minutes in a manner that suggested a luminous cloud of gas encircling the compact star. Such an accretion disk is believed to form from a trail of gas drawn off the red companion to spiral into the strong gravitational sink of the dense white dwarf, similar to the way that water in a bath swirls into the drain. X-ray astronomers believe that the great majority of x-ray stars occur in binaries that involve accretion disks at temperatures in the tens of millions of degrees (see Chapter 6).

Because the gravitational forces are so strong near the surface of a collapsed star, a great deal of energy is released as the transferred matter arrives. In a nova accretion disk, the temperature may rise to about 1 million degrees. Irregularities that transfer matter in great blobs of gas into the accretion region are believed to account for the spasmodic bursts of radiation. The larger mass ejections from the companion star would produce the recurrent and dwarf novae.

In August 1975 strong support for this scenario came from the discovery by x-ray astronomers of a transient x-ray source in the constellation Monoceros. Within a few days it became the brightest x-ray source in the sky. When the position of the x-ray nova was photographed in visible light, the photograph revealed a star that had brightened about 600 times. Next, examination of the library of old photographic plates showed that a previous eruption had occured in 1917. It was a recurrent nova. The observations make a persuasive case for the accretion model of optical recurrent novae.

A few months later an eruption was observed in a dwarf nova, RU

Pegasi. If a thermonuclear runaway had been involved, the outburst should have occurred when the white dwarf had reached its highest temperature, just on the verge of explosion, and then been consummated very quickly. In an accretion process, however, the outburst would start with the first impact of gas striking the surface, and as long as matter continued to rain down, the temperature should continue to rise. From the evidence of RU Pegasi, the nova reached its brightest condition 4 days later still. The picture is inconsistent with a catastrophic thermonuclear ignition but is very much to be expected as a result of heating by infall from an accretion disk.

The models described here can hardly be the whole story of nova eruptions. Between the class of Type I supernovae and the recurrent novae there may be many different conditions of instability involving compact stars in binary configurations that lead to a rich variety of transient outbursts.

Puffing and Pulsating Giants

The coolest stars, from red giants to red dwarfs, span a range of fifty billionfold in luminosity. All watchers of the sky are familiar with the supergiants Betelgeuse in the constellation Orion, the hunter; and Antares in Scorpius, the slayer of Orion. According to myth, as Orion sets, Scorpius rises to chase the hunter across the sky in winter and summer. Both Betelgeuse and Antares are about as large across as Jupiter's orbit. Each of these supergiants is a likely candidate for the next supernova in the Milky Way. If Betelgeuse should explode, it would provide one of the most brilliant supernova spectacles of all time, since it is only one-tenth the distance from the Earth as the supernova of A.D. 1054 in the Crab Nebula.

Even the smallest red giants, if placed at the Sun, would cover hundreds of times as much of the sky and would reach one-fourth of the way to Mercury. Of the nearby supergiants, MU Cephei is the most luminous, about 1 million times more powerful that the Sun even though its temperature is only 3000°K. To produce so much radiation, its diameter must be almost 4000 times that of the Sun. It would fill our solar system out to the distance of Uranus.

All red giants vary in an irregular or semiregular fashion. At a critical stage in a red giant's evolution, the cool, distended corona is bound only loosely to the hot core. The gaseous envelope is agitated strongly by convection and pulsation. Convection becomes violent enough to bubble great winds of matter into a thick shell of gas and dust that surrounds the star. Pulsation is like the breathing of a giant lung. It is driven by alternate ionization and recombination of hydrogen and helium atoms in the extended atmosphere. When the energy of recombining electrons and ions is converted to heat, the gas expands outward. As it swells, it cools and falls back again. During the infall the gas overheats once more, and the entire process repeats.

Most spectacular are the Mira (meaning wonderful) class of long-period variables. Even before the invention of the telescope, David Fabricius, a German theologian, noted in 1596 that Mira was visible only occasionally and disappeared for long periods of time. Today we know that Mira oscillates in brightness by a factor of 1000, with a period of 330 days. Still cooler and larger Mira variables exhibit even longer periods, up to about 600 days. Astronomers have cataloged thousands of Miras. As they pulsate, Miras expel mass prodigiously.

The very massive stars can divest as much as 80% of their mass into interstellar space when they reach the red giant stage. As they balloon in size, surface gravity gets weaker and escape velocity becomes very low. Many generate stellar winds million of times as strong as those that blow from the Sun. When the pulsations get large enough, the mechanism gets out of control and the star blows away most of its outer envelope.

Blowing Gas Bubbles in the Sky

In recent years infrared astronomers have detected evidence of graphite dust grains coalesced from carbon atoms in the outer atmospheres of red giants, and microwave spectroscopists have identified hydroxyl radicals (OH), water vapor, and carbon monoxide molecules. Because the infrared penetrates dust, it reveals the central stars that are completely shrouded in massive halos of dust and molecules. Betelgeuse, for example, has a halo about 3 arc minutes in diameter, about 1000 times the size of the Solar System.

As far back as 1867, Charles Wolf and Georges Rayet of the Paris Observatory discovered a strange star. Its spectrum showed very broad, bright emission lines that could have originated in a rapidly expanding gaseous envelope. The great line widths were attributed to the Doppler effect, meaning that the radiating gas was rushing away from the central star in all directions at a speed of about 3000 kilometers per second. As more and more Wolf-Rayet stars were discovered, it became clear that the gaseous envelope was being blown away by radiation pressure from a hot blue star at a temperature of about 100,000°K and causing the cloud to fluoresce.

About 200 Wolf–Rayet stars are known, but more important are the planetary nebulae. The name is a misnomer since they have nothing to do with planets. Nineteenth-century astronomers observed these fuzzy luminous disks and were struck by their resemblance to the rings of Saturn. We now know that they are ejected outer envelopes of red giants and that they fluoresce under the ultraviolet light of the exposed white dwarf cores. Planetary nebulae move outward at speeds of about 20 kilometers per second, much slower than is the case for Wolf-Rayet stars. As a result they are much more dense and easily photographed. Because the nebula is essentially a shell, it is relatively transparent when viewed through its center and appears brightest on the periphery, thus resembling a ring of fluorescence. The Helix nebula is one of the most beautiful examples (Figs. 58 and 59).

Although some 1500 planetary nebulae have been discovered in the Galaxy, they must be only a small fraction of the total population. As their distance from the Sun increases, they appear smaller and fainter and are increasingly difficult to detect. In the direction of the galactic center, where we could expect the greatest concentration, as we find for supernova remnants, novae, and pulsars, the obscuration of dust is particularly great. Allowing for such factors, astronomers estimate that there are as many as 50,000 planetaries in the disk of the Galaxy. They pump about 5 solar masses per year of processed star material back into interstellar space. Furthermore, about one-half the stars in the Galaxy may be expected to pass through the planetary nebula phase by the end of their evolution.

Flares Highlight the Red Dwarfs

We have been discussing a violent universe of stellar outbursts, but probably 90% of the stars we see today will continue almost unchanged into the far-distant future. Fully 75% of the stars near the Sun are red dwarfs, the senior citizens of the stellar community. Yellow dwarfs such as the Sun represent only about 4% of the stellar population. Typically, red dwarfs are about one-third the size of the Sun, and their low temperature, about 3000°K, gives them their dull red color. The pressure at the center of the star is just barely enough to support hydrogen fusion. At their modest rate of hydrogen consumption they can continue to radiate long after the more massive stars have burned out.

Occasionally even a red dwarf may act up, and when it does the result can be quite spectacular. Since these dim stars are difficult to detect at any great distance, we can sample only those within a range of a few tens of light-years. Out of this small sample about 100 flare stars have been detected. The prototype is UV Ceti in the constellation Cetus.

UV Ceti constantly flickers, but only by a few tenths of a magnitude. It exhibits a slow variation in brightness, with a period of 33 days, but more remarkable is the intense rapid flaring. Flares erupt like Old Faithful every hour on the average, and outshine the star itself by several hundred times, for about 5 minutes. The number of UV Ceti–type stars must be very large, but they are difficult to find because they are so small and dim except when flaring.

In many ways the outbursts of UV Ceti stars remind us of solar flares, but some of the red dwarf flares are 10,000 times as powerful as a great solar flare. They flare much more frequently than the Sun and strongly radiate radio waves, suggesting the injection of energetic particles into

FIGURE 58. *The ring nebula in Lyra. (Lick Observatory.)*

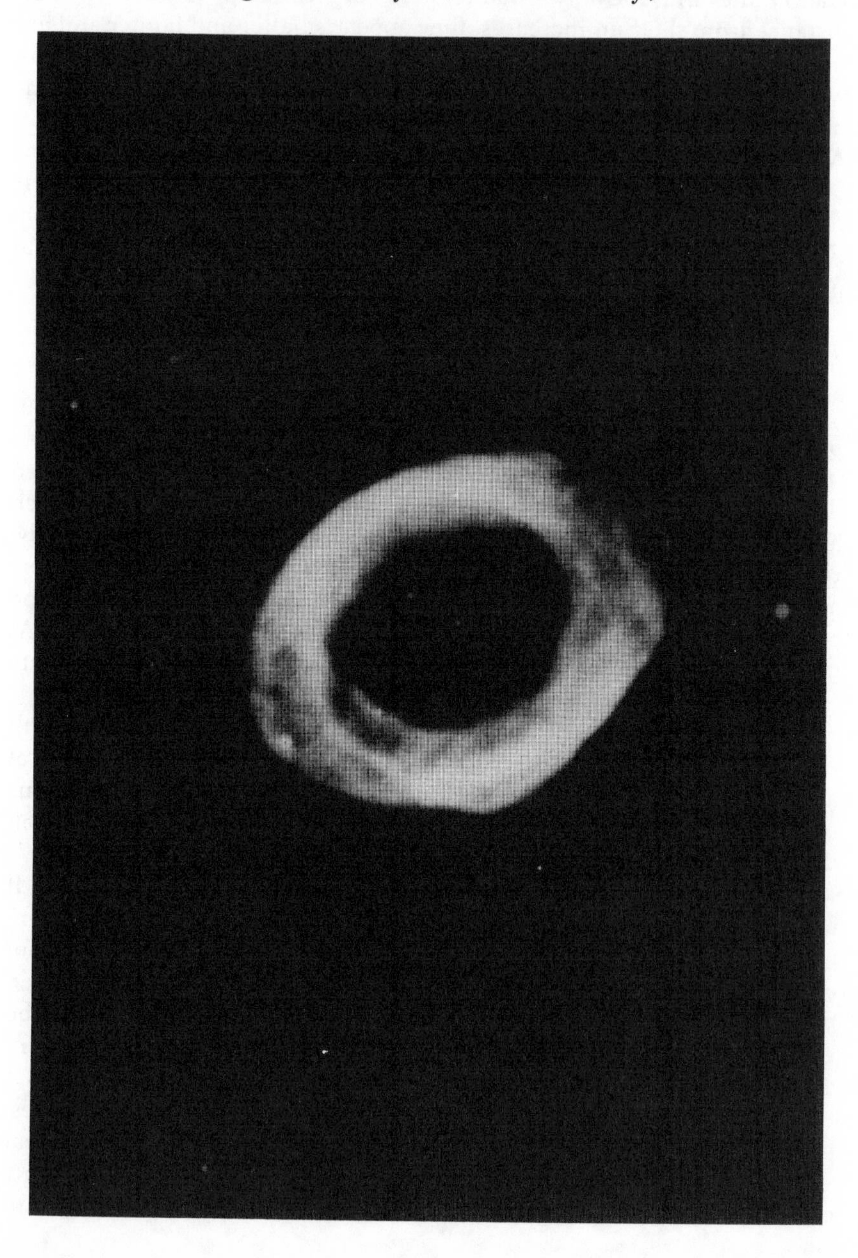

the stellar magnetic field. In fact, the flare stars were discovered with radio telescopes. The power of these stellar flares varies inversely with the size of the star; the smaller the star, the more violent the flaring. Whereas a solar flare covers about one-thousandth of the disk, a UV Ceti flare may involve almost the entire surface of the star.

One speculative model of a flare star pictures it as a highly layered structure of shells of hydrogen, each shell rotating differentially with respect to the layers immediately above and below it. This picture is reminiscent of the description in Chapter 4 of the twisted ropes of magnetic field in the Sun that are produced by differential rotation. In flare stars such magnetic field must be very much stronger and more widely distributed over

FIGURE 59. *The Helix nebula, the nearest planetary nebula to the Sun, is centered on a bright white dwarf. The circumstellar envelope shines by fluorescence under excitation by the ultraviolet light of the central star. (David Malin, Anglo-Australian Observatory.)*

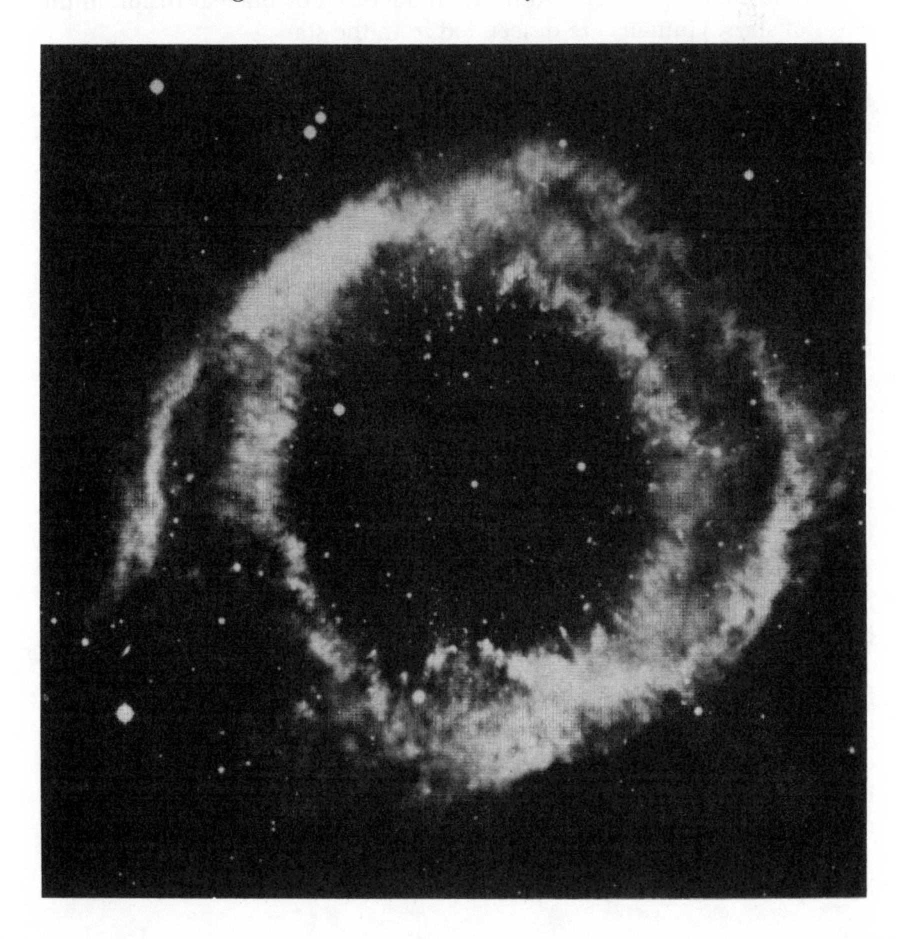

most of the surface. As in the Sun, highly ionized plasma at the surface may be contained by a network of magnetic fields until the magnetic stress reaches the breaking point and a flare erupts.

Manufacturing the Elements of the Universe

> . . . from the dust of old oblivion raked.
>
> —SHAKESPEARE,
> *Henry V*, Act II, Scene IV

Supernovae signal the death of stars but also provide the elements of life. Out of the Big Bang came most of the hydrogen and helium in the universe but hardly any of the carbon, nitrogen, oxygen, and heavier elements from which living things are made. All of the higher atomic-number species, the carbon in our tissues, the calcium in our bones, and the iron in our blood, are "star stuff" manufactured by nuclear fusion in the cores of stars. Humans are descended from the stars.

White dwarfs hoard their contents forever unless some catastrophe intervenes to expel their mass into space. When red giants blow off planetary nebulae they contribute some carbon and nitrogen to the interplanetary medium, but no substantial proportion of the cosmic abundances of heavier elements. Supernovae must play the crucial role in seeding the universe with the entire range of heavy elements of which our sun, the Earth, and we ourselves are built.

Theoretical calculations of the chemical composition of stars that are about to explode fit the cosmic abundances of the elements from carbon through iron that astronomers deduce from the spectra of the Sun and more massive stars. But how are the elements heavier than iron, such as silver, gold, lead, and uranium, created? Because they are so rare compared to iron and the lighter elements, they must have been made in a very short time compared to the billions of years of thermonuclear cooking that lead up to supernovae explosions. Theoretical and laboratory research on explosive "nucleosynthesis" has given us the answer. The seminal research in this field has been the joint work of Geoffrey Burbidge, E. Margaret Burbidge, William A. Fowler, and Fred Hoyle. Fowler received the Nobel Prize in 1985.

The great energy release of a supernova is the breeding ground of the heavy elements. Neutrons are the building blocks, and they are supplied in great quantity by the disintegration of iron. Two processes, *r* for *rapid* and *s* for *slow neutron capture*, build different sets of nuclei. Copper and lead are believed to derive from the s-process, gold and uranium from the r-process. In general, the r-process produces the neutron-rich heavy elements; the s-process synthesizes most of the neutron-poor elements between iron and bismuth.

According to this model, we would expect those elements created in stellar cores by fusion reactions via the s-process to be more abundant than those created by supernovae in the r-process. Iron coming at the endpoint of stellar fusion is more abundant than copper or lead that is formed by slow neutron capture in the cores of evolved stars. Compared to those elements, platinum and gold built by rapid neutron capture in the supernova explosion are even less abundant. Both because supernovae are so infrequent and because the r-process is so short-lived, it is not surprising that the precious metals are comparatively low in cosmic abundance. Only in a galaxy in which the supernova rate is anomalously high would the wealth of precious metals such as gold, silver, and platinum be substantially greater. Unfortunately, further synthesis converts gold to lead, reversing the ancient alchemist's dream.

If supernovae are the injection machines that fill the Galaxy with the heavy elements, red giants are the molecule factories that synthesize primitive forms of organic material. The rate of loss of material by a supergiant is typically about 1 solar mass every 100,000 years. They are the major polluters of the Milky Way. After discovering first the hydroxyl molecule (OH) and then carbon monoxide in great abundance in red

FIGURE *60. From left to right: Fred Hoyle, William A. Fowler, Geoffrey Burbidge, and Margaret Burbidge at a California Institute of Technology reunion celebrating Fowler's seventieth birthday. Their authorship of the landmark paper on nucleosynthesis is commonly referred to as "B^2 FH." (Courtesy of W. A. Fowler.)*

giants, microwave spectroscopists proceeded to find more than 1 dozen complex molecules in the coronas of red supergiants. One of these molecules is a 13-atom organic compound, $HC_{11}N$, that is comparable in molecular weight to an amino acid, a basic building block of living material.

Red giants are ideally suited to the process of molecule synthesis. Their advanced state of evolution delivers the heavy elements caused by fusion in their cores to the surface layers, where the low-temperature environment encourages molecular association. The greatest variety of organic molecules, not surprisingly, are found in carbon stars that are highly evolved red giants. Carried in the high winds that blow from the supergiants, these organic molecules are delivered to the wide reaches of the Galaxy and may play a role in its biochemical evolution.

When, on those anvils at the center of stars,
and those even more furious anvils
of the exploding supernovae,
the heavy elements were beaten together
to the atomic number of 94
and the crystalline metals with their easily lost
valence electrons arose,
their malleability and conductivity
made Assyrian goldsmithing possible,
and most of New York City.

—JOHN UPDIKE, "Ode to Crystallization."
(From *Facing Nature* by John Updike.
Copyright © 1985 by John Updike.
Reprinted by permission of, Alfred A. Knopf, Inc.
Originally appeared in *The New Yorker.*)

CHAPTER 6

Burned-Out Stars

A star is only a glowing pause in the inescapable contraction of a gas cloud to an uncertain, sometimes fantastic end.

—Kip S. Thorne,
"The Death of a Star," 1968

Fast-Spinning Neutron Stars

A supernova explosion may be the last hurrah of a dying star, but there is life after death. As the compacted core collapses to a radius of a mere 10–15 kilometers, it departs the visible universe and enters an invisible realm of radio waves, x-rays, and gamma rays. Bankrupt of all nuclear fuel for energy generation, the remnant neutron star draws on reserves of power unleashed by the gravitational energy of implosion. The tiny star spins up to millions of times the rotational frequency of its progenitor star. It then resembles an electrical dynamo that can accelerate electrons and protons to velocities approaching the speed of light. In this incarnation it can spin for millions of years, spraying the sky with sweeping beams of powerful radiation. When the Earth intercepts these beams we observe the flashing signal of a pulsar. The pulsar story that follows is yet another example of serendipity in a momentous scientific discovery.

After World War II, scientists who had developed and operated radar systems for military purposes applied their new electronic skills to radio astronomy. Before long they discovered celestial radio sources that appeared

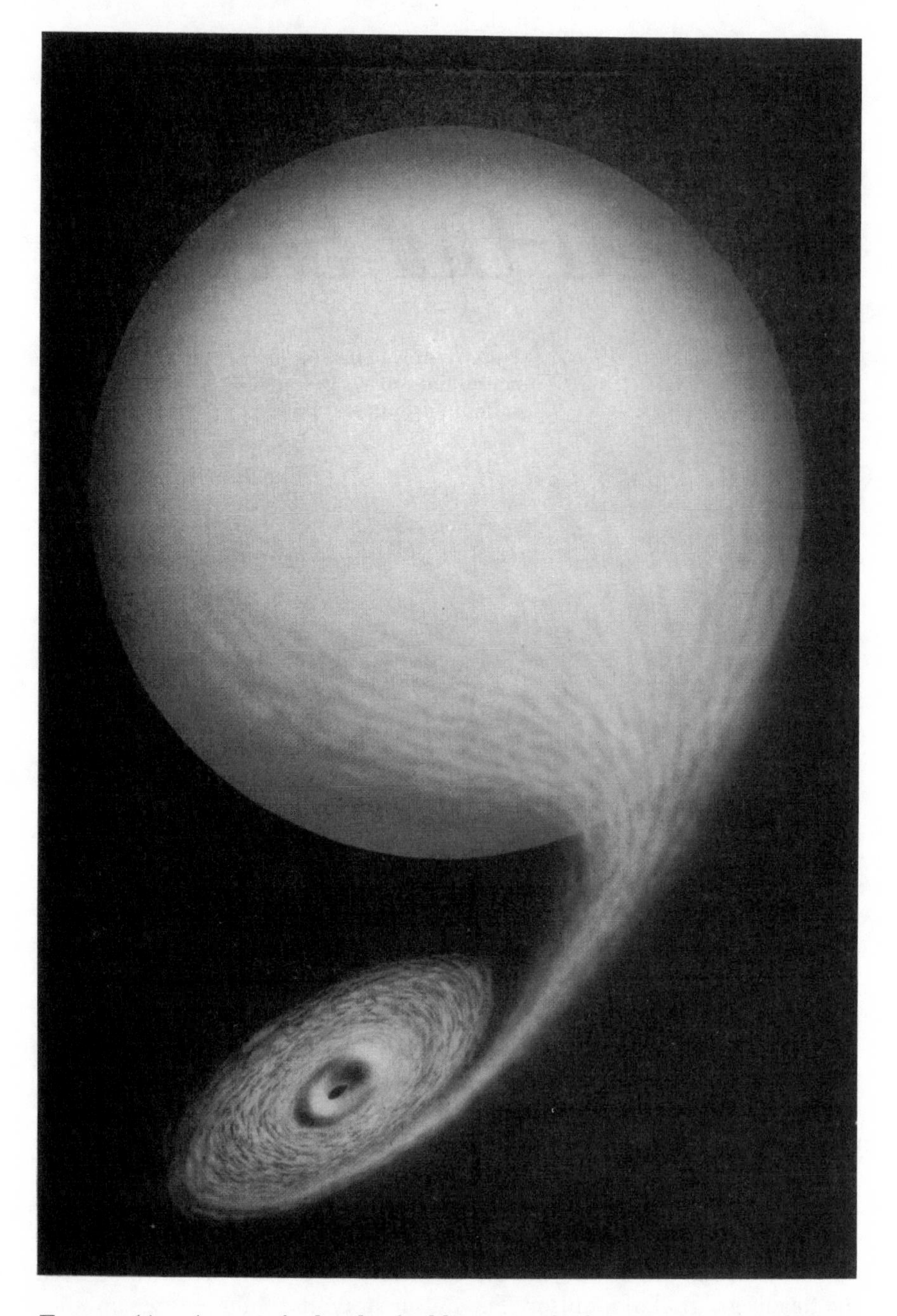

FIGURE 61. *An artist's sketch of a blue giant feeding an accretion disk surrounding a black hole. (Jon Friedman.)*

to be almost starlike. To ascertain the angular sizes of these sources, they attempted to observe how they scintillated when their radio waves passed through clouds of ionized plasma on the way to Earth. The phenomenon is analogous to the twinkling of pointlike visible stars seen through the turbulent atmosphere, in contrast to the steady images of planets. Celestial radio waves must traverse three categories of ionized gas; the interstellar gas of the Galaxy, the blobs of plasma transported through the interplanetary medium by solar wind, and the ionized blanket of the upper atmosphere—the ionosphere that reflects short-wave radio signals.

In the 1960s a radio telescope designed especially to study "radio star" scintillation was built in England near Cambridge University by Antony Hewish and his students. They confirmed all three sources of scintillation, but in a totally accidental manner, they made one of the greatest astronomical discoveries of the century. In November 1967, they discovered pulsars. A key figure in this science drama was Jocelyn Bell, a 24-year-old graduate student.

The Cambridge radio astronomers had built a radio antenna consisting of an array of 2048 dipoles covering an area the size of a football field. Each dipole was made of two parallel rods cut to match the 3.7-meter wavelength that was being observed. This long wavelength was chosen because it is easier to detect scintillation at longer wavelengths. Such an antenna is fixed in position; scanning the heavens is accomplished by the Earth's daily rotation, which carries the field of view across the sky over the local meridian, where the telescope can observe the passage of celestial radio sources for between 3 and 4 minutes. A strip of sky was recorded on 30 meters of tape each day.

It was Jocelyn Bell's task to study the tape records for evidence of fluctuating sources and to plot their positions. In examining the tape of 6 August 1967, she noticed what she described as "a bit of scruff," a patch of wavering signal about 1 centimeter long on the 120-meter chart. She realized then that "the scruff had been seen before on the same part of the records—from the same patch of sky." But the incident was dismissed as some form of local interference, and Bell was occupied with other work until near the end of October. When she resumed her study of the scintillations, she was intent on doing a better time-resolution study with a high-speed recorder if the scruffy signal reappeared. In a lecture delivered 10 years later, Bell recalled her experience of 28 November 1967:

> As the chart flowed under the pen, I could see that the signal was a series of pulses, and my suspicion that they were equally spaced was confirmed as soon as I got the chart off the recorder. They were one and one-third seconds apart. I contacted Tony Hewish who was teaching in an undergraduate laboratory in Cambridge, and his first reaction was that they must be man-

> made. This was a very sensible response in the circumstances, but due to a truly remarkable depth of ignorance, I did not see why they could not be from a star. However, he was interested enough to come out to the observatory at transit time the next day, and fortunately, (because pulsars rarely perform to order), the pulses appeared again. ("Petit Four," Eighth Texas Symposium on Relativistic Astrophysics, *Annals of the New York Academy of Sciences 302*, 1977, p. 685)

An examination of previous records showed repetitions of the pulsation on six occasions when the constellation Vulpecula, the Little Fox, passed through the field of view of the antenna. Hewish's first supposition that the signals were a local manmade effect was natural. No periodic phenomenon known to astronomers had such a fast rhythm; certainly no spin or orbital characteristic, nor any theoretical vibration. Hewish and Bell were able to rule out all possible sources of interference in the vicinity: automobile ignitions, faulty refrigerator circuits, electric blankets, and so on. Hewish checked further that no observatory was transmitting signals when Vulpecula passed overhead. Finally, he arranged for a precise time signal at a 1-second repetition rate to be superimposed on the printed record of incoming radio pulses. To his amazement, the radio pulses were precisely clocklike, to a precision of 1 part in 10 million, or 1 second in 4 months. Equally surprising was their sharpness; the pulses lasted only 0.016 second, meaning that the source was exceedingly small.

Having eliminated all possible terrestrial explanations for the source, Hewish and his colleagues now had to accept the possibility that the signals were extraterrestrial; perhaps originating on a planet circling some distant star, and generated by an intelligent source. As Hewish later reminisced, "To announce that the beacon of another civilization had been detected would obviously create a worldwide sensation."

Returning again to Jocelyn Bell's (1977) recollections:

> Just before Christmas, I went to see Tony Hewish about something and walked into a high level conference about how to present these results. We did not really believe that we had picked up signals from another civilization, but obviously, the idea had crossed our minds and we had no proof that it was entirely natural radio emission. It is an interesting problem—if one thinks one may have detected life elsewhere in the universe, how does one announce the results responsibly? [Whom] does one tell first? We did not solve the problem that afternoon and I went home that evening very cross—here I was trying to get a Ph.D. out of a new technique, and some silly lot of little green men had to choose my aerial and my frequency to communicate with us. Shortly before the lab closed for the night, I

> was analyzing a recording of a completely different part of the sky, and in amongst a strong, heavily modulated signal from the well-known radio source, Casseopeia A, I thought I saw some scruff. I rapidly checked through previous recordings of that part of the sky, and on occasions, there was scruff there. . . . This scruff, too, then showed itself to be a series of pulses, this time 1.2 seconds apart. I left the recording on Tony's desk and went off, much happier, for Christmas. It was very unlikely that two lots of little green men would choose the same improbable frequency, and at the same time to try signalling to the same planet Earth. ("Petit Four")

In short order, the number of pulsing sources found rose to four, each with a pulse rate of close to 1 per second. Because the suspicion persisted that they could be intelligent signals, they were listed by the designations LGM–1 through LGM–4, the letters standing for Little Green Men. It could be deduced from the shapes of the pulses that the sources all lay within the Milky Way.

Hewish was baffled by what the pulsars might be. The brightness of some variable stars changed with cycles as short as 1 hour or slightly less. A white dwarf in the binary system of a 1934 nova in Hercules has a cycle of 70 seconds, the fastest known until pulsars were discovered. As described in Chapter 5, in 1933 Fritz Zwicky and Walter Baade, both then at the California Institute of Technology in Pasadena, had proposed that when very massive stars die, their cores collapse into high-density stars composed entirely of neutrons, and their external parts explode into space. In 1939 J. Robert Oppenheimer and George M. Volkoff published a theory of neutron stars in the *Physical Review*. But, as Oppenheimer remarked, they never expected neutron stars to be observed: they were too small, and their radiation would peak at invisible wavelengths. With thoughts of neutron stars revived by the recent discoveries of x-ray astronomy, Hewish and Bell thought of some form of vibration in a superdense neutron star as the mechanism for generating radio pulsations, but no satisfactory physical model could be worked out.

It remained for Thomas Gold of Cornell University to provide what is now generally accepted as the correct explanation. Tommy Gold is an Austrian-born scientist who emigrated to England ahead of Hitler's occupation of his homeland. There he worked at the Admiralty Laboratories on wartime research, later collaborating with another Austrian emigré, Hermann Bondi, as well as with Fred Hoyle on the continuous creation theory of cosmology, which we shall discuss later on. He realized that if pulsars were hypothetical neutron stars, they should be spinning very rapidly. As a normally rotating star collapses, its spin rate must increase to the range of pulsation rates observed by Hewish and Bell if its angular momentum is to be conserved. The analogy is often drawn to figure skat-

ers who bring their outspread arms down close to their bodies and thus speed-up their spins. In Gold's example, a star such as the Sun, which spins at 1 revolution per month, must increase its rate of spin to about 1000 per second if it is to collapse to the theoretical 10-kilometer radius of a neutron star. Realistically, a substantial part of the star's mass would be spun off during collapse, thus carrying away much of the original angular momentum and leaving a neutron star spinning at 100 revolutions per second or less. As time went on, the spinning star would lose energy by radiation, and slow down.

When a neutron star collapses it also drags with it the original stellar magnetic field until it is concentrated one billionfold at the surface of the neutron star. In the tight grip of such a strong field, plasma at the magnetic poles would be whipped around with the spinning star. This whirling plasma could generate highly directional radio emission that would beam into space like the light of a rotating searchlight beacon atop a lighthouse. As the radio beam sweeps over the Earth, our radio telescopes record repeated flashes.

Almost 800 radio pulsars have been detected thus far, and theoretical estimates are that some tens of thousands exist in the Milky Way. Their average age may be about 3 million years, and new pulsars are probably created as frequently as supernova outbursts: about one every 30 years. Pulsar radio power varies from 10^{27} to 10^{31} ergs per second (10^{14}–10^{18} megawatts). These powers are enormously greater than any manmade radio signals. But radio power is, for some pulsars, only a small part of

FIGURE 62. Fritz Zwicky (1898–1974). *Fritz Zwicky was born in Bulgaria of Norwegian parents and grew up in Switzerland before going to the California Institute of Technology. In the early 1930s he observed that galaxies tend to cluster in groups of hundreds to thousands. He noted that individual galaxies in the Coma cluster were moving so rapidly that the cluster would be flying apart unless much more nonluminous mass than meets the eye gravitationally binds the cluster of galaxies together.*

Zwicky and Baade were the first to propose the concept of a neutron star. They advanced the view that "supernovas represent the transition into neutron stars which in their final stages consist of extremely closely-packed neutrons" ("Supernovae and Cosmic Rays," Physical Review 45, *1934, p. 138). The cataclysmic collapse of the pre-supernova was thought to release enough gravitational energy to account for all the external manifestations of stellar explosion. With Baade and other collaborators, Zwicky organized a supernova patrol that led to the discovery of most of those now known. He discovered about 100 supernovae himself. Zwicky carried the ideas of gravitational collapse beyond neutron stars to what he called "object Hades," an early intimation of black holes.*

the total. We shall see that the spectrum of a pulsar in the Crab nebula ranges all the way from radio waves to extremely high energy gamma rays. In visible light it radiates 10^{34} ergs per second, several times the luminosity of the Sun. At x-ray and gamma-ray energies, the power reaches nearly 10^{38} ergs per second, 100,000 times the Sun's power at optical wavelengths.

The Invisible Universe of X-Ray Astronomy

The 1960s witnessed an unprecedented explosion of astronomical discoveries in all parts of the spectrum from radio waves to gamma rays. From an observational standpoint, the most dramatic advances were the emergence of x-ray astronomy and the maturing of radio astronomy. Nearly all of x-ray astronomy was accomplished with the Aerobee rocket. Tailored specifically to research needs, it was a comparatively small rocket only 6 meters long and weighing less than the payload for a V-2. From its inception in 1948 to the mid-1970s, the Aerobee success rate was about 90%. Compared to the V-2, it was very well behaved; in the upper atmosphere it flew like a rigid, freely spinning top, subject only to gravitational force. Its motion could easily be analyzed to determine its orientation relative to the stars throughout a flight. With the aid of a magnetometer

FIGURE *63*. J. Robert Oppenheimer (1904–1967). *J. Robert Oppenheimer is best known to the general public as the director of the Los Alamos Scientific Laboratory when the first atomic bomb was developed there. Appalled at the fantastic death and destruction wrought by the primitive fission bombs dropped on Hiroshima and Nagasaki, he opposed the development of the far more powerful hydrogen bomb. His motives were challenged and he was censured by being barred from further classified work. A majority of American scientists expressed their confidence in him to no avail. Years later President Lyndon Johnson presented him with the prestigious Fermi Prize for his invaluable contribution to national security.*

By the physics profession Oppenheimer is remembered as an inspiring teacher of nuclear physics and a brilliant theoretician. Among his early contributions to astrophysics were explanations of the initiation and growth of a cascade of shower particles—protons, neutrons, heavier nuclei, electrons, positrons, muons, neutrinos, and gamma rays—by the impact of a single high-energy cosmic-ray particle on a molecule of the upper atmosphere.

In the mid-1930s he theorized about the formation of neutron stars in the supernova process. At the time he held little hope that they could ever be observed because of their small size and the concentration of the emission spectrum in the x-ray region. He died the year before the existence of neutron stars was confirmed by the discoveries of radio and x-ray pulsars.

that gave the rocket alignment with respect to the Earth's magnetic field and with photoelectric sensors to detect star signals, the aspect solution was readily obtained to within 1 degree of arc.

Throughout the 1950s my group at the NRL (see Chapter 3) studied solar x-ray emission. Any extrapolation of the requirements to detect sun-like x-ray sources in the Galaxy was extremely pessimistic. It would have required detectors millions of times more sensitive than those then in use. It appeared that the Sun would remain the only source of detectable stellar x-rays for the foreseeable future. In the course of our attempts to study x-rays from solar flares, however, my colleagues and I believed in 1956 that we might have detected evidence of a background of x-rays from beyond the Solar System. Spurred by these indications, we decided to attempt some primitive efforts to detect celestial x-rays more convincingly.

Within a few years other groups became interested in the possibility of x-ray astronomy. Bruno Rossi, who had a distinguished career in cosmic-ray research at the Massachusetts Institute of Technology (MIT), was joined by a young associate, Riccardo Giacconi, of the American Science and Engineering Company (AS&E) in Boston. Some of the founders of the company had been Rossi's students at MIT. He became the firm's principal consultant and chairman of the board. With support from the United States Air Force, Rossi, Giacconi, Herbert Gursky, and Franco R. Paolini initiated a program using Aerobee rockets in attempts to detect x-ray

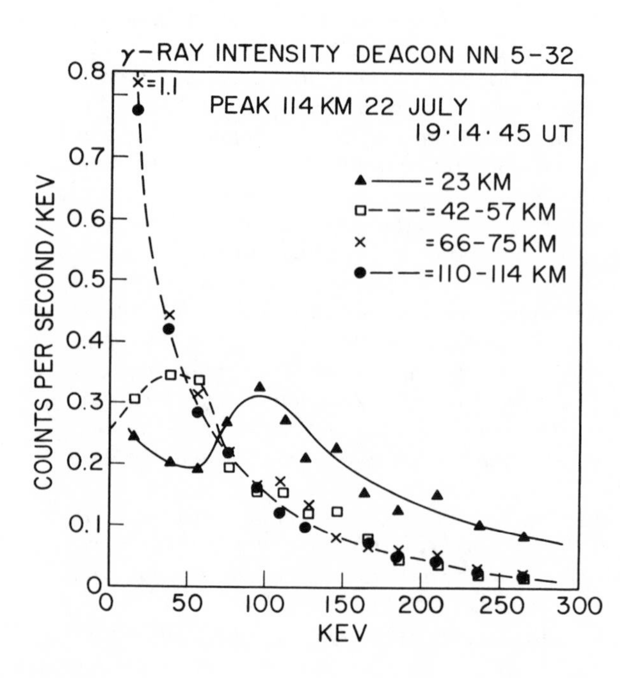

FIGURE 64. *An early suggestion of galactic x-ray emission observed during the course of 1956 IGY solar flare expedition. A scintillation counter carried on the Nike–Deacon rocket showed evidence of the cosmic-ray maximum at 23 kilometers, followed by a steady increase in softer x-ray flux with increasing altitude. Solar coronal x-rays were below the energy threshold of detection. (U.S. Naval Research Laboratory.)*

emission from the Moon. At the same time, their detectors would search for other sources of nonsolar x-rays in the hope that they might discover some unexpectedly powerful x-ray emitter.

The first attempt of the AS&E group to detect nonsolar x-rays took place on 24 October 1961. Although the rocket flew flawlessly, the scientists were disappointed as they watched the strip chart of the telemetry data; there was no indication of any interesting signals at all. They soon learned the reason for the negative results. The doors protecting the instruments on the trip up through the atmosphere had failed to open. On the second try, on 18 June 1962, almost everything ran smoothly. At an altitude of 225 kilometers the doors opened properly and the instruments scanned the sky above the x-ray–absorbing atmosphere for 350 seconds. A large intensity of x-rays, as many as 5 photons per square centimeter per second in the energy range around 4 kiloelectron volts, was detected. The first impression of the AS&E group was that they had actually succeeded in detecting lunar x-rays. On closer examination the group found that the x-rays did not come from the Moon, but rather from a direction more nearly corresponding to the center of the Galaxy. The Moon experiment had failed, but celestial x-rays had been detected and x-ray astronomy was primed for a remarkable era of discovery.

FIGURE 65. *Positive evidence of x-ray emission from the general direction of the galactic center was obtained in 1962 with proportional counters flown aboard an Aerobee rocket. Subsequent observations showed that 60% of the intensity came from a single x-ray star, Scorpius X-1, and the remainder from the combined emission of a large collection of stars along the galactic plane. (Giacconi, Gursky, Paolini and Rossi, American Science and Engineering Co., Phys. Rev. Letters, Vol. 9, p. 439, 1962.)*

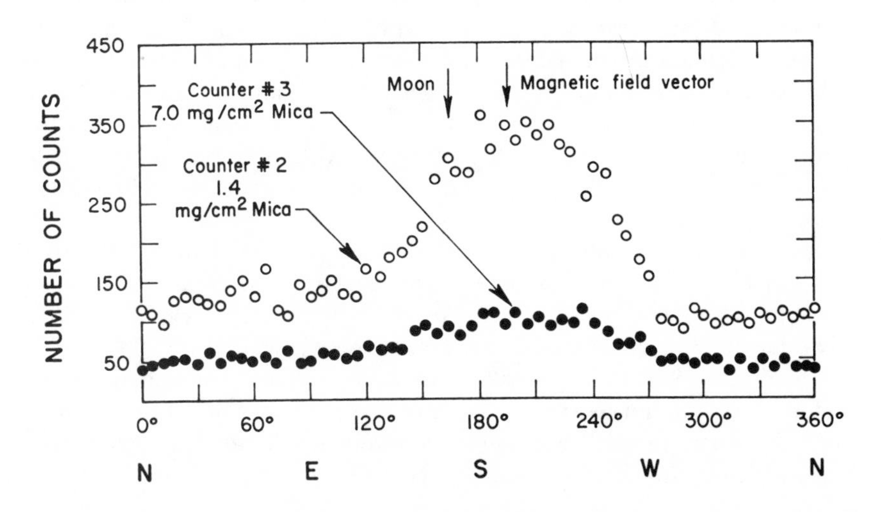

Within a few months of the time of the AS&E group's discovery of celestial x-rays, my group at the NRL launched its own detector, about 10 times as sensitive as that flown by AS&E, which scanned across an intense x-ray source in the constellation Scorpius, about 20 degrees above the galactic center. The spatial resolution of the detector was sufficient to enable us to suggest that the object was starlike. The detector also scanned past the Crab nebula and revealed an x-ray flux about 7 times lower than that of the Scorpius source. Lastly, the instrument picked up diffuse radiation from all over the sky, which could have been galactic or extragalactic in origin. Comparison of these results with those of the AS&E group showed excellent agreement. The important advance made by the NRL measurement was its relatively precise positioning of the strong source that we called Scorpius XR-1 and of the source in the Crab nebula. (The designation of x-ray sources later dropped the letter *R*.)

The discovery of the bright x-ray source, Sco X-1, was an astonishing result, but more attention was initially paid to the diffuse background. This was surprisingly intense, and it was suggested that perhaps it could be used to test one of the popular theories of cosmology at that time, the steady-state, or continuous creation, model developed by Hermann Bondi, Thomas Gold, and Fred Hoyle. In this concept, matter is created in the form of neutrons that subsequently decay into protons and electrons. The rate of creation of neutrons is sufficient to fill in the space left by expansion of the universe and thus maintain a constant density in intergalactic space. Accordingly, the universe today is no different than in the past, nor will it be any different in the future (see Chapter 7).

The decay of neutrons into electrons and protons contributes very high energy to these particles so that the resultant intergalactic matter (IGM) is hot, on the order of 300 million degrees, and should emit x-rays. Fred Hoyle, in a rough calculation, estimated that the intensity observed in the rocket experiments was sufficient to confirm the concept of continuous creation and the decay of neutrons into hot electrons and protons. Subsequent calculations carried out by Robert J. Gould and Geoffrey R. Burbidge at the University of California at San Diego came up with a different result. The detected flux was short by a factor of 70 of confirming the continuous creation model of the universe. Although this did not absolutely contradict the theory, it contributed to a general decline in its acceptance.

X-ray astronomy caught on very quickly. By the end of the 1960s, it claimed 170 specialists. Though not the largest community of astronomers, the number involved in x-ray astronomy grew more rapidly than in any other field of astronomy in the 1960s and early 1970s.

The discovery of x-ray sources immediately led many scientists, myself included, to attempt to explain the physical processes of x-ray emission. Some scientists postulated *synchrotron radiation;* like the high-speed particles whirling around the circular racetracks of the physicists' accelera-

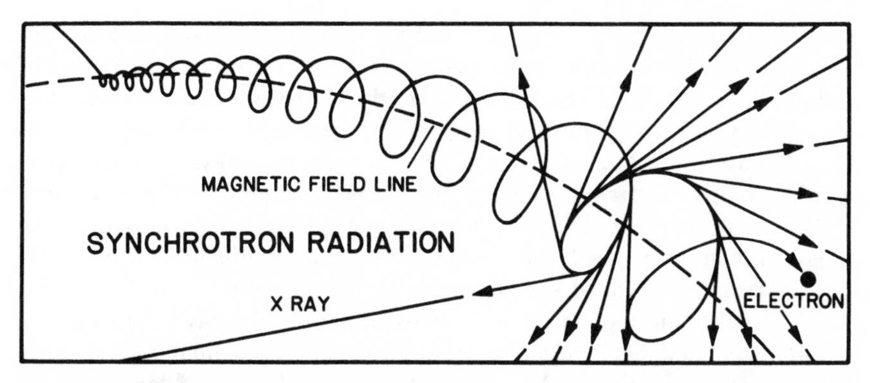

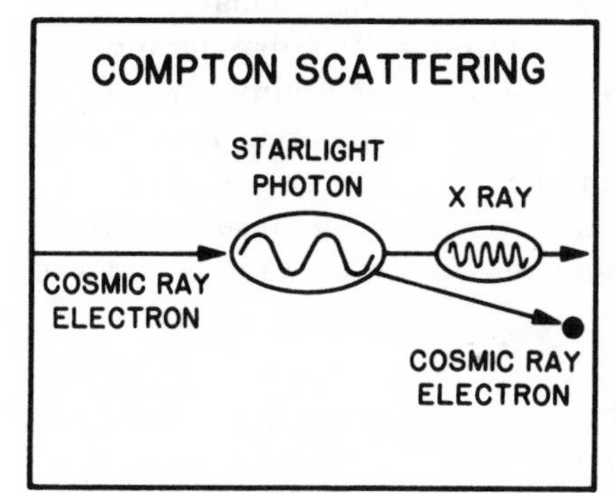

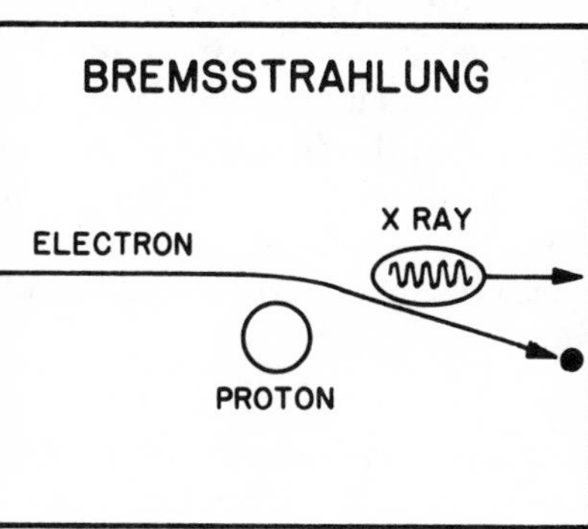

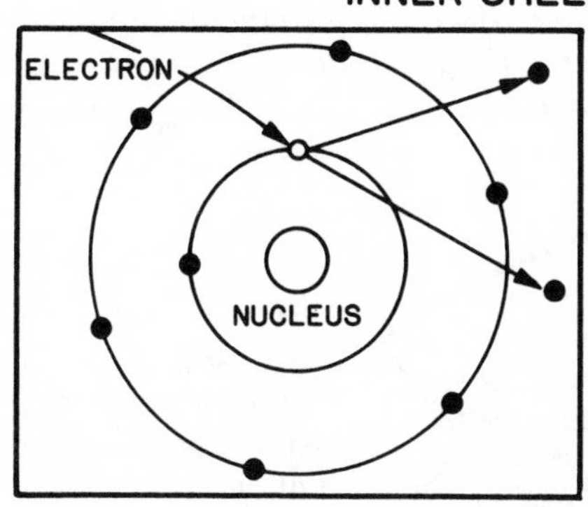

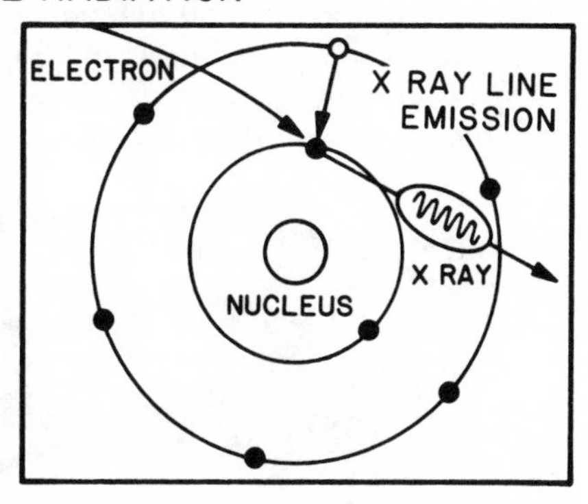

FIGURE 66. *The x-ray emission processes. Electrons spiraling about magnetic field lines radiate a continuous spectrum of synchrotron x-rays along the direction of motion. In Compton scattering, a relativistic electron collides with a visible light photon and boosts its energy to the x-ray range. Brehmsstrahlung x-rays are produced when an electron passes in near collision with an atomic nucleus. Atomic transitions produce x-ray line emission when an electron or an x-ray photon ionizes an inner shell of an atom and a higher-shell electron falls into the vacant place in the inner shell.*

tors, the charged particles in celestial x-ray sources must spiral around magnetic field lines and in the process emit a continuous spectrum of radiation from visible light to x-rays. Another process is the *inverse Compton effect*, in which relativistic electrons (traveling at nearly the speed of light) carom off photons that recoil at very high energies; x-ray energies. A third possible source of x-rays is the process of *bremsstrahlung* in a very hot plasma. Very hot electrons curl past the positively charged nuclei of the plasma at high speeds, emitting x-rays as their paths bend. This is the process responsible for the x-rays from the solar corona.

My first inclination was to seek an explanation of x-ray sources in the neutron star model. The interest in neutron stars as potential x-ray sources was stimulated by the work of Hong-Yee Chiu of the Goddard Institute for Space Studies. I learned of Chiu's work from a 1964 monograph that he had published, entitled *Supernovae, Neutrinos and Neutron Stars*. Chiu argued that the core temperature of a neutron star theoretically must be about 1 billion degrees. Energy transmitted from the core would heat the surface to a temperature of about 10 million degrees. Because the surface layer, presumably iron, would be of high density, it would radiate intense

FIGURE 67. *A cross section of a neutron star. A neutron star is a cosmic physics laboratory for the most exotic forms of matter. Whereas atoms are mostly empty space, a neutron star is a gigantic nucleus of 10^{57} neutrons. Its core density reaches about 1 billion tons per cubic inch. At its surface is a gaseous atmosphere only a few meters thick and supported on a crystalline crust 1 million trillion times as rigid as steel. The gravitational field at the surface is 100 billion times as strong as the gravitational field of the Earth. Its interior is superfluid and superconducting at a temperature of more than 1 million degrees kelvin.*

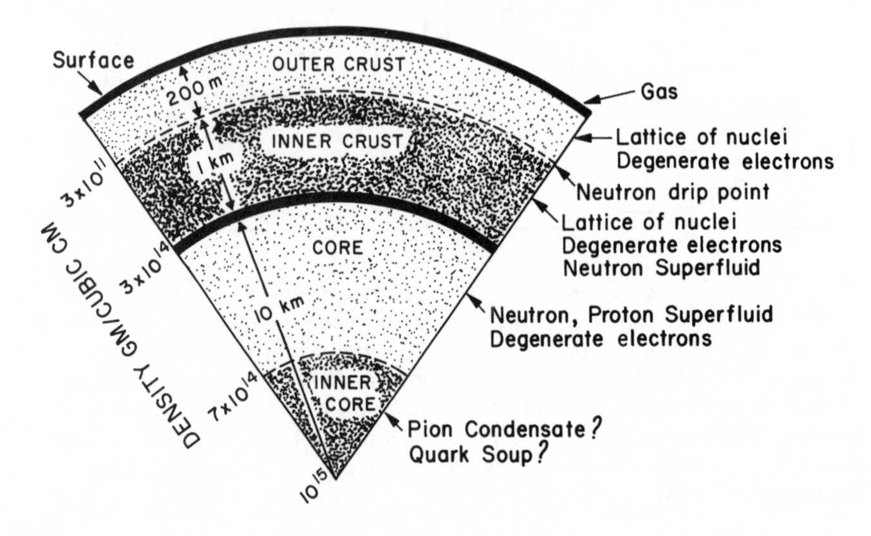

x-rays with peak emission at about 3 angstroms (4 kiloelectron volts). After its initial formation the neutron star would remain very hot for as long as 1000 years, by which time it would have cooled to about 1 million degrees and its radiation spectrum would have shifted to longer wavelength x-rays. In optical wavelengths, it would shine relatively weakly.

Although the Crab's nebular emission in radio waves and visible light had been satisfactorily explained as synchroton radiation, it did not extrapolate to a predicted strong flux of x-rays. Furthermore, synchrotron radiation exhausts the energy of the electrons that produces it. The higher the energy of the electrons, the faster they decay in the production of radiation. To explain the x-ray emissions from the Crab nebula would require electrons with energies as high as 10^{14} electron volts. A nebula of charged particles would last for only a few years at these energies. Some process was necessary for continuous acceleration of the electrons, but at that time, no conceivable mechanism was theoretically obvious.

If one of Chiu's neutron stars existed at the center of the Crab nebula as a result of a supernova explosion observed in 1054, however, that neutron star would be less than 1000 years old as we see it now and, theoretically, would still be very hot. A neutron star at a surface temperature of 10 million degrees at the distance of the Crab nebula could produce just about the intensity of x-rays observed by the rocket experiments.

In early 1964 we grabbed an opportunity for testing the neutron star hypothesis with a rocket experiment. Our radio astronomers systematically predicted opportunities for observing celestial occultations of radio sources by the Moon. It turned out that the Crab nebula would be eclipsed by the Moon in July 1964, an event that occurs once every decade. By using the Moon as a screen to cover the nebula gradually, we could determine the angular size of the x-ray source. If a neutron star was a source of x-ray emission at the center of the Crab, the occultation would produce an abrupt disappearance of x-rays in a fraction of 1 second. If the x-ray emission were to disappear gradually, it would indicate that the x-ray source was an extended nebula. At that time, the technology required for carrying out such an occultation experiment was not fully developed. It was essential that a rocket stabilizer be available that could hold the x-ray detector locked onto the Moon during the course of the occultation. The experiment, if successful, could provide persuasive evidence of the existence of a neutron star.

For rocket studies of the Sun, devices called *attitude control systems* had been designed early in the 1950s. These electromechanical devices aimed the rocket's instruments at the Sun by tracking it with phototubes. To perform a similar function on celestial sources at night is much more difficult because no particular star in the night sky stands out so prominently. NASA, through its Goddard Space Flight Center, designed a gyro-controlled stabilizer for nighttime rocket astronomy. The first control units were tested in 1959; by 1964 the system was still far from reliable. Four-

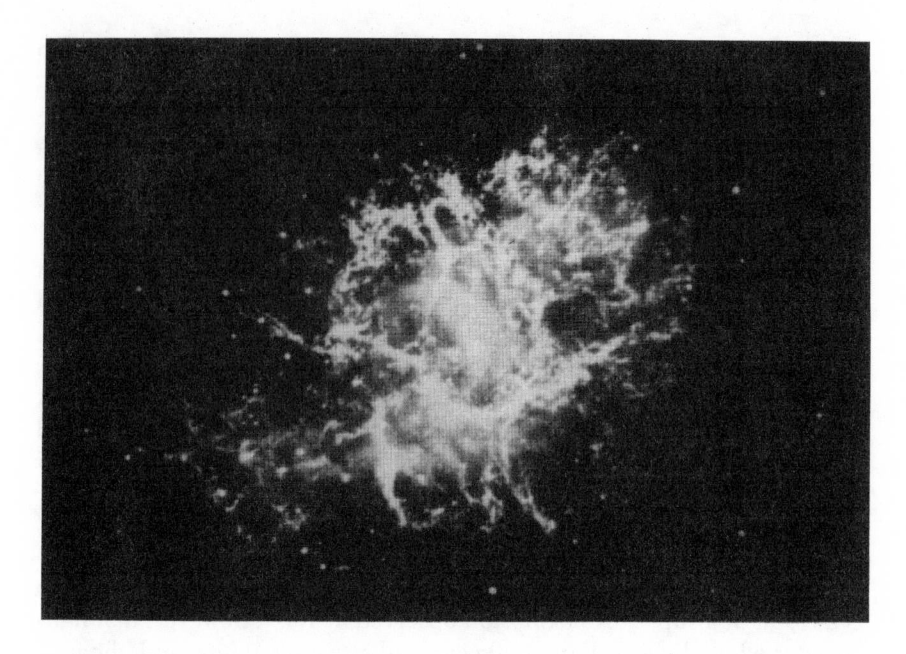

Figure 68. *The Crab nebula is a rapidly expanding web of tangled filaments of gas, the debris of a supernova explosion that was observed by oriental astronomers in* A.D. *1054. At the center of the nebula is a neutron star pulsar spinning at 33 revolutions per second. (Lick Observatory.)*

Figure 69. *Herbert Friedman with instrument section of Aerobee rocket that was used to observe the lunar occultation of x-rays from the Crab nebula on 7 July 1964. The rocket's gyro stabilizer kept the two proportional counters locked on the nebula as the Moon passed in front of it. (U.S. Naval Research Laboratory.)*

teen tests had been conducted by Goddard, and only three had operated within acceptable tolerances. In spite of this poor record, I felt that the once-in-10-years opportunity was too important to pass up, and I decided to accept the risks of conducting the occultation experiment with the available attitude control system. I was encouraged by my good friend Iosif Shklovsky, the eminent Soviet astrophysicist, to attempt the observation. In a letter to me in the Spring of 1964, he commented that his Soviet colleagues were not yet up to conducting such an experiment but he was confident that we at the NRL could carry it off.

Additional problems involved timing, which was crucial if the detectors were to arrive at an altitude above the absorbing atmosphere in coincidence with the occultation of the nebula. Because of its liquid fuel, carrying out precisely timed launchings with the Aerobee required lengthy preparation and countdown; rarely were rockets launched precisely at the designated time. But in spite of a host of problems, the experiment worked almost perfectly. The rocket took off 7 July 1964, at precisely the proper time, and the attitude control system, for the first time, functioned exactly as desired. Elation over the success of the operation, however, was tempered by disappointment over the scientific result. A neutron star was not identified. Instead, the occultation of the center of the nebula revealed a steadily decreasing intensity that proved that the source was broadly distributed rather than concentrated in a neutron star. The actual width of the x-ray–emitting portion of the nebula was as much as 1 arc minute, or about 1 light-year. This was enormous compared to the expected 20-kilometer diameter of a neutron star. If an x-ray–emitting neutron star existed in the Crab nebula, it could not have contributed much more than 5% of the total x-ray flux or it would have been detected.

Within a short time the AS&E scientists also demonstrated that the x-ray source, Sco X-1, could not be a hot neutron star. They were able, in August and October of 1964, to measure the spectrum of the object. They concluded that it did not produce its x-rays from a hot dense body; its spectrum more closely resembled a synchrotron or thin gas thermal bremsstrahlung mechanism. Before another year had passed, the x-ray astronomy group at the Livermore Radiation Laboratory made an even more precise measurement and concluded that the x-rays from Sco X-1 resulted from thermal bremsstrahlung in a dilute plasma at a temperature of about 60 million degrees.

Our X-Ray Galaxy

Realizing that two x-ray sources were hardly a foundation on which to base a theory of the process of x-ray emission, the x-ray astronomy groups began to undertake a more systematic effort to produce a substantial catalog of x-ray sources. Experience in all fields of astronomy in different

wavelength ranges has clearly demonstrated that every subfield of astronomy is represented by a diversity of classes of objects.

The procedure for surveying the sky for additional sources, which was adopted by the NRL group, was a simple and natural way to use the Aerobee rocket. It was deliberately given a very slow spin rate that enabled its instruments to scan almost the entire celestial hemisphere in a single flight slowly enough that individual sources could be identified in flight. By 1967 the number of known x-ray sources exceeded 30. Most were detected by NRL; some were discovered by the groups at AS&E, Lockheed Corporation, Los Alamos National Laboratory, and the Lawrence Livermore National Laboratory.

Even with this limited collection of sources, it was possible to make some general estimates of the numbers of x-ray sources in the Galaxy and of the Galaxy's total x-ray luminosity. It appeared that most of the sources clustered toward the galactic center and the galactic plane. We figured that the total number of detectable x-ray sources in the Milky Way would exceed 1000 and that the luminosity would be about 7×10^{39} ergs per

FIGURE 70. Iosif Samuilovich Shklovsky (1916–1985). *Iosif Samuilovich Shklovsky described himself as a youthful portrait artist by calling, who came to science by accident. His earliest theoretical work in 1945 produced a modern theory of the ionization of the solar corona. In 1949 he turned his attention to problems of galactic astrophysics and proposed that the hydrogen 21-centimeter radiation of interstellar gas should be easily detectable. He further anticipated the development of the rich field of microwave spectroscopy of galactic molecules.*

In 1953 Shklovsky theorized that both the optical and radio emission of the Crab nebula were synchrotron radiation of relativistic electrons circulating in the nebular magnetic field. Shklovsky recognized that the synchrotron spectrum of supernova remnants could extend as far as the x-ray region. With Vitaly Ginsburg he applied the synchrotron hypothesis to extragalactic objects as well, and established an entirely new description of the universe in which magnetic fields and high-energy particles dominate the radiation processes rather than thermal emission of stars.

Shklovsky, a citizen of the U.S.S.R., was a forthright defender of human rights and had a wry sense of humor that often got him into trouble with the authorities. When his own travel was restricted for spurious reasons such as alleged bad health, colleagues encountering a robust Shklovsky would remark that they heard he was ill. To which he would reply, "Yes, I have diabetes. Too much Sakharov!" (Sakhar *is Russian for sugar and Shklovsky was one of the staunch defenders of Sakharov in the Soviet Academy of Sciences.) Although elected a corresponding member of the Academy in the mid-1960s, he was never advanced to full membership. One prominent Soviet scientist commented in explanation: "Fifty percent of what Shklovsky does is brilliant but no one can tell which fifty percent it is."*

second, or about one-thousandth of the optical power radiated by the Galaxy. An extensive all-sky catalog (HEAO-1, 1984) lists about 1000 sources, but high-resolution studies of even small fields reveal hundreds of weaker sources.

Although most of the objects did lie close to the galactic plane and therefore appeared to be members of the Milky Way, a few sources lay at high latitudes toward the galactic poles and possibly were extragalactic x-ray sources. Of these, we soon discovered that a giant galaxy M87 and the quasar 3C-273 were among the high-latitude x-ray sources. At their known distances, the x-ray intensities must have been more than 1 billion times greater than from the Crab nebula. Another great surprise was that several x-ray sources seemed to be strongly variable. In particular, Cygnus X-1 decreased in brightness by a factor of 4 over the course of the year. The NRL group speculated that the variability might have been due to the x-ray star being eclipsed by a companion in a binary system. This hypothesis was not substantiated for Cygnus X-1, but before long, results obtained by the *Uhuru* x-ray satellite, p. 213, confirmed that the eclipsing variability characteristic was very common to x-ray emitters.

Clocking the Crab Pulsar

After the disappointing failure to identify a neutron star in the Crab nebula by the occultation observation, a connection between pulsars and supernova remnants was very clearly demonstrated in November 1968, when the most remarkable pulsar of all was discovered in the Crab nebula. Two young radio astronomers, David H. Staelin and Edward C. Reifenstein, observing at the National Radio Astronomy Observatory in Green Bank, West Virginia, detected pulses from the Crab but could not ascertain a period. They were confused by what appeared to be two pulsars in the same field. Within a matter of weeks, the puzzle was resolved by John M. Comela, with the large 1000-foot radio astronomy telescope at Arecibo, Puerto Rico. There were, indeed, two pulsars, one with the shortest period yet observed, 0.033 second, and the other with close to the longest period, 3.7 seconds. Before long, the fast pulsar was identified with the center of the nebula (5 arc seconds in diameter), and the slow pulsar was positioned 1.5 degrees away.

Although the radio pulsars discovered by the Cambridge astronomers seemed more precise than the best quartz clocks—stable to 10^{-14} second per second—the Crab pulsar soon showed a measurable slow drift. The observations at Arecibo revealed that the Crab pulsar period increased by 36 nanoseconds per day. It soon became clear that all radio pulsars slow down. This observation confirmed the spinning neutron star model in which the energy of radiation must be drawn from the store of rotational kinetic energy.

Discovery of the Crab pulsar was a signal for optical astronomers to turn their telescopes on the central star that Walter Baade, many years earlier, had suspected to be the remnant of the supernova of A.D. 1054. The spectrum observed in 1942 was highly unusual. From its ultraviolet brightness, it appeared to be very hot. There were no emission or absorption lines; an altogether strange situation. William J. Cocke, Michael J. Disney, and Donald J. Taylor at the Steward Observatory in Arizona, were almost immediately successful in detecting optical pulsations with a period of 0.0333 second.

The identification of the pulsar as Baade's central star was established at Kitt Peak Observatory by means of stroboscopic photography. Light from the telescope was imaged by a television camera, and the beam was interrupted by a rotating shutter at exactly the pulsar frequency. If the shutter was in phase with the pulsations, the entire field of pulsar plus surrounding stars was photographed. When the shutter was out of phase, however, the light of the pulsar disappeared but the remaining star field was unaffected. There could hardly have been a more dramatic demonstration of the optical pulsing. A star brighter than the Sun had been caught flashing on and off 30 times per second.

Within a few months my colleagues (Edward T. Byram, Talbot A. Chubb, and Gilbert Fritz) and I launched an Aerobee rocket to search for x-ray pulsations. We succeeded in observing the same pulse pattern that had been found in radio and visible wavelengths, but the x-ray pulsed power was 10,000 times as strong as the radio and about 6% of the total nebular x-ray emission. It was now clear why we had failed to find the neutron star in our 1964 occultation observation. Six percent was somewhat below our limit of sensitivity for distinguishing a sharp break in the occultation light curve.

Immediately following the discovery of the x-ray pulsations, Tommy Gold chaired a symposium on pulsars at the National Academy of Sciences. He had deduced from the slowing down of the radio pulsar how much energy was being lost from the spinning neutron star. It was far more than could be attributed to the radio emission. When I reported that the total x-ray power was more than 10,000 times the radio power, the answer was clear. Gold's model of a spinning neutron star radiating energy at the expense of angular momentum was fully demonstrated.

The full message of the Crab pulsar had now been satisfactorily deciphered. A supernova had left a collapsed core, a spinning neutron star; the gravitational energy released by collapse powered the spin. It is currently spinning away 10,000 trillion trillion kilowatts, and the store of rotational energy will keep the pulsar mechanism working for thousands of years to come. A similar pulsar powered by leftover spin is located in the region of the Vela X nebula and was discovered in 1972. Its period is also very short, 0.089 second. Unlike the Crab, its soft x-ray emission is thus far undetected, and the pulse profile changes substantially with energy. Its pulsations, like the Crab's, reach to very high gamma-ray energies,

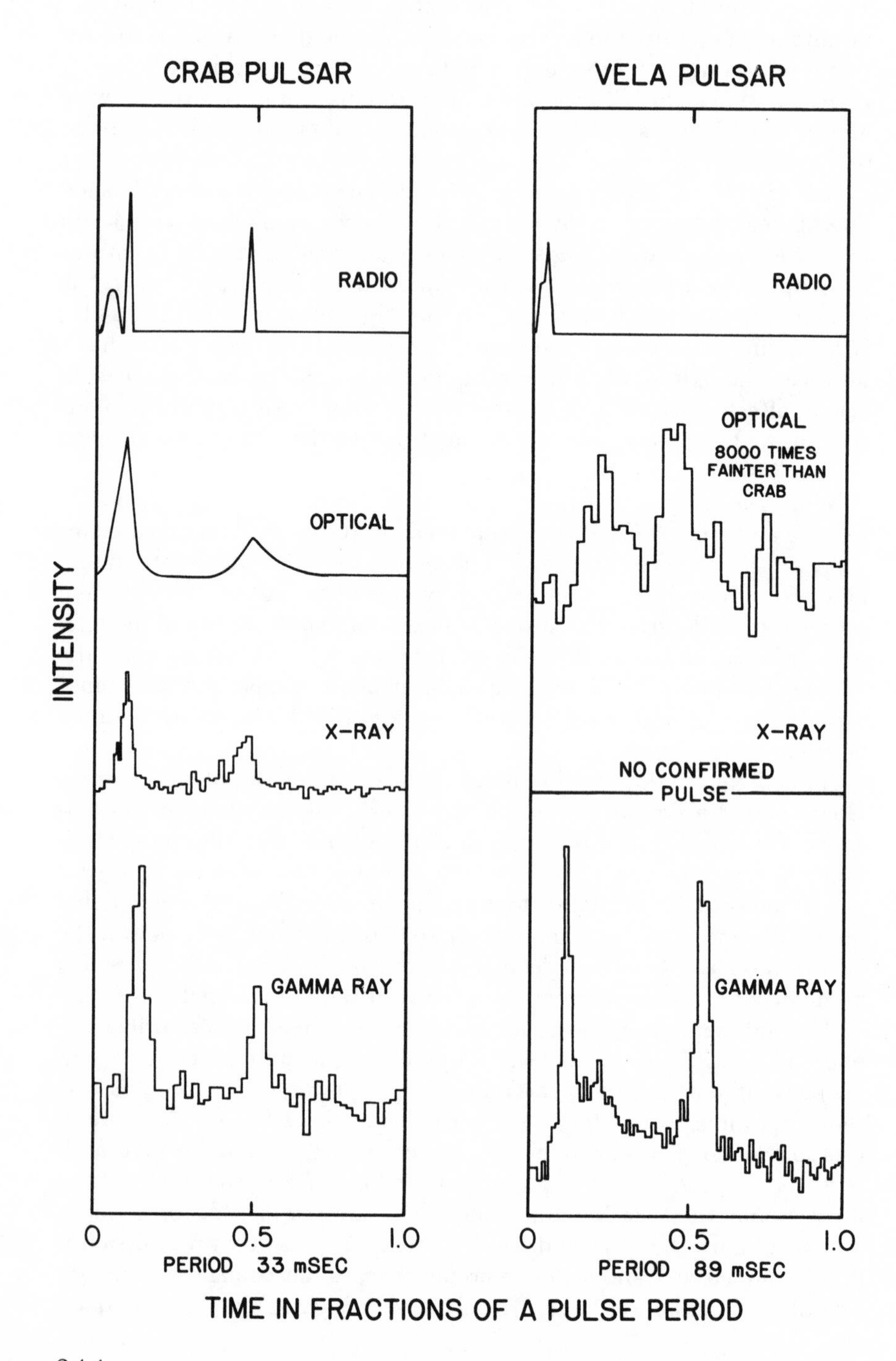
CRAB PULSAR
VELA PULSAR
RADIO
RADIO
OPTICAL
OPTICAL
8000 TIMES
FAINTER THAN
CRAB
INTENSITY
X-RAY
X-RAY
NO CONFIRMED
PULSE
GAMMA RAY
GAMMA RAY
0
0.5
1.0
PERIOD 33 mSEC
PERIOD 89 mSEC
TIME IN FRACTIONS OF A PULSE PERIOD

but they are out of phase with the radio pulses. The only two examples of wide-spectrum radio to gamma-ray pulsars are so different that it is clear we have much to learn about the physical mechanism involved.

The Unremarkable Visible Likeness of Sco X-1

The nature of x-rays from the Crab could be guessed at with some confidence from the time of its first detection because the optical nebula and its connection with an ancient supernova was fairly well understood. But the mystery of Sco X-1, the strongest x-ray source, was deeply puzzling as long as no clue whatsoever to an optical counterpart was available. Since the track of the Moon never crosses Sco X-1, the occultation measurement of position could not be applied. But a *tour de force* of instrumentation to obtain an accurate position and identification with a very faint blue star was carried off by the AS&E group and visiting Japanese scientist Minoru Oda. On 8 March 1966 they flew a device known as a *modulation collimator*. It consisted of a set of planes of wire grids. As an object is viewed by the proportional counter detector through these grids, the scanning motion produces a vernier pattern that makes it possible to determine not only the size of the source but its precise position in the sky. The observation indicated an upper limit of 20 arc seconds for the angular size of the source, which ruled out the possibility that it was a supernova remnant. If it were expanding as fast as the Crab nebula, for example, it could not have been older than 50 years and it would be extremely unlikely that such a recent supernova explosion in the Galaxy could have gone unobserved.

Optical astronomers searched photographic plates of the region of the sky around Sco X-1 but found no plausible candidates. A more likely possibility was that the source was actually pointlike and, if this were the case, the optical object would shine very faintly like a thirteenth-magnitude blue star in the visible region. Very quickly, following the x-ray positioning of Sco X-1, there began a campaign to find an optical counterpart. In Japan, astronomers using a 74-inch telescope identified a blue star of 12.6 magnitude within 1 arc minute of the Sco X-1 position. One week later, observers with the great 200-inch telescope at Mount Palomar ver-

FIGURE 71. *The pulse profiles of neutron star pulsars in the Crab and Vela supernova remnants. The Crab synchrotron radiation spectrum shows a strong resemblance in the pattern of main pulse and interpulse in all wavelength regions. The Vela pulsar shows no interpulse in radio waves, very faint evidence of pulsation in visible light, no confirmed x-ray pulsation, but strong pulsation in gamma rays. Models of pulsar-emission mechanisms are still uncertain and no satisfactory explanations have been found for these differences.*

ified that result. It now appeared that the brightest x-ray object in the sky was associated with a visible star of very minor optical significance. The historical record of optical observations suggested that the source resembled an old nova with characteristic fluctuations in brightness. Its distance was estimated to be 250 parsecs, which gave a value of 10^{36} ergs per second for its x-ray luminosity. This power was approximately equal to that from the Crab nebula.

The similarity of the faint blue star to an old nova led to much theoretical speculation about the possible mechanism of x-ray emission. Most old novas are members of binary systems in which a compact object, such as a white dwarf, is coupled gravitationally to a giant star. In theory the gas that is tidally sucked from the large partner would be heated to very high temperature because its kinetic energy would increase in proportion to the mass of the compact partner, and inversely with its radius. The power produced by this accretion process could provide more than 10 times the energy per nucleon if it fell onto a white dwarf or hypothetical neutron star with the mass of the Sun, as is achieved in nuclear burning in the core of the Sun. Shklovsky particularly preferred a model in which the accreting object could be a neutron star.

The Uhuru *Era*

The first x-ray astronomy satellite was launched into orbit around the equator from a modified oil rig off the coast of Kenya. Appropriately, it was named *Uhuru*, meaning "freedom" in Swahili, in honor of its launch on Kenyan Independence Day, 12 December 1970. The *Uhuru* mission was conceived and carried out under the leadership of Riccardo Giacconi while he was affiliated with the AS&E group. In somewhat over 2 years of operating life, it produced a catalog of 339 discrete sources and lead to the discovery of previously unknown periodic variations in intensity of a number of objects, proving that they belonged to double-star systems and that they were powered by gravitational accretion processes.

The payload of the *Uhuru* satellite was not much heavier than one for rocket astronomy. It weighed only 64 kilograms, and its basic detection system consisted of a pair of proportional counters of moderately large area. Once it began to operate, it immediately demonstrated several advantages of the "long-playing" satellite. Because its instruments made repeated observations of objects and could accumulate counts from successive scans, they added up to much greater sensitivity for the detection of weak sources. It was able to pick up x-ray stars one-tenth as bright as those that had been detected from stabilized rockets. Although position location was still crude, it was sufficiently improved that a substantial number of sources could be identified with visible counterparts.

The first great surprise came when *Uhuru's* instruments picked up a source previously discovered in 1967 and known as Centaurus X-3. It was found to exhibit extremely regular periodicities. At a pulse period of 4.8 seconds, it resembled a pulsar. But it also showed a secondary period of 2.087 days, which was clearly identified with a binary system orbital periodicity in which the x-rays from a compact object were eclipsed every time it passed behind the companion star. Almost immediately, ground-based astronomers identified one of the components as a blue supergiant.

Uhuru next discovered Hercules X-1, which had periodicities of 1.24 seconds and 1.7 days. The optical counterpart was a blue star with about twice the mass of the Sun. The same orbital period was observed in optical wavelengths as was detected in x-rays. Although the basic binary nature of both Cen X-3 and Her X-1 was clearly evident, it came as a surprise that the pulse period of Her X-1 decreased slightly as time went on. If it had been a pulsar operating like the one in the Crab nebula, its pulse period would have slowly increased because of the enormous amount of energy being radiated. The opposite behavior could be understood if a compact neutron star accreted matter from a stellar wind streaming from the surface of the blue companion. The infalling material would not strike directly on the neutron star, however. Because of its angular momentum, the flow toward the neutron star follows a spiral path to form an accretion disk. Gas in the spinning disk loses energy by friction with surrounding gas and falls inward from the disk onto the neutron star at the rate of about one-billionth of 1 solar mass per year.

In the strong magnetic field of the spinning neutron star, infalling plasma is channeled to the surface at the magnetic poles, where two hot mounds of x-ray–emitting plasma build up to a temperature of about 100 million degrees. As the star rotates, the poles are alternately visible from the Earth, thus producing the pulsed radiation. Unlike the neutron star in the Crab nebula, which slows down because it transforms spin energy to radiation, Her X-1 gains energy from the conversion of the potential energy of infalling gas. The transfer of angular momentum from the accretion disk accelerates the spin. Several old x-ray pulsars have now been discovered that spin at nearly 1000 revolutions per second, so-called millisecond pulsars.

Besides the fundamental periodicities of the neutron star's spin and orbital revolution, Her X-1 exhibits additional variations on the basic theme. During each 1.7-day orbital cycle the x-ray intensity dips for about 6 hours when the neutron star passes behind its companion and out of view from Earth. A 35-day variation is related to a wobble of the spin axis like the precession of a spinning top. Of the two beams emerging from the magnetic poles of the neutron star, one swings almost directly toward the Earth for 9 days and produces the strong pulse. The intermediate pulse is weaker and lasts for fewer days because it doesn't align as well with the Earth. The fast-pulse profile is also modulated by precession. When the average brightness is high in the 35-day cycle, the 1.24-second pulse appears alone. However, when both beams of the neutron star barely brush

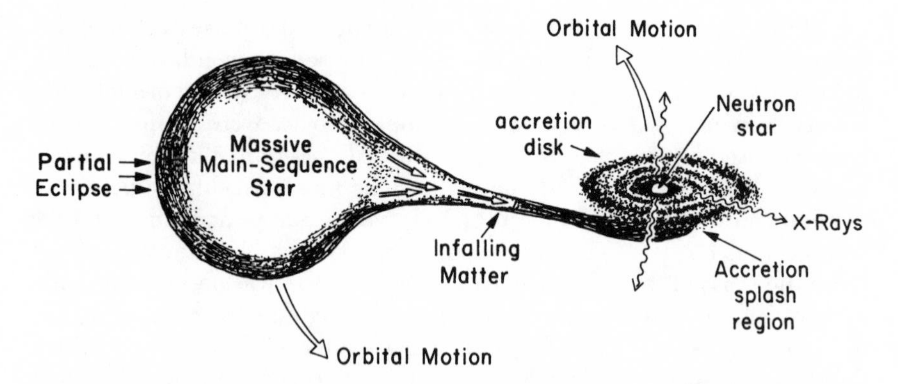

FIGURE 72. *A model of an x-ray binary consisting of a main-sequence companion star coupled to a neutron star. Tidal forces draw a flow of gas that spirals into an accretion disk surrounding the neutron star. As the gas flows toward the neutron star along its magnetic field, it grows so hot that intense x-rays are emitted. Depending on the perspective of the system as seen from Earth, the x-ray flux will exhibit the periodicity of the spinning neutron star and eclipses of flux by the companion star and by the accretion stream during the orbital period.*

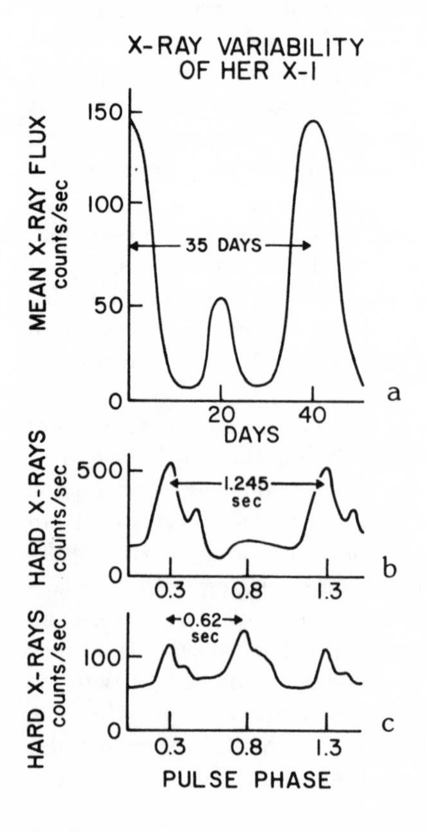

FIGURE 73. *Influence of neutron star precession on Her X-1 pulsation patterns. (a) Mean variation of x-rays induced by 35-day precession. (b and c) Modulation of fast-spin pulsation by precession. X-rays are received from only one pole of the neutron star in (b). Interpulse appears in (c) when x-rays are received from both poles each spin period.*

the Earth, the average flux is weak and the pulse pattern becomes double in each spin cycle.

Like Cyg X-3, Her X-1 has become an object of special study for ultra-high energy gamma rays and cosmic rays. Cerenkov telescopes on Mount Hopkins and Mount Haleakala in Hawaii have revealed 1-trillion-electron volt photons. At Los Alamos an extensive cosmic-ray air shower array has shown evidence of bursts of from 10^{14}- to 10^{15}-electron volt–photons from the direction of Hercules. All three sets of observations reveal the signature of the 1.24-second neutron star period. There seem to be slight differences in the x-ray and gamma-ray periodicities that are mystifying. Much longer sets of observations will be necessary to clarify what is going on.

The evolution of a binary-star system can exhibit a rich variety of dynamic transformations that involve mass transfer back and forth in relatively short phases. Often, in a matter of mere millions of years, the originally more-massive star may shrink to compact size, whereas the overflow of mass to the companion causes it to bloat to supergiant size. When the compact size eventually collapses to a neutron star, a reversed flow of gas from the giant companion creates an encircling accretion disk and inaugurates a phase of strong x-ray emission.

As the spin of the neutron star is whipped by the torque of accreting material, it can generate a blast of intense radiation that blows away the atmosphere of the companion. Recent observations by Princeton astronomers working at Arecibo have revealed eclipses of a rapidly spinning radio pulsar in a close 9-hour orbit about a companion that weighs a mere 2% of 1 solar mass even though it is larger than the Sun. In this bizarre configuration the companion star appears to be evaporating under the pressure of radiation from the neutron star, so that it will completely disappear in a few hundred million years.

Although it was the first, and still is the brightest, x-ray source thus far detected, Sco X-1 remained a mystery object until the advent of *Uhuru*. Without evidence of a neutron star, the accretion model advocated early on by Shklovsky and others was highly speculative; no strong x-ray pulsations were detected such as were observed by *Uhuru* from Cen X-3 and Her X-1. The source of pulsed x-ray generation in those binary sources was clearly a rapidly spinning magnetic neutron star. In 1975 regular variations in the Doppler shifts of its spectral lines and in its optical brightness were recognized as evidence that Sco X-1 had a companion star that circled it with a period of 0.787 days. The clue to the presence of an accretion disk was found in 1981, when simultaneous x-ray and optical variations were observed with no time delay. Such coincidence could be understood if the x-rays from the neutron star generated light very close by; for example, on the inner portion of an accretion disk rather than on the more separated optical companion star, as is the case with the other x-ray binaries. Another surprise developed in 1985, when it was discovered that Sco X-1 produced quasi-periodic oscillations in its x-ray intensity. Variations in brightness of several percentages are not perfectly periodic; the frequency drifts over the range between 6 and 50 cycles per second. Unlike the precise pulsars with their remarkable stabilities, Sco X-1 is a sloppy clock.

The phenomenon of quasi-periodic oscillations has been observed by the European *EXOSAT*, the Japanese *HAKUCHO*, and NASA's *HEAO-1* in a number of galactic bulge sources, but the process is not well understood. A beat frequency mechanism seems most promising, in which the

FIGURE 74. *About once a day a very intense flash of gamma radiation may be detected from an unpredictable direction in the sky. Bursts as short as .001 seconds and as long as 80 seconds have been observed; most last from 1 to 10 seconds. The burst shown here appeared 5 March 1979 and was 10 times as intense as any other on record. It flashed to peak in less than 0.0002 second. As it decayed the intensity oscillated with a period of about 8 seconds.*

The shortness of the burst suggests a neutron star since the dimension of the source cannot exceed 0.0002 light-second, or about 40 miles. The oscillation implies a spin period of 8 seconds, too fast for any type of star other than a neutron star. Many burst spectra show a line at about 420 kiloelectron volts that is consistent with electron–positron annihilation radiation, which is normally at 511 kiloelectron volts but has been redshifted by the intense gravitational field at the surface of a neutron star. All of the observed gamma-ray bursters are believed to lie in the Milky Way; none has recurred at precisely the same location or has been identified, but new evidence places them far beyond a known object.

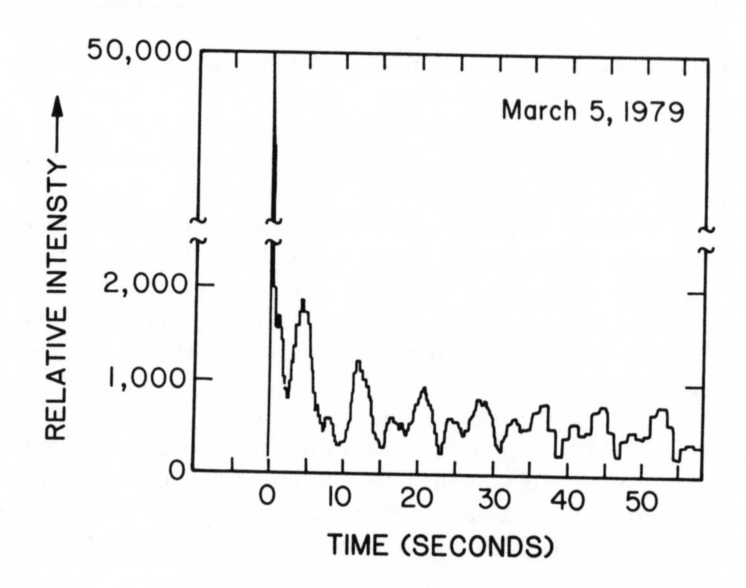

the quasi-periodicity reflects the difference in frequency of the spinning neutron star and the orbital frequency of clumped plasma in the inner accretion disk.

X-Ray Transients

We have discussed variable x-ray sources marked by regular periodicities; that is, pulsars and eclipsing binaries. Another important class is *transients:* sources that brighten spectacularly over a period of days and then fade away over several months. At the other end of the scale are *bursters*, which flash abruptly and then gradually disappear after from 10 seconds to 1 minute. Where nothing had been observed before, a burster may suddenly radiate x-rays with the power of 1 million suns.

The first reports of x-ray bursts came from observations with the *Astronomical Netherlands Satellite (ANS)* in 1973 and the American *Vela* satellites designed to monitor nuclear weapons tests in space. The *ANS* bursts seemed to originate in a globular star cluster, NGC 6624. Many more were soon detected from objects not associated with star clusters. In some cases, optical bursts had been seen in coincidence with x-ray bursts. A particularly remarkable object known as the *rapid burster*, discovered by Walter Lewin and his co-workers at the Massachusetts Institute of Technology, repeats its flashes like staccato machine-gun fire. After a very large burst it may cease for a while, like a relaxation oscillator that needs to recharge whatever reservoir is involved. The popular model for bursters involves accretion onto an old neutron star with a weak magnetic field. The gas builds up over the surface of the neutron star until its density and temperature reach that required for thermonuclear ignition. Bursters are nature's nuclear warriors. So enormous is the power gained from gravitational accretion that a pea dropped onto a neutron star could release the energy of a Hiroshima bomb.

Black Holes: Out of This World

If the incredible crushing-down of stellar mass to make a neutron star is a stunning concept, the idea of a black hole is truly "out of this world." Collapse beyond neutron star density must go all the way to infinite density, which is a physical "singularity" that defies detailed description. Einstein's theory is applicable only until density is about 10^{88} tons per cubic inch, which is the density that the Sun would reach if it were crushed to one-millionth of the size of a proton. Higher density is beyond physical understanding.

In 1783 John Michell, director of the Thornhill Observatory in Yorkshire, England, calculated that an object would need to have a speed greater than one five-hundredth the speed of light to escape the Sun.

From that number, Michell next argued that a star 590 times as large as the Sun and of the same average density would prevent even light from escaping: "All light emitted from such a body would be made to return to it by its own power of gravity" (Paper read before the Royal Society of London, 1783). Pierre Simon de Laplace, the great French astronomer and mathematician, came to the same conclusion in 1796. John Wheeler, a Princeton physicist, coined the term *black hole* in 1960. Michell's and Laplace's massive star fits that name, but black holes of stellar mass are, according to the modern physics of general relatively, highly collapsed regions of spacetime. Gravity is the weakest of all physical forces in the world at large; it took the entire Earth to cause Newton's apple to fall. But in the neighborhood of a black hole, the concentration of gravity becomes overwhelming and all-devouring. Nothing, not even light, can escape.

Rockets that reach speeds of 11.2 kilometers per second (7 miles per second, or 27,000 miles per hour) escape the Earth's gravity and travel to the planets and beyond. For all the objects in the Solar System, the velocity of escape is quite small compared to the velocity of light. On the Sun, surface gravity is 28 times as great as on the Earth, so that a person who weighs 150 pounds on Earth would weigh almost 4200 pounds on the Sun. If we imagine the Sun being steadily crushed to smaller and smaller volume, an object on its surface would move closer to its center. According to Newton's law of gravitation, gravitational pull increases in inverse proportion to the square of the distance from the center. Were the Sun reduced one hundred thousandfold to the size of a neutron star, its surface gravity would increase 10 billion times. Our 150-pound person standing on the neutron star would weigh 50 trillion pounds and have an escape velocity of about 200,000 kilometers per second, 6% of the speed of light. If the neutron star were shrunk to a radius of 3 kilometers, escape velocity would reach the speed of light and the star would become a black hole.

Only a few decades ago neutron stars and black holes seemed almost beyond belief. I recall Oppenheimer's reaction when I reported the detection of x-ray stars to a colloquium at Princeton University in 1963 and suggested a link with neutron stars. He was obviously thrilled at the prospect that his theoretical neutron star, which had seemed hopelessly beyond the possibility of detection when he first proposed a theoretical model 25 years earlier, could now become the subject of detailed observation. Soon radio and x-ray pulsars were discovered, and the case for neutron stars became incontrovertible.

Encouraged by the consistency between the theory and observations for simple symmetrical collapse to a neutron star, theorists were now prepared to confront the question of ultimate collapse. They described a black hole as a very simple object, bound purely by a self-sustaining gravitational field and retaining no causal history of the material source from which it had evolved. No imprint remained of any distortions carried along in physical asymmetries of magnetic field or of the complex structure of the evolved stellar core on the way to total collapse. Only three characteristics sufficed to characterize the collapsed object in its final state: mass, angular momentum, and electric charge. The gravitational field could be described precisely, and its periphery, the "event horizon," was as smooth and featureless as a soap bubble. Wheeler used the graphic description, "Black holes have no hair" ("Our Universe: The Known and the Unknown," *American Scholar 37*, 1968, p. 248).

In the early debate about white dwarfs (see Chapter 5) Chandrasekhar implied that total collapse of a star is a perfectly natural consequence of the principle of relativity, and Eddington countered with: "There should be a law of nature to prevent the star from behaving in this absurd way" The large majority of astrophysicists now accept the inevitability of collapse to a black hole, but the concept is still a very disturbing one. Our most promising avenue to observational confirmation is to probe the accretion environment with x-ray telescopes. Thus far only three x-ray stars match the theoretical requirements.

The most likely candidate black hole is Cygnus X-1. It was detected in 1964 in one of the early NRL rocket surveys, and when observed again 1 year later, its x-ray flux had decreased to one-fourth. Such variability was startling in an object whose x-ray power was 1000 times the total optical power of the Sun. When it was observed more carefully by *Uhuru* in December 1970, its intensity fluctuated so rapidly that the *Uhuru* instruments had difficulty following the variations. A rocket measurement shortly afterward revealed fluctuations as fast as one-thousandth of 1 second. From the rapidity of these fluctuations, it was evident that the x-rays came from a compact object smaller than a white dwarf.

A star cannot change its brightness in less time than it takes light to cross its diameter. For example, if the Sun were suddenly to blink off, an observer on Earth would first see the center of the disk grow dark. Light from the edge would travel an additional 800,000 kilometers (the radius of the Sun) toward Earth before a ring of light around the limb would disappear. The additional transit time takes 2.7 seconds, and the Sun would appear to fade from center to edge in that time. Thus, when Cygnus X-1 exhibited millisecond bursts, it meant that the x-ray–emitting region could be no more than 300 kilometers across (10^{-3} second $\times$ velocity of light $=$ 300 kilometers).

The rocket experiment also located the position of the source with sufficient accuracy that the only object in the field was a blue giant star, HD 226868, 30 times as massive as the Sun. Such blue giants were not known to emit detectable x-rays by themselves, so there must have been some secondary process involved. The wavelengths of its spectral lines increased and decreased in oscillatory fashion with a period of 5.6 days; this was clearly a Doppler effect. From this behavior, it was possible to conclude that HD 226868 and an invisible companion star orbit around a common center of mass and are separated from each other by a distance less than that between Mercury and the Sun.

If the mass of one member of a binary pair is known, the companion mass can be determined from the orbital period according to Kepler's law of planetary motion. For example, if the Moon had the density of marshmallow, it would orbit the Earth much farther out and very slowly. Were it made of lead, its orbit would shrink close in and its period would become much shorter. From the orbital characteristics of Cyg X-1, astronomers estimated that the companion had about 9 times the mass of the Sun. In theory, the mass of a neutron star cannot exceed 3 solar masses; greater mass must collapse to a black hole. The evidence strongly suggested, therefore, that the x-ray–emitting compact object in Cygnus X-1 was a black hole.

Gravitational Redshift and Time Dilation

When massive stars exhaust their nuclear fuel and implode, the end stage cannot be described in physical detail. Physicists refer to a "singularity" of infinite density and infinitesimal volume in which all of the collapsed mass is entombed. In this singular state, time has no meaning; there is no beginning and no end. Within the collapsing system, the implosion takes perhaps one ten-thousandth of 1 second, but to a remote observer on a distant planet, the infalling star never reaches its singular state. The reasons for these different perceptions are gravitational redshift and time dilation, the stretching of time according to Einstein's relativity theory.

The prediction that time slows down in a gravitational field has been proven by direct experimentation. In the early 1960s Harvard University physicists Robert V. Pound and Glen A. Rebka, Jr. made use of the very precise frequencies of certain gamma-ray lines emitted by radioactive nuclei; a form of emission known as the *Mössbauer effect.* They carried their radioactivity clock to the top of a tower and measured the difference in frequency of the gamma-ray lines between the high point and the basement level. The lower level, being closer to the center of the Earth, was in a more intense gravitational field. A slowdown of time with height was clearly detected, in precise agreement with the prediction of Einstein's general relativity theory.

Now suppose that with a telescope we could observe clocks carried by hypothetical astronauts who are plunging toward a black hole. We would note that their clocks are slowing down as the astronauts come closer to the hole. Eventually, at the boundary, their clocks stop completely. The astronauts' own falls slow down, and they appear to come to a halt at the same time that their clocks stop. From our point of view, it takes the astronauts an infinite length of time to cross the horizon. The astronauts, if they could communicate with us instantaneously, would tell us that we were all wrong. Their clocks, heartbeats, and other vital signs would seem to be behaving entirely normally. According to their own measures, they cross the black hole horizon in a split second.

If we could catch sight of the collapse of a star to a black hole, we would see its light redshifted in proportion to the slowing down of time. Atoms radiate visible frequencies like ultraprecise clocks and, just as was observed with gamma rays in the Harvard University experiment, visible light is redshifted as the strength of the gravitational field increases. Accordingly, the light of a collapsing star should turn progressively redder until it passes through the visible to the infrared. When collapse has reached to within one-third the radius of the black hole, the escaping light is doubled in wavelength. At a distance of only one-millionth of the black hole radius, light is redshifted one thousandfold. At one-trillionth of the radius, the redshift reaches 1 million, and infinite redshift is almost an infinitesimal distance away.

As the implosion of the star freezes in time when clocked by a distant observer, the passage from visible to invisible, because of redshift, will still be very fast. A star 10 times as massive as the Sun would fade from view in a few millionths of a second. If the collapsing object were as massive as 1 million stars, it would still disappear in about 0.25 second. If an entire galaxy of stars collapsed to a black hole, it would fade away in a matter of days. Obviously, the chance of observing any such event is extremely small.

The universe seen from the surface of a collapsing star would appear very strange indeed; almost an *Alice in Wonderland* experience. In *Through the Looking Glass,* Alice and the Red Queen converse: " 'What sort of things do you remember best?' Alice ventured to ask. 'Oh, things that happened the week after next,' the Queen replied in a careless tone."

To observers on the collapsing star, all external light would be blueshifted and everything happening in the outside world would be speeded up. When the star reached its black hole radius, the hypothetical observers would see the entire future flash before their eyes. If, for example, the universe is closed and oscillating from Big Bang expansion to eventual halt and then collapse, the observers would see all the galaxies streak away and then rush back until the density of the collapsed universe matched that of the black hole. Only then would outside time match time inside the black hole.

In addition to the redshift, the curvature of spacetime near the black hole contributes to the blackout of the collapsing star. As the star falls inward, the warping of spacetime becomes more and more severe. Light rays must follow the curvature of space and bend away from the classical straight-line trajectories of flat space in our familiar world (see Chapter 7). Those rays that pass a black hole at large distances are only slightly deflected. Nearer to the black hole, the bending becomes more pronounced. If a searchlight beam is aimed straight at the black hole, its light will disappear into the hole. As the beam is directed off the hole, it will quickly reach a distance at which it executes a circular orbit. This distance defines the *photon sphere.* Astronauts orbiting in the photon sphere and looking straight ahead will see the backs of their heads. If light deviates ever so slightly from the photon sphere, it must spiral in or out, away from the circular orbit.

Imagine, now, that astronauts carrying powerful searchlights, are on the surface of a collapsing star. What the outside observer sees is fairly normal until the astronauts reach the photon sphere, where relativity causes strange things to happen. If they point the searchlight beams directly upward, the light rays escape straight away into space. As soon as they point the beams away from the vertical, spacetime curvature bends the beams toward the photon sphere. The more the searchlights are tipped away from vertical, the greater the deflection of their light. Finally, with the beams aimed at a sufficient angle, the light rays are bent so far that they return to the star. Accordingly, we can define an imaginary *exit cone* within which light escapes the star forever; outside the cone, light returns to the star, never to escape again. As the star continues its collapse, the exit cone narrows. Finally, the curvature of spacetime becomes so great that the exit cone closes down completely. No matter which way the searchlights are pointed, light cannot escape: The escape velocity has become greater than the speed of light. At the instant that the cone closes completely, the astronauts cross the *event horizon;* that is, the edge of their world. No longer can they communicate with the outside universe. To the outside observer, the moment of crossing the event horizon marks the disappearance of the collapsing star. Thus, the curvature of spacetime acts together with the gravitational redshift to cause the collapsing star to vanish into the black hole.

Black holes come in all sizes. Although our discussion thus far has dealt with black holes the mass of several Suns, physicists suspect that there are baby black holes no more massive than the Earth and giant black holes that contain the mass of hundreds of millions of Suns. If the Earth were squeezed down to a radius of 1 centimeter, it would become a black hole.

Once a black hole is born it can never disappear. Black holes can only grow larger. A single black hole can never divide, but two black holes can coalesce into one. We suspect that in the nucleus of an active galaxy, violent energy generation is associated with a giant black hole. Wandering stars that fall into the gravitational grip of the black hole are torn apart, and their wreckage swallowed up. The gas sucked toward the black hole is compressed and heated to enormous temperatures, and the energy-release process is about 100 times as efficient as in thermonuclear energy conversion. Super black holes could grow to hundreds of millions of solar masses in about 1 billion years.

The tidal force exerted by a black hole depends on the relative average densities of the black hole and of the object that is under its gravitational spell. As a rough rule, the body under the influence of the gravitational field of the black hole will be torn apart if its average density is less than the density of the black hole. A black hole with the mass of the Sun has an average density of 10^{16} grams per cubic centimeter (60 billion tons per cubic inch). A black hole of 100 million solar masses, such as might lurk in the center of an active galaxy, is no more dense than water, about 1 gram per cubic centimeter. A normal star of solar mass also has an average density comparable to water. If it approached the 10^8–solar mass galactic nucleus, it would be torn apart. A black hole of mass equal to a small galaxy, about 3 billion solar masses, would have the average density of air. Stars approaching such extended black holes would fall straight in, without disruption. If a space ship entered a black hole of about 1 million solar masses, its astronauts would experience only slight tidal pull. Were the ship to enter a super black hole of billions of solar masses, its passengers would hardly be able to feel the crossing of the event horizon.

Thus far we have discussed black holes of masses equal to the Sun and up to masses of the order of galaxies, but it is also possible that black holes exist of much smaller mass than the Sun. These miniature black holes could not be produced by collapse in the manner of a massive star. The Earth, for example, could never collapse to a black hole because it has no potential source of gravity sufficient to crush the matter of which it is made. Yet Stephen Hawking, a British physicist, has proposed that baby black holes of a wide range of masses could have formed in the epoch of the Big Bang ("Particle Creation by Black Holes," *Communications in Mathematical Physics 43*, 1975, p. 199).

In its early moments, the universe was very dense and chaotic. Black holes may have separated out of this medium. They might be no bigger than an atom yet contain a mass of 10^{20} grams (100 trillion tons). Even smaller black holes the size of nucleons would have masses of about 10^{15} grams (1 billion tons). If these subsolar-mass black holes exist, they may pervade the universe. Some scientists have speculated about the possibility of capturing a baby black hole in Earth orbit. By feeding it waste matter, an abundance of energy could be generated and transmitted back to Earth. Others have conjectured about a baby black hole at the center of the Sun, or about a black hole collision with the Earth. A tongue-in-cheek suggestion has been made that the great Siberian crater was produced in 1908 by impact with a baby black hole. A baby black hole of atomic size, however, would drill through the Earth with hardly any noticeable effect. Its 10^{20} grams might increase by no more than 1 gram as it passed through.

Are Black Holes Forever?

When quantum effects are considered, a black hole, according to Stephen Hawking, must emit quanta of radiation as if it were a black body at a finite temperature. But the emission temperature is inversely proportional to the mass of the black hole; for 1 solar mass the temperature is only one-millionth of 1 degree kelvin, and essentially unobservable. As it radiates, the emission gradually uses up the mass of the black hole, but at such a minuscule rate that the lifetime of a stellar mass black hole is greater than 10^{75} years. If baby black holes as small as 10^{-5} grams were created in the Big Bang, they would have a lifespan of about the same as the universe's. In the final stage of evaporation, temperature would skyrocket with the release of bursts of gamma rays observable in our eon of time. The search for a background of such baby black hole gamma radiation has revealed no evidence thus far.

The stars may dissolve and the
fountain of light
May sink into ne'er ending
chaos and night.

—Percy Bysshe Shelley,
"The Irishman's Song," 1809

CHAPTER 7

Island Universes

The fires that arch this dusky dot-
Yon myriad-worlded way-
The vast sun-clusters' gather'd blaze,
World-isles in lonely skies,
Whole heavens within themselves, amaze
Our brief humanities.

—Alfred Lord Tennyson,
"Epilogue," 1850

Through the Veil of the Milky Way

On a clear summer night the band of the Milky Way arches across the sky, its stars crowded together so closely that it resembles a continuous glow of light. Our Sun and about 100 billion other stars fill the disk of the Milky Way. When we look within the disk, the view is so dense with stars and dust that it is impossible to see to the center or the edge. When we turn to the sky above and below the plane, we see that the stars have thinned out dramatically but fuzzy patches of light are discernible; these could be nebulous glows or dense collections of stars in distant galaxies such as the Milky Way.

One thousand years ago Arabian astronomers took note of these nebulous patches, but without telescopes no detail could be made out. In the mid-eighteenth century Thomas Wright and Immanuel Kant advanced the concept of "island universes" to explain the wispy nebulous patches seen against a foreground Milky Way. At about the same time, William Herschel constructed his so-called 40-foot telescope with 48-inch mirror and systematically searched the sky for nebulae, but he still failed to resolve stars. His work, carried on for 42 years after his death by his son John, lead to the publication of a catalog of over 5000 nebulae. At first John Herschel thought the nebulae were all composed of stars, but by the end of the century he had deep doubts that they were stellar aggregations. One hundred years later the doubts still persisted, and hardly anybody thought there were galaxies beyond the Milky Way. Not a scrap of positive evidence existed. Not until the 1920s did scientists recognize the great universe of galaxies.

To progress from describing stars to modeling galaxies requires an enormous scale change in perspective. To understand the luminosity of stars requires the physics of atoms only 10^{-8} centimeters in size. To model galaxies as large as 10^{22} centimeters in diameter involves a scale change of 10^{30} times. In the atomic domain everything is governed by electromagnetic forces; on the galactic scale gravity is the overwhelming force. Even the most massive and brilliant stars do not exceed about 50 million times the mass of the Earth, but galaxies contain billions, and in rare cases even trillions, of stars bound together by their mutual gravitation. Furthermore, we now suspect that they contain more invisible mass than luminous stars; mass that reveals itself only by the dynamical effects of gravitation.

Throughout the universe galaxies tend to assume three basic forms: spiral, elliptical, and irregular. Within those categories there exists considerable diversity, but the threefold distinction is quite clear and useful to understanding their evolution. The classification still follows Edwin Hubble's original descriptions, which were based on the observations carried out with the 100-inch Mount Wilson telescope from the end of World War I to just prior to World War II.

The spirals resemble enormous whirlpools that have been flattened into disks that spin around a central bulge. Within the disk, curving trails of stars, gas, and dust unwind from the core, and with increasing distance unravel in shreds and forks. Some spirals are wound very tightly, others very loosely. Our Milky Way is an intermediate variety. One variety of spiral is called a *barred spiral* because it exhibits an extended bar across its nucleus. The spiral arms take off from the ends of the bar.

Elliptical galaxies are more-or less egg-shaped and range from nearly spherical to rather flattened ovals. They have used up almost all of the gas and dust essential for the formation of new stars and are populated only by old, dying stars. A supergiant elliptical galaxy may be three times as large across as the Milky Way and contain as many as 10 trillion stars. Dwarf ellipticals still measure about 1 billion billion miles across.

Irregularly shaped galaxies, as the name implies, are neither disklike nor ellipsoidal, but chaotic. All are rich in large clouds of glowing ionized gas and are in the process of breeding young blue stars. The Magellanic Clouds are almost contact neighbors of the Milky Way. They have indistinct shapes and rank as dwarf ellipticals.

As is the case with stars, there are small galaxies and giant galaxies in each class, quiet galaxies and explosive types such as the radio galaxies, Seyfert galaxies, and quasars. On the smallest scale there are binary pairs of galaxies that orbit each other, and there are clusters of only 1 or 2 dozen galaxies, such as the "local group" of which the Milky Way is a member. On the largest scales there are giant clusters of thousands of galaxies, such as Coma Berenices, and still-larger associations of clusters into superclusters. Relatively speaking, the distance between galaxies in clusters, on the average, is only about one-tenth the separation of stars within galaxies, and collisions are rather common.

The superclusters appear as gigantic strands of galaxies elongated over hundreds of millions of light-years. They resemble tangled skeins that on the very largest scales form a spongy structure. Startling evidence has emerged of voids as large as 250 million light-years in diameter; more than 1 billion Milky Ways could fit into such enormous volumes. Within these spaces there seem to be no galaxies at all. Astronomers pursuing these studies are beginning to suspect that the superclusters occupy no more than perhaps 5% of the volume of the universe, and that the rest of the universe is empty. Some theorists speculate about the possibility that vast galaxies of dark matter fill the emptiness.

Ever since Hubble classified galaxies according to shape, astronomers have sought to understand the basis for the evolution of those shapes. Most early theories sought for simple progression from one shape to another. For example, galaxies might begin as ellipticals, then develop spiral structure, and finally break down into the chaos of irregulars. Equally good arguments could be made for a reverse sequence. Beginning with galaxies in disordered chaos, they might gradually wind up with spirals and finally consolidate into the simple symmetry of ellipticals. Both evolutionary models were based on the premise that a galaxy's shape somehow reflected its age. When astronomers learned how to determine the age of a star, they were able to tag the age of the galaxy in which the star was located. In all types of galaxies they found stars several billion years old, and thus they were forced to conclude that all galaxies were of roughly the same age and, that neither ellipticals nor irregulars were younger than spirals.

What did emerge from this examination of content rather than shape was the discovery that ellipticals use up their gas and dust in one great burst of star formation very early on, so that thereafter they harbor old stars almost exclusively. Spirals had managed to conserve a large proportion of gas and dust and still create new stars more-or-less prolifically. Irregulars are the most active stellar nurseries of all. Just how the shapes relate to these simple differences is believed to be a matter of the initial conditions of the amount of mass and angular momentum in the condensing gas cloud, and of interactions with nearby galaxies.

ASTRONOMICAL DISCOVERY

At Home in the Milky Way

A broad and ample road, whose dust is gold,
And pavement stars, as stars to thee appear
Seen in the galaxy, that milky way
Which nightly as a circling zone thou seest
Powder'd with stars.

—JOHN MILTON,
"Paradise Lost," Book VII, 1667

When we see the sparkling band of the Milky Way trailing across a clear night sky, we now recognize that we are looking from inside a cross-section of a very large spiral galaxy, the dimensions of which strain the imagination. It takes a ray of light 100,000 years to travel from one edge of the galactic rim to the opposite side. Overall the Galaxy resembles a thin disk with immense spiral arms wrapped around a central bulge. Enveloping this entire structure is a very faint spherical halo. We live in the tranquil suburbs, about 30,000 light-years from the violent activity of the center.

Filling the disk are more than 100 billion stars, wide lanes of gas and dust, giant interstellar clouds, high-energy cosmic rays, wispy webs of magnetic fields, and radiation of all wavelengths from gamma rays to radio waves. The bulk of the galactic mass, however, is now believed to reside not in the disk, but in the surrounding galactic corona. Trillions of invisible small stars, typically about one-fiftieth of the mass of the Sun and just barely able to support fusion, may fill the great, enveloping halo. Neutrinos, if they have nonzero mass, are yet another form of invisible mass that might make up a substantial part of the Galaxy, but neutrino mass has still not been demonstrated.

The entire pinwheel turns majestically about the galactic hub. At a distance of 10,000 light-years from the center, the orbital period is about 50 million years. The Sun, at 30,000 light-years, takes 240 million years per revolution. Still further out, at about 50,000 light-years, the time it takes to make one turn around the Galaxy reaches about 400 million years. Over billions of years the entire pattern might be expected to wind up tightly, as the inner portions accumulate more laps than the outer regions, but it doesn't appear to work that way. The spiral form is believed to exist in a very slowly rotating density wave pattern. Stars, gas, and dust circle through and crowd up in the density waves, where some stars die and new ones are born to highlight the structure.

Between 10 and 15% of the mass of the Milky Way disk consists of interstellar gas and dust. In the dustiest galaxies this figure may be as high as 25%. When we look through the disk toward the galactic center at optical wavelengths, only 1 photon in 100 billion gets through the dust clouds. The view of the galactic center is so totally obscured that a star as bright as the Sun would be invisible to us, and the energy machine that drives the violent activity in the core may only be guessed at. At infrared and radio wavelengths as well as in the x-ray and gamma-ray spectrum, radiation gets through more-or-less unhindered, and the murky clouds are effectively dispelled. Studies in these extremes of the spectrum are leading astronomers toward a solution of the mystery at the heart of the Galaxy. Some of the things they have learned are that perhaps 1 million solar masses of hot gas have been expelled at great speed from the galactic center relatively recently and that as many as 5 million solar masses are crowded within just a few light-years of the nucleus.

Viewing our own Galaxy from within the disk places us at a great disadvantage, but by looking perpendicular to the disk through its thin dimension we can have an almost unhindered look at the rest of the universe of galaxies. Nearby, at a distance of only 2 million light-years, is the magnificent Andromeda galaxy, tilted about 12 degrees to our line of sight. It is virtually a twin of the Milky Way, but is filled with about 300 billion stars. Its spiral arms, composed of complex bright and dark lanes, are moderately tightly wound around a bright nucleus. The stars are crowded within the thickness of the disk, no more than about 250 light-years from the equatorial galactic plane. Sparkling, hot blue giant stars are strung out like diamonds along the spiral arms, embedded in the dense banks of gas and dust from which they recently formed. Here, also, are the supermassive giants undergoing the explosive death throes of their evolution, as described in Chapter 5.

Some 40 years ago Walter Baade identified two populations of stars in galaxies. Stars in the disks of spirals and in irregular galaxies such as the Magellanic Clouds belonged to *Population I.* Those in the halo and nuclear bulge were labeled *Population II.* Elliptical galaxies appeared to be composed entirely of Population II stars.

Population I includes very young, hot, luminous stars that are loosely grouped into "stellar associations" of recently formed stars. Less conspicuous are the "open clusters" of smaller, more compact groups of a variety of ages; these are comparatively stable. Chemical abundances of the elements resemble the Sun's, about 99% hydrogen and helium by mass and 1% of everything else.

Baade's Population II originally included only red giants, dwarfs, and globular star clusters, which are all about as old as the Galaxy. A globular cluster may contain over 100,000 stars that are very tightly packed and very old, and include the oldest stars in the Galaxy.

Baade's idea of two populations, although a very simple concept, was surprisingly successful at first in unraveling the complex interrelations of age, stellar evolution, element building, and dynamics. Before long, however, the scheme was recognized as an oversimplification with many exceptions.

The simplest view of the formation of the Galaxy is that it condensed

FIGURE 75. *At a distance of 2.2 million light-years, the Andromeda galaxy (M31) is a giant spiral, almost a twin of the Milky Way. Recent studies of stellar velocities in the central bulge suggest the presence of a black hole of 1 million solar masses. (Lick Observatory.)*

out of a 100-billion–solar mass cloud of hydrogen and helium a few hundred million years after the Big Bang. Within the slowly rotating spherical mass, local regions of higher density began to coalesce early to form the first-generation stars in globular clusters. We see these original clusters still contracting and still relentlessly following their original orbits in the halo, forever preserving the evidence of the primordial shape of the Galaxy.

As collapse accelerated in the central region, the nuclear bulge began to take shape and the stellar nuclear hydrogen reactors began to generate energy. Stars raced through the central region in elliptical orbits and settled into the egg-shaped nuclear bulge. For elliptical galaxies this may be the end point of their formation. In a spiral, however, considerable amounts of hydrogen–helium gas and dust were apparently left over to form the thin, extended disk. The first generation of Population I stars was born from the primitive hydrogen and helium piled in the arms. As these early stars evolved and exploded, later generations were processed from the resulting "dirty mix" of interstellar gas filled with heavy elements of the present normal cosmic abundance.

Interstellar Gas

Most of the gas in the Galaxy is cold, neutral atomic hydrogen. In 1944, while the Netherlands was still under Nazi occupation, a young Dutch astronomer, Hendrik van de Hulst, predicted that the Galaxy would be aglow with a characteristic radio wavelength of neutral hydrogen at 21 centimeters. Van de Hulst, a student of the distinguished astronomer Marcel G. J. Minnaert at the University of Utrecht, chose to continue his theoretical research at the University of Leiden when Minnaert was placed in a detention camp.

At Leiden, van de Hulst came under the influence of Jan Oort, who had received an inkling of the galactic radio-emission discoveries of Grote Reber in America through a smuggled issue of the *Astrophysical Journal.* As van de Hulst strove to understand Reber's observations, he concluded that stars and interstellar dust could not account for the radio emission. Instead he derived a theory of microwave line emission at a characteristic wavelength of 21 centimeters that related to the orientation of the spin axis of the electron in the hydrogen atom.

The hydrogen atom consists of a central proton orbited by an electron, both spinning like the Earth and the Sun. According to quantum theory, the spin is either clockwise or anticlockwise and is represented by a spin vector oriented along the spin axis. The electron and proton vectors may point in the same direction (parallel) or in opposite directions (antiparallel). In the parallel orientation the atom has very slightly more energy than when the spins are antiparallel. Because the atom always seeks its lowest energy state, an electron with parallel spin must eventually flip

over to the antiparallel position. When this occurs a photon of radio energy exactly equal to the energy difference of the two atomic states is emitted.

The hydrogen atom, on average, remains in the higher-energy parallel-spin state for about 10 million years before it flips its spin. Occasional atomic collisions tend to flip the antiparallel spins back to parallel so that at any time about three-fourths of the atoms are in the parallel-spin orientation and the remainder are antiparallel.

Van de Hulst first thought that the 21-centimeters radiation would only be observed by some incredible happenstance. Detection would require a unique transition in the most abundant species of galactic gas, and at an ideal wavelength. Still, his notebook contained the remark, "if now the whole sky isn't gleaming at this wavelength!" (*Nederlands Tijdschrift Naturkunde 11*, 1945, p. 210). True enough, in subsequent years it was found to be impossible to direct an antenna toward any position close to the galactic plane without hearing the song of neutral hydrogen at 21 centimeters. The Galaxy contains so many hydrogen atoms that about 10^{54} flips take place each second.

Van de Hulst's prediction wasn't communicated widely to the scientific community because of wartime difficulties. Almost at the same time, a similar theoretical conclusion was reached by Iosif Shklovsky in the Soviet

FIGURE 76. Jan Hendrik Oort (1900–). *Jan Hendrik Oort is the dean of twentieth-century astronomy. He was born in Frankener, the Netherlands, the son of a doctor of medicine and the grandson of a professor of Hebrew who was one of the principals in the preparation of the Leiden Dutch version of the Bible. He rose to the position of Director of the Leiden Observatory in 1945, and his inspiration is largely responsible for the remarkable achievements of Dutch astronomers over the past half-century.*

In Oort's student years the prevailing view was that the Sun was close to the center of the Galaxy. His early studies of the skewed distribution of stellar velocities led him to conclude that the Sun was indeed far from a center about which the stars rotated. After H. C. van de Hulst suggested on theoretical grounds that the radiation of galactic hydrogen atoms at a wavelength of 21 centimeters should be very intense, Oort and his colleagues proceeded to map out the spiral pattern of the gas clouds of the Milky Way and its central bulge. He later detected the violent expulsion of gas from the central region, hinting at the existence of a massive compact galactic nucleus.

In recent years some scientists have seized on Oort's hypothesis that all comets escape from a vast swarm that surrounds the outer reaches of the solar system to explain an apparent 27-million-year periodicity in biological extinctions. Among various mechanisms proposed for periodic peaks in cometary bombardment of the Earth is the shaking-out of comets and asteroids by a binary companion star of the Sun that sweeps through the comet cloud every 27 million years.

Union. It was not until 1951 that sufficiently sensitive instruments were developed to confirm the theories. The hydrogen signal was first recorded by Harold Ewen and Edward Purcell at Harvard University, and quickly confirmed in the Netherlands and Australia.

Astronomers observed the 21-centimeters emission in all directons in order to determine the number of hydrogen atoms in the line of sight. Because the atoms in the disk of the Galaxy are rotating with orbital velocities that can be calculated for each radial distance from the galactic center, it is possible to observe the Doppler shift in wavelength of the radio photons and thereby identify where in the spiral arms the hydrogen is located. For example, atoms near the galactic center orbit more rapidly and their Doppler shift is larger, whereas those farther out move proportionately more slowly, with a smaller Doppler shift.

By mapping the Doppler-shifted hydrogen-line radiation, radio astronomers have sought to unravel the spiral-arm structure of the Galaxy. They find that the hydrogen atoms do indeed concentrate in "spiral

FIGURE 77. *Fan-shaped streamers and billowing clouds of gas and dust fill the Orion nebula, a nursery of young stars. The nebula is about 15 light-years across, and at the distance of 1600 light-years looks as large as the Moon. Stars now being born in the dark clouds are completely invisible in the optical range, but longer-wavelength infrared shines through the dust cocoons that envelope the stars. (Lick Observatory.)*

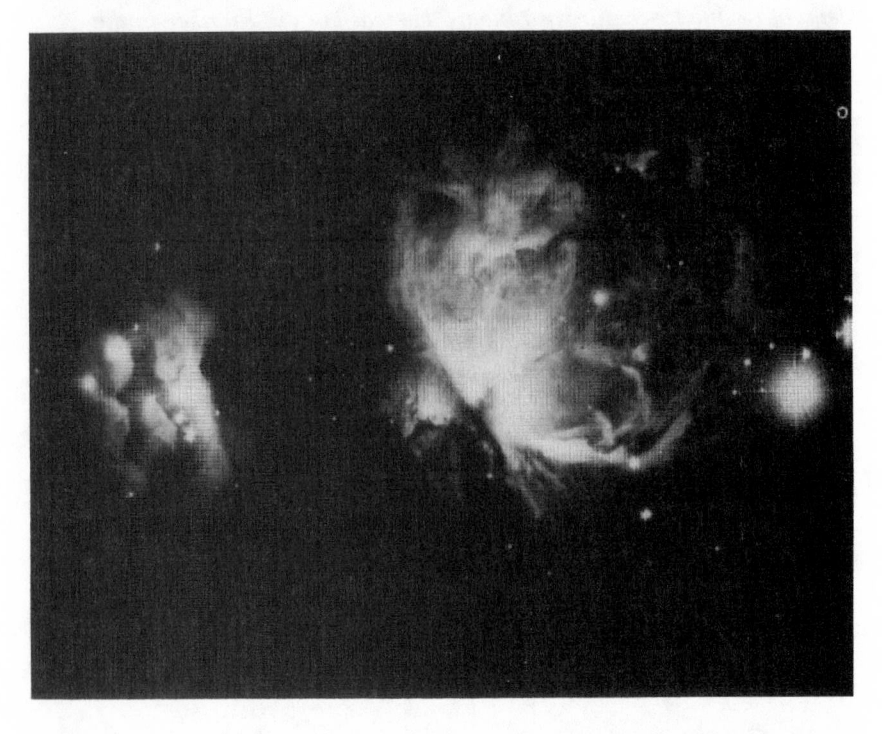

arms," although the pattern is somewhat indistinct. At least in the parts of the Galaxy as far out as the Sun, a four-armed pattern has been deduced. Closer in, the pattern becomes more complex and is difficult to distinguish with precision. Hydrogen in the spiral arms is 10 times more concentrated than it is in the disk. Furthermore, hydrogen in the disk of the Galaxy appears to be rushing outward through the arms at speeds of from 30 to 100 miles per second. Jan Oort estimated that as much as 1 solar mass of hydrogen per year flows out of the nucleus into the arms.

When two hydrogen atoms join to form a hydrogen molecule, the new molecule does not emit radio waves. It can be seen only in absorption in the extreme ultraviolet spectrum and when illuminated from behind by a bright ultraviolet star. To obtain the tell-tale spectrum requires that the spectrograph be carried above the atmosphere on a rocket or spacecraft.

Hydrogen molecules do not form by having two atoms simply collide in free space. Instead, the atoms must attach to dust grains and migrate along the surfaces until they meet and stick together. This process is called *grain catalysis.* The grains may be silicates, graphite, or perhaps other substances, typically about 0.5 micron in diameter. Subsequently the molecules pop off the dust into interstellar space. The galactic disk contains about as much mass in the form of interstellar hydrogen atoms as in hydrogen molecules, but taken all together the gas amounts to only a few percentages of the mass of the stars.

Several thousand giant molecular clouds, consisting mainly of molecular hydrogen, are estimated to exist in the galactic disk. They measure hundreds of light-years across and weigh as much as 1 million suns. They are the largest gravitationally bound objects in the Galaxy and are the principal reservoirs out of which material is supplied for the birth of new stars. The largest clouds are those responsible for the most massive star formation. They concentrate in the spiral arms. Nearby Orion is a well-studied example of a stellar nursery embedded in a giant molecular cloud. Only a small portion of the cloud has thus far condensed into hot stars that excite the nebular fluorescence.

Outside of the giant clouds, the interstellar medium contains very few molecules. Inside the clouds however, most of the atoms have joined to form molecules. Some 50 different molecular species have now been discovered by radio microwave spectroscopy. Most of these are diatomic, such as molecular hydrogen, carbon monoxide (CO), and cyanogen (CN); or triatomic, such as ammonia (NH_3), water (H_2O), and hydrogen sulphide (H_2S). But some are much more complex and have as many as 13 atoms per molecule. (The enormous amount of formaldehyde observed led one wag to comment that God must be dead because space is filled with embalming fluid. A happier "spiritual" thought is that the total amount of ethyl alcohol in Sagittarius is equivalent to about 50,000 trillion trillion fifths at 100 proof, prompting another wit to remark, "that is why our galaxy must be a barred spiral.")

ASTRONOMICAL DISCOVERY

The Beast at the Center of the Galaxy

> *Ever eating, never cloying,*
> *All devouring, all destroying,*
> *Never finding full Repast,*
> *Till I eat the World at last.*
>
> —JONATHAN SWIFT,
> "On Time," 1727

Although the center of our galaxy is completely invisible behind dense clouds of dust, radio, infrared, x-ray, and gamma-ray telescopes can easily penetrate the dust barrier to reveal a highly active nuclear region.

Most closely identified with the exact center of rotation of the Galaxy is an extremely bright and compact radio source known as Sagittarius A* (Sag A*). Although it is no more than about 20 times the radius of the Earth's orbit, it is about 10,000 times as powerful as the strongest known radio pulsar. Some binary x-ray sources are also radio emitters, but these are typically 100,000 times weaker than Sag A*. On the basis of all the evidence, Sag A* is a unique radio source in the Galaxy; it is several orders of magnitude brighter than typical stellar radio sources. It could be explained by a modest rate of accretion onto a 1-million–solar mass black hole, and it is sufficiently compact to fit that description.

It is difficult to pinpoint the nucleus of the Galaxy given the poor resolution of gamma-ray telescopes, but there is evidence of a very puzzling and remarkably powerful gamma-ray source that suggests a black hole at the galactic center. In 1970 a detector carried on a balloon was launched from Australia by a team of astronomers from Rice University, led by Robert Haymes. Radiation was observed at the energy characteristic of the annihilation of colliding electrons and positrons, 0.51 million electron volts. In 1957 improved instrumentation flown by a Bell Telephone Laboratories group led by Marvin Leventhal gave unambiguous confirmation of the gamma-ray line.

It is believed that energetic gamma-ray photons emitted from the outskirts of a black hole can create electrons of ordinary matter and positrons of antimatter. When these particles, in turn, encounter each other, they disappear with the release of a pair of gamma rays of the characteristic 0.51 million electron volts energy. The intensity observed was a startling 100,000–1 million times the total power of the Sun at all wavelengths. A billion tons of positron antimatter may be produced and annihilated each second. Nowhere else in the Galaxy has a similar source been detected.

Surprisingly, subsequent observations from balloons and the *HEAO-3* satellite showed that the gamma-ray line varied in intensity over a 6-month period, indicating that it could not be bigger than about 6 light-months in diameter. Two closely spaced balloon flights revealed that the line intensity changed substantially in only 10 days. But the source has not been seen since 1979. From an exhaustive analysis of all the observations of both the line and the adjoining continuum radiation, Goddard Space Flight Center's Reuven Ramaty and Richard E. Lingenfelter of the University of California, La Jolla, concluded that the source could be a black hole of fewer than 100 solar masses, within 1 degree of the galactic center but not necessarily coincident with it.

When the gamma-ray observations were first reported, Donald Lynden-Bell, a distinguished theorist then at the California Institute of Technology, referred to the source as "The beast at the galactic center." True, the gamma-ray telescope has poor resolution, only about 4 degrees, but even that rough coincidence of such a remarkable source with such a unique position in the Galaxy is very highly suggestive.

Soon after the first balloon observation, x-rays from Sagittarius were detected by *Uhuru* and, toward the end of the decade, the *Einstein Observatory* nailed down the position of the source to within 1 arc minute of the galactic center. Furthermore, the x-rays varied on a time scale of about 3 years, implying a source only 3 light-years in diameter. We can confidently expect that the next round of observations from forthcoming x-ray and gamma-ray observatory spacecraft will help greatly to resolve the mystery of the source.

In the near-infrared we see evidence of a dense crowding of cool red giants at about 1 million stars per cubic light-year, or about 100 million times the density of stars around the Sun. Longer infrared wavelengths that give evidence of the concentration of ionized gas lead to estimates of several million solar masses of diffuse mass as well as to a point source of comparable mass. The ionized gas most likely originates outside an inner region of 10-light-years diamater that is largely devoid of gas and dust. Surrounding this void is a doughnut-shaped, turbulent cloud of at least 10,000 solar masses of neutral gas orbiting the galactic center. A group of infrared astronomers at the University of California in Berkeley led by Nobel Laureate Charles Townes, speculate that the doughnut cloud is the result of a gigantic explosion that may have occurred only a few hundred thousand years ago, when a large blob of gas was consumed by the black hole. Townes and his colleagues believe that the black hole must contain about 4 million solar masses, even larger than the mass estimated from the gamma-ray emission.

Worlds in Collision

Galaxies used to be thought of as "island universes" that evolved in magnificent isolation. Even if there were a rare collision, it was believed that the 100 billion stars within a galaxy were so widely spaced that the galaxies would pass through each other with almost no chance of individual stars bumping into each other.

The idea that galaxies in collision could merge as a result of tidal friction was proposed by Swedish astronomer Erik Holmberg in 1940. He pointed to the comparatively small mean separation of galaxies, from about 10 to 100 times their diameters, and the relatively high probability of close encounters. Then he conjectured that a pair of approaching galaxies would begin to revolve about each other and the resulting tidal forces would dissipate enough energy to induce their orbits to shrink. Eventually the galaxies would coalesce. During the 1950s Fritz Zwicky became intrigued with evidence of faint streams of stars tens of thousands of light-years long that seemed to bridge galaxies in close proximity. At Mount Palomar in the 1960s, Halton Arp identified hundreds of interacting galaxies that exhibited bridges, plumes, tails, antennas, and other bizarre connnections. But over those 2 decades there was little serious theorizing about interactions between galaxies.

The collision theory was debated earnestly in 1951, when radio astronomy made its first detection of a powerful extragalactic source. F. Graham Smith of Cambridge University had set his two war-surplus radar dishes 1000 feet apart and linked them to make an interferometer. When he targeted the second strongest radio source in the sky, Cygnus A, he was immediately able to position it with such precision that Walter Baade and Rudolph Minkowski at Mount Wilson could search out its optical counterpart. The object appeared on their photographic plates as a faint nebulous patch that contained two adjoining condensations. From the redshift, the distance was estimated to be 700 million light-years.

FIGURE 78. Walter Baade (1893–1960). *Walter Baade was born in Germany and educated at Göttingen University. He observed at the Hamburg Observatory for 11 years before coming to Mount Wilson Observatory in 1931. While most of his colleagues were engaged in war research during World War II and the lights of Los Angeles were dimmed, Baade, who was technically an enemy alien, had prime use of the 100-inch telescope and accomplished a remarkable body of research. He identified two different populations of stars in the central bulge of Andromeda. Population I was made up of young blue stars and Population II were old red stars. In the spiral arms outside the bulge he found mainly Population I Stars. Globular clusters that fill a spherical halo of the Milky Way were made up of older Population II stars dating back to the time of formation of the galaxy. Later, with Rudolph Minkowski, Baade used the 200-inch telescope to identify optically several of the first bright sources discovered by the pioneers of radio astronomy.*

When Baade discovered that the pulsating Cepheid variables in the disk of the Milky Way were 4 times as bright as those in the halo, he was able to correct an embarrassing astronomical error that made the Galaxy appear to be only one-half as old as the Earth. His new distance scale doubled the size of the universe.

To everybody's surprise, here was a distant extragalactic source that radiated about 700 billion trillion kilowatts of radio power, millions of times the radio output of the Milky Way. Ten billion Crab nebulae would barely match one Cygnus A. When Stanley Hey discovered the source in 1946, it was thought to be a relatively nearby object inside the Milky Way. Astronomers called it a *radio star* because it scintillated like a twinkling star. Now it was clear that it looked starlike because it was so distant. Astronomers realized that a new era was dawning in astronomy. Soon it would become possible to probe farther back into the early universe with radio interferometers than with optical telescopes.

To Walter Baade, the optical appearance of two concentrated objects inside a fuzzy halo strongly suggested a pair of galaxies colliding and producing the intense radio emission in the process. His partner, Rudolph Minkowski, expressed great skepticism. Soon the astronomical community was split into two camps, pro and con. The press concocted exciting stories of radio waves broadcasting the thunder of worlds in collision, and the BBC aired a recording of the astronomer's taped signals.

The following story recounted by Robinson, Schild, and Schucking (*Quasi-Stellar Sources and Gravitational Collapse*, 1965, p. xi) tells us something of the strong personalities of the two great astronomers who led the debates over colliding galaxies:

> To listen to Cristoforo Columbo sitting in a dimly lit, favorite tavern with a glass of port recounting his adventures and

FIGURE 79. Rudolph Minkowski (1895–1976). *After serving in the German army in World War I Rudolph Minkowski conducted spectroscopic research at the State Physical Institute in Hamburg and later collaborated with Walter Baade at the Hamburg Observatory. When the Nazis came to power, anti-Semitic persecution forced him to leave Germany, and he came to Mount Wilson in 1935. There he specialized in studies of supernova remnants and established their ages and energy contents. Later, at the Palomar Observatory, he supervised the mapping of the northern sky with the 48-inch Schmidt telescope.*

With Walter Baade, he identified many radio sources, including supernova remnants and active galaxies. Cygnus A, originally thought to be a superheated thermal radio source produced by the collision of two galaxies, was soon proven to be a synchrotron radiation source. Minkowski placed Cygnus A at the enormous distance of almost 700 million light-years, making its energy content in high-energy electrons and magnetic fields equivalent to the total conversion of 1 million solar masses. He also solved the mystery of Casseopeia A, the brightest radio source in the sky. His analysis of rapidly moving filamentary wisps and an expanding nebula showed that they originated 300 years earlier at the center of a supernova explosion that was not observed because it occurred at a distance of over 10,000 light-years and was strongly obscured by interstellar dust.

describing vistas of far lands never seen before must have made an exciting, unforgettable evening. But one might doubt if this could ever match the strange fascination of an evening with the late Walter Baade from the Mount Wilson and Palomar Observatories. For more than a quarter of a century he had worked with the biggest optical telescopes on earth. . . . Baade saw the mysteries of the universe as the greatest of all detective stories in which he was one of the principal sleuths . . . he told us one evening the story of Cygnus A.

"In 1951, at a seminar talk that Minkowski gave in Pasadena on the theories of radio sources, I got mad. I had just published the theory of colliding galaxies in clusters and identified the Cygnus A source with such a pair in collision. Nobody would believe there were extragalactic radio sources. Minkowski reviewed all the other theories first; and then at the end of the seminar, as if he were lifting a hideous bug with a pair of pincers, he presented my theory. He said something like: 'We all know this situation: people make a theory and then astonishingly they find the evidence for it. Baade and Spitzer invented the collision theory; and now Baade finds the evidence for it in Cygnus A.'

"I was angry (said Baade) and I said to him, 'I bet a thousand dollars that Cygnus A is a collision.' Minkowski said he could not afford that, he had just bought a house. Then I suggested a case of whiskey, but he would not agree to that either. We finally settled for a bottle and agreed on the evidence of a collision—emission lines of high excitation. I forgot about the thing until several months later. Minkowski walked into my office and asked, 'Which brand?' He showed me the spectrum of Cygnus A. It had neon V (quadrupally ionized neon) in emission and 3727 A (ionized oxygen) and many other emission lines. I said to Minkowski, 'I would like a bottle of Hudson Bay's Best Procurable,' that is the strong stuff the fur hunters drink in Laborador.

"But that was not everything. For me a bottle was a quart, so what Minkowski brought me was a hip flask. I did not drink it. I took the flask home as a trophy.—Two days later it was Monday. Minkowsky visited me in order to show me something—he saw the bottle and emptied it."

Baade chuckled: "Isn't it a shame that you get no returns when the horse you bet on is a dead-sure thing?" (By permission. Copyright © 1965 by The University of Chicago. All rights reserved.)

"That was six years ago," the editors noted, "and the horse is dead. Most of the experts would agree now that Minkowski had a right to consume the whiskey because Baade had not won his bet."

Baade's collision model proposed that violent motions generated within the interacting galaxies dissipated in strong frictional heating. The resulting radio emission would be characterized by a thermal spectrum. Instead it was soon found that Cygnus A was a synchrotron source of the type produced by electrons moving with nearly the speed of light in a magnetic field. With an early type of interferometer, R. C. Jennison and M. K. Das Gupta at Jodrell Bank discovered in 1953 that the radio emission from Cygnus A did not emanate from Baade's fuzzy galaxy, but from two diffuse patches on either side of it. For the next 2 decades very little attention was given to collision models, and radio astronomy began to reveal a class of active galactic nuclei out of which there emerged complex structures of jets and great radio lobes invisible to the optical astronomers.

With very long baseline interferometry (VLBI) that used telescopes as far apart as California and West Germany, the image of Cygnus A shown in Fig. 80 was constructed. A pair of tightly aligned jets of gas squirt out of the core of the Galaxy in opposite directions to a distance of 1 million light-years and terminate in vast lobes of radio-emitting plasma. Lines of magnetic force are carried outward with the jets, and electrons moving at close to the speed of light travel in tight spirals about the direction of the magnetic field, producing the polarized radio waves characteristic of synchrotron radiation. As the jets bore into the ambient gas to great distances, their energies run down until the ram pressure of the intergalactic gas dominates. The jets then expand into puffs and bubbles blown about by shock waves in vast plasma clouds. Within each of the lobes of Cygnus A, two hot spots appear that dominate the radio emission. The energy content of the lobes is enormous, roughly equivalent to the energy that would be liberated by the conversion of the mass of 100,000 stars.

All of this description of Cygnus A argues against Baade's early hypothesis of galaxies in collision, but more recent, refined observations of the central region in both optical and radio wavelengths reveal a complex structure that revives the collision theory.

Almost all radio galaxies, even those of rather low overall luminosity, exhibit radio-emitting jets. How the jets are produced is still a deep mystery. The most popular speculation is that jets are linked to a supermassive rotating black hole and its accretion disk. In the best-resolved cases the jets appear to be oriented perpendicular to the plane of the Galaxy, and remain steady for as long as 100 million years. In clusters of galaxies the intracluster gas is far more concentrated than in the intergalactic medium, and the resistance to galactic velocities of several thousand kilometers per second can bend back the jets into wishbone shapes and extended tails (see Fig. 81).

The most important questions for theoretical models are: What mechanisms make jets? And, what feeds the central energy source that transfers mass and energy to the radio lobes via the jets? If the central object is a black hole, what feeds the black hole? One proposal by Jeremiah

FIGURE *80. A radio image of the distant galaxy Cygnus A made with the VLA. Early radio telescopes could not resolve any structure in this brightest extragalactic radio source. Optically it appears to be a peculiar giant elliptical galaxy (see* inset*) crossed by a dust rift that suggests an earlier collision of two galaxies. With the VLA, at a wavelength of 6 centimeters the image resolves into a bright central "dot" straddled by a pair of widely separated, billowing lobes filled with wispy forms. Two fine jets connect the bright dot at the center of the lobes. Near the edge of each lobe is a bright concentrated source of emission that must be fed by the jets with energy supplied from a central "engine" believed to be a massive black hole in the nucleus of the optical galaxy. (VLA: Courtesy NRAO/AUI, observers R. A. Perley, J. W. Dreher, J. J. Cowan;* inset: *Hale Observatory.)*

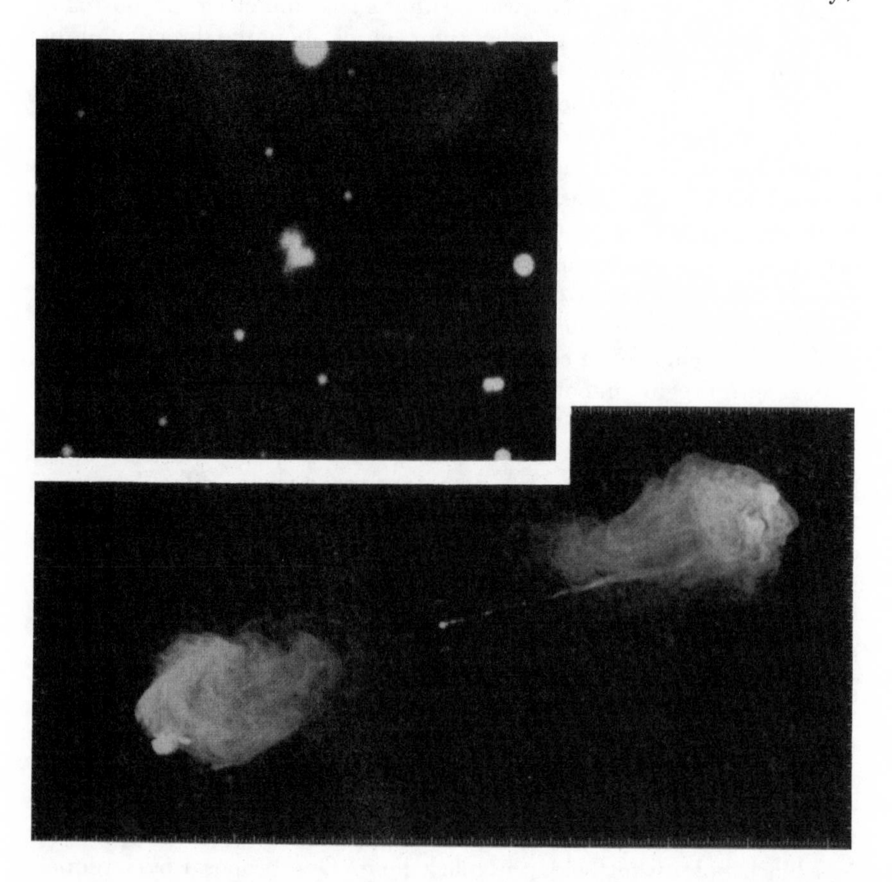

Ostriker of Princeton University is a process that he terms *cannibalism.* The central black hole of a "cannibal" galaxy can consume a "missionary" galaxy that comes within gravitational capture range. If the missionary galaxy itself has a black hole in its core, it will settle into a binary orbit around the cannibal galaxy. As the orbit shrinks, an enormous amount of gravitational energy will be released.

Many schemes have been proposed to explain the jets, but they all smack of speculation without substantial verification. Certain basic ingredients must be present, such as a deep gravitational well, an ample gas supply, and sufficient spin to squirt gas streams in opposite directions. Jets have such a wide range of characteristics that perhaps no single mechanism can explain them all. Progress will surely come from future advances in observational capabilities.

SS 433, a Jetting Wobbler

Astronomy seems to produce one incredible surprise after another with astonishing frequency, but the saga of SS 433 must rank high on the list

FIGURE 81. *A twin-tail radio galaxy, 3C-75, observed at 20-centimeters wavelength by the VLA. The jets emerge from twin nuclei of the central galaxy and appear to wrap around each other. (Courtesy NRAO/AUI, observers: F. N. Owen, C. P. O'Dea, and M. Inoue.)*

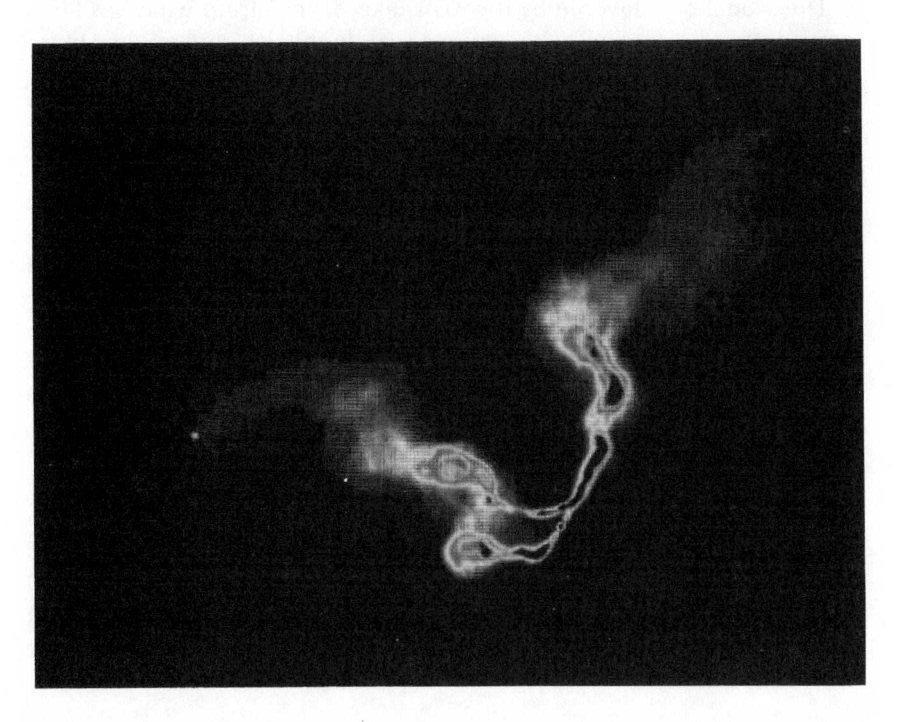

of mystery stories of the last decade. This strange object gets its name by being number 443 on a list of stars with emission lines in their spectra. *SS* stands for C. Bruce Stephenson and Nicholas Sanduleak, who compiled the list. It lies very close to the galactic plane at a distance of about 15,000 light-years. Not only do the emission lines appear at unfamiliar wavelengths, they vary in wavelength by large amounts over the course of several weeks. Detailed spectral studies in 1978 by Bruce Margon of the University of Washington revealed three sets of emission lines, of which one remained essentially constant whereas the other two appeared to be oppositely Doppler-shifted, one to the blue and one to the red.

How could the source be coming and going at the same time? If the Doppler interpretation were correct, the velocities of approach and recession were an amazing 27% of the speed of light, about 100 times faster than any star had ever been observed to move. If a star were to attain even a few percentages of that speed, it would rapidly escape from the Galaxy. The source breaks all records for a Doppler blueshift anywhere in the universe.

If all of this startling evidence were not enough, there was the additional feature that everything changed substantially from night to night. Observations over the course of 1 year showed that the Doppler velocities varied forward and back over a range from zero to a maximum of 80,000 kilometers per second, with a period of 164 days. The enormous velocities were changing by about 1000 kilometers per second each day.

The model now favored by theorists takes its cue from what has been learned of radio galaxies with symmetrical double jets and applies those ideas to a stellar system. It involves a binary-star system consisting of a relatively normal star and either a neutron star or a black hole. In some unknown manner the compact object produces two oppositely directed twin jets that stream outward at more than one-fourth of the speed of light. The 164-day period must derive from a wobble of the axis of the jets.

SS 433 demonstrates that jets are not an exclusive property of distant galaxies. We now have a jetting object within our own galaxy that can be studied from relatively close up, in which all the phenomenology of galactic jets is mimicked but the action is greatly accelerated and is observable on time scales of weeks and months rather than hundreds of millions of years.

SS 433 is also a variable radio and x-ray source, but without any obvious regularity. It appears to lie within an old supernova remnant. Searches for other stars that exhibit similar characteristics have thus far failed to turn up any candidates. The jets in SS 433 appear to be about as long as the distance from the Sun to Earth, whereas galactic jets are often 10 billion times longer. Still, it seems that the basic physical mechanism that operates in the nucleus of a galaxy of the mass of 100 million suns may also drive a star system of only a few solar masses.

Collision Modeling with Computers

The interacting galaxies that Zwicky observed in the early 1950s did not receive much theoretical consideration because the prevailing notions were that gravitation could not produce jets, filaments, or drawn-out wisps.

FIGURE 82. *The Whirlpool galaxy (M51) appears to be a binary pair joined by tidal forces. (Lick Observatory.)*

The 1970s brought a revolution in concepts of interacting galaxies. Alar and Juri Toomre at the Massachusetts Institute of Technology were able, with the aid of computers, to model galactic collisions that reproduced structures like those observed. Now there is substantial evidence of mergers involving quasars, and the former *Infrared Astronomy Satellite (IRAS)* found many inconspicuous galaxies that exhibit great bursts of star formation, apparently as a result of recent galactic collisions. These "starburst" galaxies radiate the greatest part of their energy in the infrared.

Following its launch in 1983, *IRAS* scanned the sky for about 10 months over infrared wavelengths from 12 to 100 microns, a range that characterizes objects at temperatures of 30°K to 300°K. To the surprise of most astronomers, the *IRAS* results indicated that colliding galaxies emitted as much as 99% of their radiation in this range. In contrast, normal galaxies emit less than one-half their luminous power in the same band. The mechanism for production of starbursts is now believed to be the supersonic collisions of galactic gas clouds. Millions of bright newborn stars at the centers of these collisions radiate normally in the visible and ultraviolet, but the surrounding gas and dust absorbs and transforms the shorter wavelengths to heat radiation.

Earlier in this chapter we mentioned that stars in elliptical galaxies were thought to have formed rapidly, out of infalling gas clouds. Those in spirals could have condensed more slowly, after settling into a rotating disk. In such scenarios the shape of the galaxy is determined early on and never evolves from elliptical to spiral and *vice versa.* The Toomre brothers rejected these ideas and proposed instead that ellipticals are the jumbled star piles left from collisions of spiral galaxies well after the Big Bang. Although it was viewed at first with skepticism, there is growing support for their collision hypothesis.

Quasars

Twinkle, twinkle, quasi-star
Biggest puzzle from afar
How unlike the other ones
Brighter than a billion suns
Twinkle, twinkle, quasi-star
How I wonder what you are.

—George Gamow,
"Quasar," 1964

The first identification of a "radio star" (3C-48) was made by Allan R. Sandage, who presented his work to the American Astronomical Society at a Christmas meeting in 1960. Its spectrum was a blue continuum that suggested synchrotron radiation and was marked by broad emission lines in a pattern that was undecipherable. Although it appeared to be starlike, the spectrum "fingerprint" gave no clue to what it was made of. The answer to the puzzle eluded astronomers for more than 2 years and then shocked them into disbelief when the meaning of the spectrum was finally understood.

At the time of the discovery of 3C-48, primitive radio telescopes had found several sources that covered small areas of the sky and appeared to be discrete, but telescope resolution was too poor to permit identification with known optical objects. A breakthrough came when one of the strongest radio sources, 3C-273, was occulted by the Moon and its coordinates were established with very high accuracy by Cyril Hazard and his associates, in Australia. To their surprise they found that the source appeared to be double, with its two components separated by 20 seconds of arc. Nothing appeared on star plates to match the stronger component, but the weaker one coincided with a faint, thirteenth-magnitude blue star that hardly seemed to merit any special study.

Hazard passed his information on to Maarten Schmidt at the Palomar Observatory, who pointed the 200-inch telescope at the star. Schmidt found nothing particularly interesting about the star itself, but next to it, at the position of the stronger radio component, there appeared a faint bluish jet. A star with a jet? Obviously, it had to be something very strange.

The mystery deepened when Schmidt obtained a puzzling spectrum of the object. Instead of a continuous spectrum with dark absorption lines such as one would except from a star, there were broad, bright emission lines, as though the source were a nebula. Nor was there any sense to the pattern of the emission lines, which didn't match those of any known substance. Schmidt was obsessed with an intuition that the pattern of lines was very familiar. After 6 weeks of puzzlement, his mental block suddenly vanished and he realized that it was the simple pattern of the hydrogen atom, but redshifted by an enormous 16%. That shift corresponded to a recessional velocity of 45,000 kilometers per second, 15% of the speed of light. Schmidt went home that night still staggered by the implications of his discovery and commented to his wife, "Something really incredible happened to me today."

The significance of these numbers was fantastic. According to Hubble's law, 3C-273 was among the most-distant objects observed, some 2 billion light-years away, and even though it appeared starlike, its luminescence was 100 times that of the entire Milky Way. Schmidt quickly communicated his startling information to his close associate, Jesse Greenstein, who immediately made the same connection with a redshifted hydrogen spectrum for 3C-48. Only this time, the interpretation was even more incredible: a redshift of 37%, nearly one-third the speed of light and a distance of 3.7 billion light-years. The large variations in brightness of 3C-48 in less than 1 day implied dimensions of only a matter of light-days. At the enormous distances indicated by the redshifts, these objects would be radiating from 100 to 1000 times as much light as

FIGURE *83. Quasar 3C-273 at 2700 million light-years looks like a blurred star with a faint jet about 1 million–light-years long. It radiates about 100 times as much visible power as all the stars in the Milky Way, and even more in x-rays and gamma rays. Because its x-ray luminosity varies substantially on a time scale of 1000 minutes, its diameter must be on the order of the distance traveled by light in that time—1000 light-minutes—which is about the diameter of the Solar System. The energy machine that can produce such enormous power is believed to be a black hole containing the mass of 100 million suns. (Hale Observatory.)*

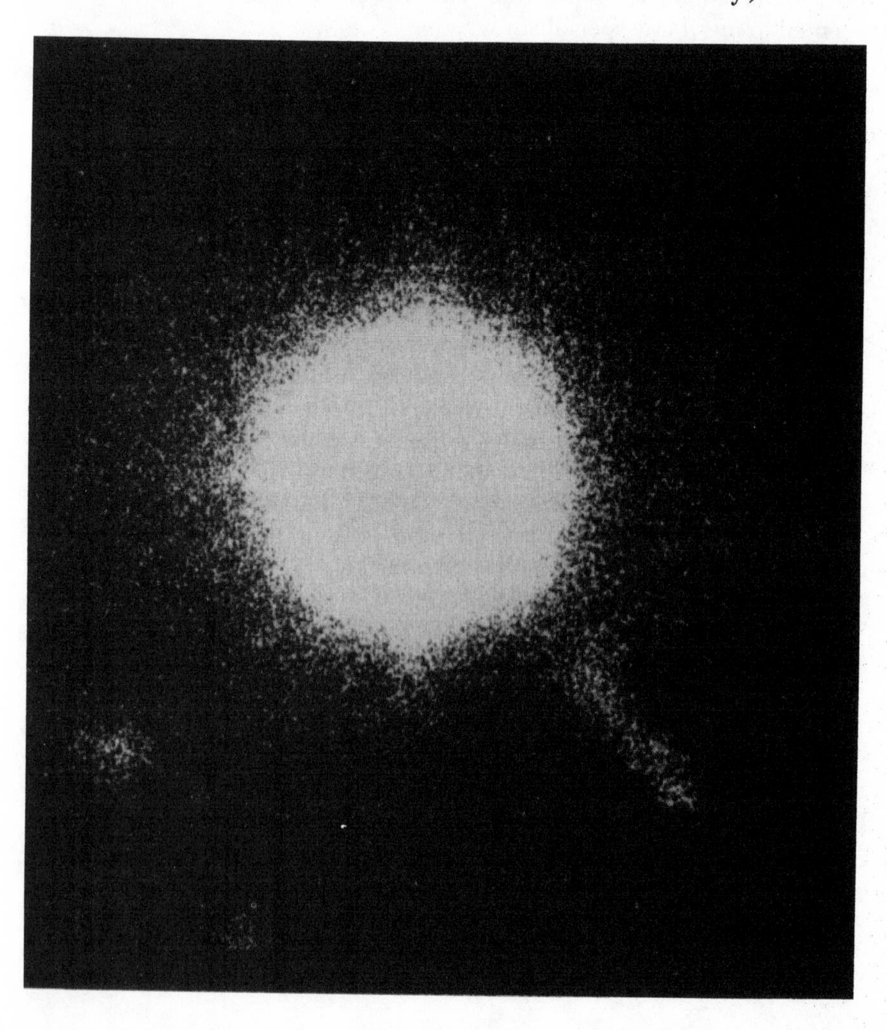

the entire Milky Way from volumes less than one-hundred-thousandth of its size.

Baffled by what they might be, astronomers at first called them *quasi-stellar objects* and then shortened the name to *quasar*. In the quarter-century since their discovery, more than 5000 of these sources have been cataloged. Detailed surveys of small regions of the sky indicate about 20 quasars per square degree, down to twentieth magnitude. Over the full 40,000 square degrees of sky, future searches could therefore turn up 1 million more. There is still no unanimous agreement on how they get their energy, although there is a general consensus that the answer must be a supermassive black hole.

Great controversy raged for many years over whether quasars were truly cosmological objects. Astrophysicists sought explanations that would bring the quasars in close and make their luminosities much more normal. James Terrell of the Los Alamos Scientific Laboratory proposed that they could be bodies ejected from the nucleus of our own galaxy in the last few million years and that they are now well past us and racing away. Placing the quasars at a distance of only millions of light-years rather than billions would reduce the light estimates by millions of times. But if locally produced, why weren't some flung out in our direction and therefore blueshifted? No blueshifted quasar has ever been detected. Another serious objection is that it would be impossible for our galaxy to eject 1 million quasars at such high velocities because the energy required exceeds that contained in all the mass of the galactic core region.

Another suggestion was that the redshift was evidence of a gravitational effect of Einstein's law of general relativity, such as we discussed in Chapter 6. As light expends energy to make its way out of a strong gravitational field, its wavelength must grow longer. It is essential that a very large mass be concentrated in a very small volume. This explanation can be rejected by a rather sophisticated spectroscopic argument. Some of the spectral emission lines in quasars can only originate in very tenuous gas at a density of no more than about 10,000 atoms per cubic centimeter. At such low density the gas is a very inefficient radiator of emission lines. To compensate for the weak emission process, the gas volume must be extremely large, which is directly opposed to the need for great compactness to produce a large gravitational redshift. The only way to resolve the issue would be to make the quasar mass very large: at least 100 times the mass of our galaxy. Unfortunately for this model, the black hole radius would then be about 6 light-years, and it would be impossible to account for observations of light fluctuations as short as weeks or even days (light-travel time across the black hole would take 6 years).

A strong statistical argument against any local hypothesis for the location of quasars was that discrete, already identified radio galaxies and quasars accounted for about one-fourth of the total radio flux that reaches us from the entire sky. To avoid the dilemma that all sources in the universe would contribute much more radio flux than we receive, the known sources must either represent a large fraction of the observable universe or our location must be central to an unusually large concentration of objects. There is no evidence for the latter situation.

The clincher for cosmological distances has come with the detection in recent years of fuzzy images of normal galaxies at the same positions as quasars and with identical redshifts. With the use of CCD cameras it has been possible to discern quasars in the midst of barred spiral galaxies, and a surprisingly large number appear to have smaller companion galaxies. Are galaxy collisions important to the evolution of quasars?

As time went on and the evidence for putting quasars at cosmological distances steadily increased, contrary opinions almost completely disappeared, but the puzzle of the remarkable energy output remained. At the time of the discovery of quasars, astrophysical concepts of energy generation did not go beyond conversion of mass by nuclear fusion. But nuclear reactions at best convert only a fraction of 1% of the mass to energy, and the power radiated by quasars would quickly consume more mass than could be confined in the small volumes that they appeared to occupy.

Among the early speculations was the possibility of a chain reaction of supernovae. In the densely packed conglomerations of stars that race around in the environment of a galactic nucleus, the debris and shock wave from one supernova could accelerate the evolution to a supernova condition in a nearby star already close to the end of its life. The impact would throw the star out of its orbit and into collision with another nearby star, and so on. If even one collision occurred per day, the luminosity would be sufficient to account for quasars. If the supernovae produced pulsars, the pulsar radiation alone could match the energy production requirement.

Another idea, put forth by Philip Morrison, was that the quasar machine could be a giant pulsar. He proposed a spinning ball of about 100 million solar masses with a rotation period of about 1 year. With increasing age his "spinar" would shrink and spin up, becoming steadily brighter. Glitches such as occur in neutron star pulsars might explain the variability in light output.

These proposals and many more have fallen by the wayside for good reasons, and a consensus has steadily formed around the concept of a supermassive, spinning black hole surrounded by a huge, doughnut-shaped whirlpool of inwardly spiraling gas. One-hundred million solar masses could lurk in the bottomless pit of the black hole, which to generate its power must cannibalize about 10 solar masses per year of stars that wander by. The spinning gas in the accretion disk moves faster and gets hotter as it spirals inward, reaching temperatures in the hundreds of millions of degrees so that it radiates at all wavelengths, including x-rays. At the inner edge of the whirlpool velocities reach thousands of kilometers per second. A strong axial magnetic field along which energetic particles are accelerated to relativistic velocities in streaming jets is held steady for millions of years by the spin of the black hole.

Studies of the optical variability of many quasars reveal substantial changes in brightness during the course of 1 month. The optical diameter must therefore be one light-month at most. In x-rays the variability can occur in a matter of minutes, implying a diameter of light-minutes (the Sun–Earth distance is only 8 light-minutes). Because x-rays are generated in the hottest, innermost part of the accretion disk, the x-ray diameter must be smaller than the optical diameter.

Quasars are observed with a wide range of redshifts, from as small as 0.1 to slightly larger than 4, although the largest and smallest values are comparatively rare. At the greatest redshifts the velocity of recession exceeds 90% of the speed of light. Most of the redshifts cluster at recessional velocities that correspond to an epoch of the early universe. It seems that there must have been many more quasars then than either earlier on or later. The quasar epoch possibly represents only 10% of the history of the universe. Did they evolve, die, and become something else, possibly the dormant phase of galactic nuclei? We still do not know.

To the Heart of Darkness

Radio galaxies, Seyfert galaxies, and quasars all seem to belong to a broad class with violently disturbed central regions referred to as AGNs, for Active Galactic Nuclei. In the 1940s Carl Seyfert discovered the class of galaxies that bear his name. Only a few dozen of all galaxies belong to this group, and they are marked by extremely bright and broad emission lines in the spectra of their nuclear regions. One of the most striking Seyferts is NGC 1068, a spiral galaxy with a nucleus that is extremely bright in radio, infrared, visible, and x-ray emission. At its center hot, turbulent gas clouds move with internal velocities as high as thousands of kilometers per second.

The giant elliptical galaxy M-87 in the center of the Virgo cluster of galaxies (see Fig. 84), is an example of a powerful radio galaxy with a young jet, only about 4000 light-years in length, projecting from its nucleus. The jet strongly resembles, in miniature, the 160,000-light-year–long jet in the quasar 3C-273. Although the jet appears to be one-sided in visible light, there is good evidence in radio images of a counterjet in the opposite direction. The light from the jet and the nucleus is synchrotron radiation at all wavelengths, from visible to x-rays. Along the jet is a series of knots that may result from constrictions in the magnetic field or densifications produced by shock waves.

Spectroscopic studies of the Doppler shifts in the central region of M-87 give evidence of an enormous mass whose gravitational attraction speeds up stars to much higher velocities than expected in a normal galaxy. The mass deduced from the star's velocities is about 5 billion solar masses,

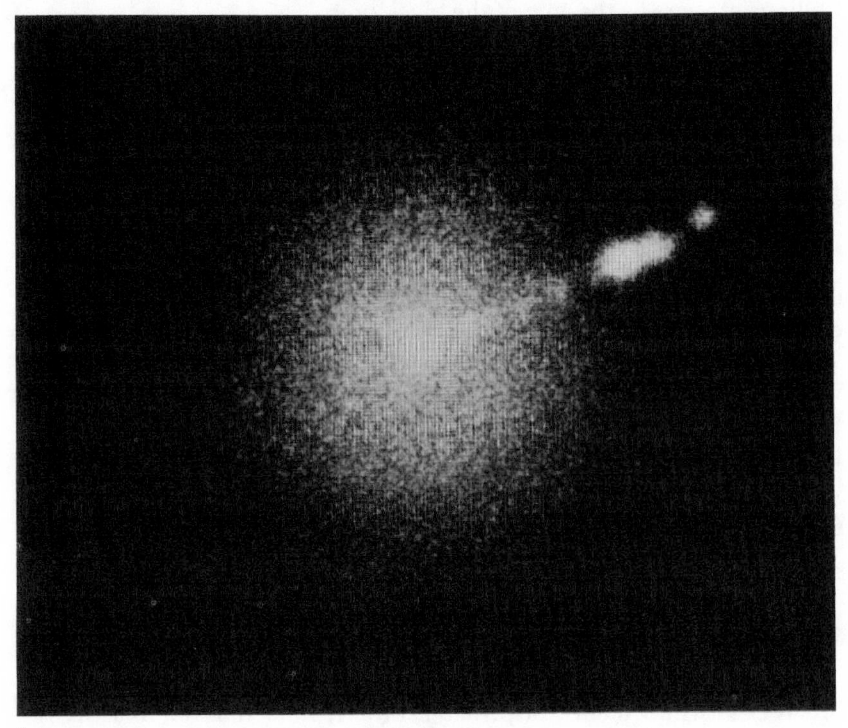

(a)

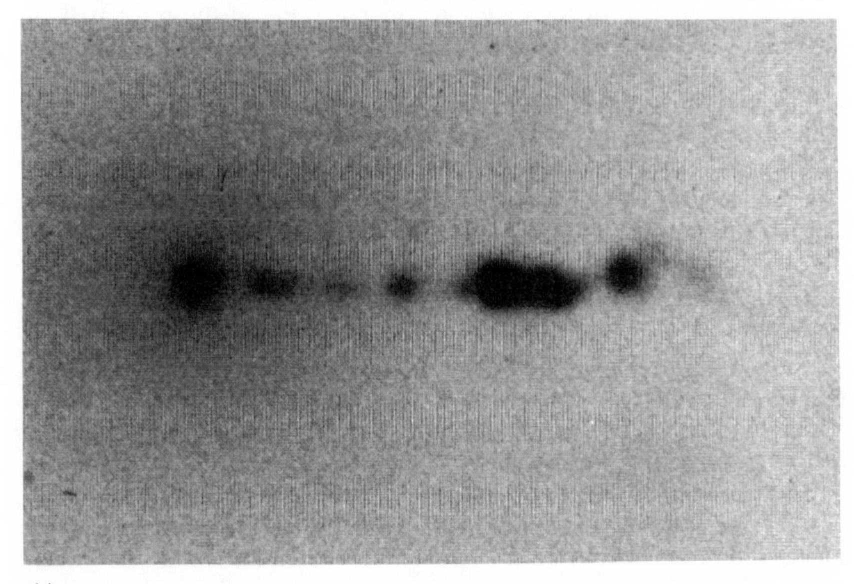

(b)

FIGURE *84. M-87 (Virgo A) is a giant elliptical galaxy with a prominent jet emerging from its core. In a normal optical exposure (a) the galaxy appears about 40,000 light-years across and the 5000–light-year jet is hidden inside the glare. When the exposure is adjusted to enhance the image of the jet relative to the galaxy, it appears as in (b). The jet is marked by knots. It runs straight for the first 3000 light-years to the brightest knot, then wiggles for another 2000 light-years through two more knots. The bluish light is polarized, evidence of synchrotron radiation. In 6-centimeter radio waves, the VLA image (c) shows extensions of the jet into wide lobes filled with filamentary structure.*

The density of stars increases so rapidly toward the core, and their velocities are so high, as to suggest the powerful gravitational pull of a black hole of some 5 billion solar masses. However, unlike quasars, which show bright, pointlike x-ray centers, there is no comparable x-ray concentration in M-87. (Courtesy of Hale Observatory Halton Arp; Canada–France–Hawaii telescopes, observers J.–L. Nieto, G. Lelievre, G. Wlérick, B. Servan; and NRAO/AUI, observers F. N. Owen, D. C. Hines.)

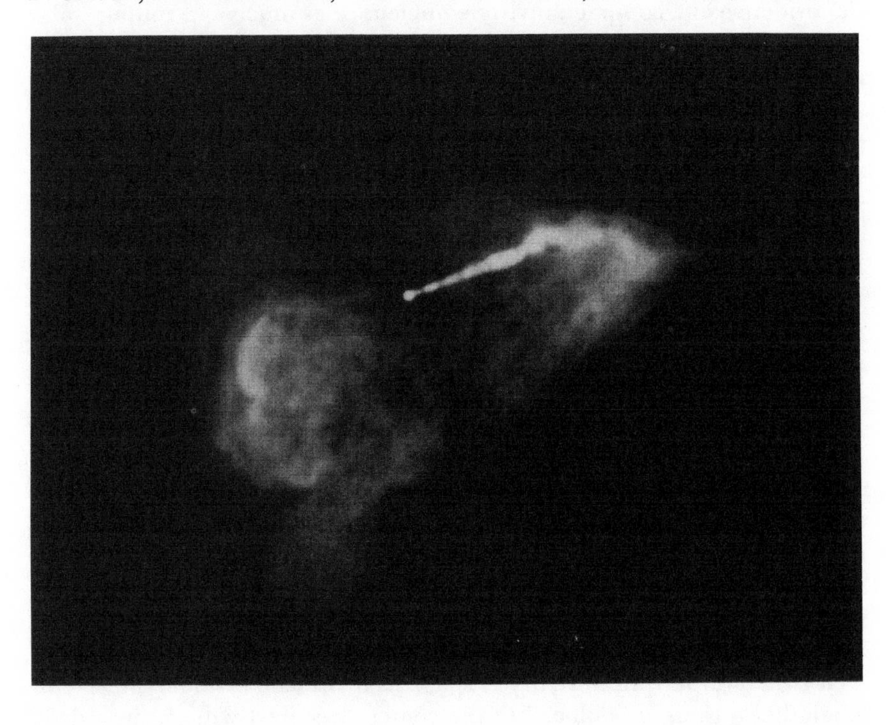

and it must be packed into a volume only some 10 light-years in diameter. It is difficult to conceive of so much mass being squeezed into such a small volume without having it collapse to a black hole about 15 billion kilometers in radius.

Of course all of these scenarios are speculative and there is no positive proof that a black hole is involved, but the circumstantial evidence is highly persuasive. Although we are not certain how any of the active galaxies work, there is a strong general resemblance among them that suggests a possible evolutionary sequence from quasars to radio galaxies. The jets in the quasar 3C-273 and the giant radio galaxy M-87 appear remarkably similar except for scale. In many ways the tight, violently active nucleus of a Seyfert galaxy looks like it could be a phase in the afterlife of a quasar. Each AGN requires a "monster" at the center and an ample supply of stellar food to feed its high metabolism.

Recently, surprising evidence has emerged that even a sedate spiral galaxy such as Andromeda, a twin of the Milky Way, may hide a 50 million–solar mass black hole in its core; this is more than 100 times the mass of the suspected black hole at the heart of our galaxy. Douglas O. Richstone of the University of Michigan and Alan Dressler of the Carnegie Institution observed that, as in M-87, the stars near the Andromeda nucleus are moving too fast to be bound by the visible mass. Also, the tiny elliptical galaxy M-32, just below the bulge of Andromeda, showed similar evidence of a compact, invisible nucleus of as much as 5 million solar masses. Richstone speculates that Andromeda may have been a Seyfert galaxy in its youth, or perhaps a smallish quasar.

Weighing Light

In both Newton's theory of gravitation and Einstein's law of general relativity, a ray of light should bend as it passes a large mass; however, Einstein predicted twice as great a bending as Newton. For Newton, space was flat and the familiar Euclidian geometry that governs our experience on Earth was assumed to hold throughout the universe. According to Newton, a ray of light would deviate toward a massive object such as the Sun because light had weight and would be attracted by gravity. The universe of Einstein offered a fundamentally new concept of gravity. Matter was no longer viewed as an entity in space exerting a pull of gravity on other objects. Instead, the presence of matter created a warp in four-dimensional spacetime. The more massive an object, the greater would be the spacetime curvature in its vicinity. Light would travel along geodesics—the shortest distances between points in spacetime—on curved paths.

It is helpful to think of curved space as a stretched rubber sheet. If a small, heavy weight is placed in the center, the sheet will sag in a deep dimple. A ball rolled toward the edge of the depression will dip toward the center and then pull out at an angle to its original direction. Should the ball roll in only slightly off-center, it will be drawn to the well and spiral down. Spacetime surrounding the Sun is curved in a similar way, and light rays bend like the path of the rolling ball.

When Einstein published his theory in 1915, he pointed out that a total eclipse of the Sun would provide an ideal opportunity to observe the bending of light. As a ray of starlight grazes the limb of the Sun, a viewer on Earth, sighting along the ray path of the arriving beam, would see the star displaced by 1.75 arc seconds. Under normal conditions the brilliant glare of the Sun would blot out the stars. Only with the light of the disk eclipsed would it be possible to observe stars up to the edge of the Sun.

Eddington received a copy of Einstein's publication by way of the neutral Netherlands in the midst of World War I and became very excited by the possibility of testing Einstein's theory. He was a conscientious objector, impatient for the war to be over and for scientists to resume their normal free intercourse. With the approaching end of the war, it appeared that the first favorable opportunity would be the eclipse of 29 May 1919. The Sun would then be in the midst of a field of exceptionally bright stars, the Hyades, and well positioned for the observations. Nearby, but sufficiently separated to provide an undisplaced reference point, is the bright star Aldebaran.

During the final months of the war British astronomers pushed preparations for their eclipse expeditions. In popular language they described their mission as an experiment to weigh light. A.C.D. Crommelin and C. R. Davidson of the Royal Observatory, Greenwich, went to Sobral in northern Brazil; Arthur Eddington and E.T. Cottingham set up camp on a race course on Principe Island in the Gulf of Guinea. The entire occasion was filled with special drama. From Germany, which up to only 1 year earlier had been Britain's bloody enemy, had come the theory that only Eddington and perhaps a very few others were able to appreciate. The prediction contradicted Newton's laws, which had seemed so undeniably established, and now Newton's countrymen were taking their telescopes to far parts of the Earth, anticipating that they would prove Einstein right and Newton wrong.

On the morning of May 29, there was a fierce thunderstorm on Principe. The Hyades, known from ancient times as the "rain makers," appeared to be living up to their reputation. Fortunately, by the time totality arrived, the sky was clear enough to expose a number of photographic plates. Sadly, only 2 produced measurable star images. Things seemed much better at Sobral. Even though the sky was at first overcast, it cleared for 5 or 6 minutes of totality, and 26 plates were exposed. Although only 7 turned out to be useful, they were sufficient.

To confirm a shift in position of a star near the Sun, it is necessary to photograph the same area of the sky at a later time of year, with the Sun out of range. The Sobral team returned in July for that purpose, Eddington again accompanied the expedition and had the results of the measurements he had made earlier. When all was done, Eddington found an average shift that agreed with the prediction of relativity. He wrote, "The evidence of the Principe plates is just about sufficient to allow the possibility of the 'half deflection' and the Sobral plates exclude it with practical certainty." Eddington rushed to inform Einstein of the verification of his theory. It is said that Einstein casually remarked, "I knew the theory was correct." When asked, "What if the theory had proved wrong?" he replied, "I would have felt sorry for the good Lord! The theory is of course all right." Eddington's reaction was much more emotional. He referred to the episode of verification as "the greatest moment of my life" (J. Mehra, ed., *Physicist's Conception of Nature*, 1973, p. 131; J. Newman, ed., *World of Mathematics*, 1956, p. 1101).

The notion that our most elementary concepts of space and time had proved to be erroneous caught the public imagination. Einstein, 40 years old at the time, suddenly emerged as a world figure.

In 1930, in an entertaining after-dinner toast to Einstein, who was present, George Bernard Shaw made the following remarks:

> Religion is always right. Religion solves every problem and thereby abolishes problems from the universe. Religion gives us certainty, stability, peace and the absolute. It protects us against progress which we all dread. Science is the very opposite. Science is always wrong. It never solves a problem without raising ten more problems. Copernicus proved that Ptolemy was wrong. Kepler proved that Copernicus was wrong. Galileo proved that Aristotle was wrong. But at that point, the sequence broke down because science then came up for the first time against that incalculable phenomenon, an Englishman. As an Englishman, Newton was able to combine a prodigious mental faculty with the credulities and delusions that would disgrace a rabbit. As an Englishman he postulated a rectilinear universe because the English always use the word 'square' to denote honesty, truthfulness, in short; rectitude. Newton knew that the universe consisted of bodies in motion and that none of them moved in straight lines nor ever could. But an Englishman was not daunted by the facts. To explain why all the lines in his rectilinear universe were bent he invented a force called gravitation and then erected a complex British universe and established it as a religion which was devoutly believed in for 300 years. The book of this Newtonian religion was not that Oriental magic thing, the Bible. It was that British and matter of fact thing, a Bradshaw [railway timetable]. It gives the stations of all the heavenly bodies, their distances, the rates at which they are traveling and the hour at which they reach

> eclipsing points or crash into the earth. Every item is precise, ascertained, absolute and English.
>
> Three hundred years after its establishment, a young professor rises calmly in the middle of Europe and says to our astronomers, 'Gentlemen, if you will observe the next eclipse of the sun carefully, you will be able to explain what is wrong with the perihelion of Mercury.' The civilized Newtonian world replies that if the dreadful thing is true, if the eclipse makes good the blasphemy, the next thing the young professor will do is to question the existence of gravity. The young professor smiles and says that gravitation is a very useful hypothesis and gives fairly close results in most cases but that personally he can do without it. He is asked to explain how, if there is no gravitation, the heavenly bodies do not move in straight lines and run clear out of the universe. He replies that no explanation is needed because the universe is not rectilinear and exclusively British; it is curvilinear. The Newtonian universe thereupon drops dead and is supplanted by the Einstein universe. Einstein has not challenged the facts of science but the axioms of science and science has surrendered to the challenge. (Blanche Patch, *Thirty Years with GBS*, 1951)

The 1919 observations supported Einstein's theory to an accuracy of about 15%, but other theories of general relativity were subsequently proposed that predicted the bending of light to somewhat different degrees and made it important to measure the amount of bending to still-greater accuracy. The precision of measurement was dramatically improved in 1974–1975, when the radio astronomers at the National Radio Astronomy Observatory (NRAO) at Green Bank, West Virginia, observed very powerful radio quasars rather than optical stars. The quasars pass behind the Sun during the course of the year as do ordinary stars, and they can be followed up to the edge of the Sun without being lost in a glare of solar radio emission, which is very weak compared to the optical disk. No eclipse is therefore needed. The NRAO team chose to observe the quasar 0116+08, which passes behind the Sun every April 11. By using a two-telescope interferometer with a spacing of 20 miles, they could measure the deflection of the radio rays to an accuracy of 1%. A half-century of faith in Einstein was rewarded; his theory was precisely confirmed and competing theories were discarded.

Gravitational Lensing

For years after the original detection of the bending of light, scientists were intrigued by the possibility of gravitational lensing. Could a star focus the light of another star situated at a greater distance? Einstein himself worked on the theory of such lensing and concluded that the effect would be too small to be detected. Only if the lensing star and the source behind it were very closely aligned with the observer would the effect be appreciable. For two stars, that coincidence is so rare that the probability of observation is infinitesimal.

Although Einstein's calculation for the lensing effect of a star was discouraging, only 1 year later Fritz Zwicky suggested that the prospects for lensing were very much greater with distant galaxies of very large mass. These are comparatively abundant, so that the probability of observing a lens effect with galaxies, he concluded, was much better than with stars. His paper received very little attention, however, and for many years afterward few astronomers gave much further thought to gravitational lenses.

In the 1960s theoretical physicists returned to the problem and investigated it in much greater detail. The essence of the lensing effect is that the curvature of space in the vicinity of a massive object will deflect a ray of light in direct proportion to the mass and in inverse proportion to the distance at which the light passes the lensing object. They found a multitude of interesting possibilities, depending on the arrangement of the elements in the gravitational lens system. For example, the lens might produce an image in the form of a thin ring, two narrow crescents, or two small elliptical spots.

An observational breakthrough occurred when Dennis Walsh, Robert Carswell, and Ray Weymann, observing with the 2.1-meter telescope at Kitt Peak, found an odd pair of quasars near the Big Dipper. They lay on an almost north–south axis and were separated by only 6 seconds of arc (about the angular size of a baseball at a distance of 2 miles). The components were named 0957+561A and 0957+561B and were regarded as independent quasars. The spectra of the two objects, however, had lines shifted toward the red by exactly the same amount, corresponding to a distance of about 3 billion light-years. In all respects their spectral details were identical, the only difference being that quasar A was brighter by a constant factor at every frequency.

With no more than 1500 known quasars uniformly distributed on the sky—about 1 quasar for every 30 square degrees of sky—the possibility that two quasars having such a small separation and exhibiting identical redshifts and spectra could occur by chance was so small that astronomers had to conclude that they were twins of some sort. Most likely they were an apparent pair, a double image of one quasar. Suppose that between the quasar and the Earth, in the line of sight, there existed a massive compact body. Light passing on one side would be bent so that the quasar appeared displaced outward on that side. Light passing by the other side would be bent in the opposite sense to produce another outwardly displaced image. Hence the double image.

Further studies of the A and B images revealed that the imaging body lay within a large cluster of galaxies at about one-half the distance to the quasar. It appears that the principal lens element is a giant elliptical galaxy, with a mass of more than 1 trillion suns, but it is only the brightest of some 60 galaxies in the cluster, each working its own deflection of the light rays. As a consequence, the A and B components are not symmetrically aligned with respect to the main lensing galaxy. Furthermore, there are some fine spectral differences that come about because the light of the B image traverses a large portion of the main lensing galaxy and A's light passes more to one side.

Quasars often exhibit dramatic changes in brightness, sometimes in less than 1 day. The brightness of twin B was soon observed to change over a span of months, whereas A did not show corresponding variations. How is it possible for two images of the same quasar to vary out of step in brightness? The answer may lie in the different ray paths to the A and B images; the A component may take a shorter time to reach us than the light that forms the B image. Following a variation in source brightness, A would change first to be followed at a later time by a variation in B.

Some six credible examples of gravitational lenses have now been identified. Indirectly, the discovery of more gravitational lensing galaxies will begin to tell us something about the total mass of the universe. Because the required coincidence in line of sight between the source quasar, the lensing galaxy, and the observer is so statistically rare, even a few examples would raise our estimates of the abundance of faint galaxies. And with further observational refinements, gravitational lenses could provide a probe of the mass distribution within distant galaxies.

Astronomers quickly seized on the explanation of gravitational lensing to produce identical twin images of a single quasar, but Arp and Hazard were stumped by the very large separation of the two images: 157 arc seconds, more than 20 times greater than any previously identified pair. So extreme was this displacement that it seemed impossible to identify a sufficiently massive intervening object. If the lensing were produced by a supermassive cluster of galaxies, it would need to be visible between the B and C images, but no such luminous mass has been detected. Another possibility was a black hole, but it would have to contain the mass of a quadrillion (10^{15}) suns: 10 million times as great as the black holes believed to lurk in the nuclei of quasars. Here the puzzle rested until the early part of 1986, when Bohdan Paczynski, an emigré Polish astronomer at Princeton University, returned to the gravitational lens explanation, but with a startling new twist. He proposed that the massive lens was a "cosmic string," a relic of the earliest instant of the Big Bang, when such bizarre structures are theorized to have frozen out of the unstructured chaos. "There is a crack in everything God has made" (Ralph Waldo Emerson, *Essays*, First Series: *Compensation*, 1841).

According to theory, a cosmic string is a thread of energy almost infinitesimally thin (10^{-30} cm) and incredibly massive (10^{22} grams per centimeter) stretching across the entire universe. Such a thin crack in the universe would be invisible but so massive that its gravitational effect would be enormous. Light rays passing through the distorted space surrounding a string could be deflected by as much as several arc minutes: enough to explain the spacing between 1146 + 111 B and C.

Wild excitement flashed through the astronomical community, and front-page stories appeared in the press when Paczynski's Princeton colleague, Edwin Turner, and his associates reported in May that new observations with the 4-meter telescope at Kitt Peak confirmed that not only were both redshifts precisely 1.012, identical to 1 part in 1000, but their spectroscopic similarity was as close as for any previously discovered gravitational lens. Almost immediately, however, the enthusiasm was chilled when John Huchra at the Multiple Mirror Telescope in Arizona observed distinct differences in the ultraviolet emission spectra of carbon and iron in the two images. These wavelengths were shorter than the visible light observed by Turner. New evidence of differences in the near infrared quickly followed from observations at the European Southern Observatory by Peter Shaver and S. Cristian, who concluded that the lens hypothesis was not supportable and that the two objects, for all their similarity, must be two different quasars.

The story does not end here. Is there no case left for the cosmic-string lens? It is being argued that the differences in ultraviolet and infrared spectra do not necessarily rule out the lensing explanation. The two images are so far apart that the two sets of light rays travel paths that may differ by thousands of light-years. Their respective spectra may change sufficiently in 1000 years to account for the differences. So the lensing hypothesis may still survive.

How will astronomers seek to resolve the uncertainty? If the object is a black hole of such enormous mass, it should appear as a hole in the cosmic microwave background radiation at least 0.1 arc second across. Radio astronomers at the VLA in New Mexico are seeking such evidence at the very limit of their resolution. Other astronomers believe that a string may also reveal itself by effecting a detectable change in the temperature of the 3-degree background radiation across the string. Furthermore, if the string is there, it should produce multiple images of other quasars in the field, but more have yet to be found. As this story is being written, there are rumors of supportive evidence for other lensed objects in the field. The situation is now highly confused, but nonetheless wonderfully exciting.

Cosmic string is being proposed as the explanation for much more than multiple quasar images. In the simple context of an expanding universe the Milky Way and several hundred neighboring galaxies should be flying apart, but a detailed study of their motions shows a more complex pattern. The galaxies appear to be drawn toward a central position by a phenomenally powerful but invisible gravitational source that observers have named "the great attractor." Our galaxy is moving toward the dark attractor at more than 1 million miles per hour.

Astrophysicists speculate that a huge loop of cosmic string, 100,000 light-years in diameter, may drift through space 150 million light-years from Earth, pulling the galaxies in its wake. If these ideas are right, a cosmic string may have had much to do with the initiation of galaxy formation in the early universe. Furthermore, the galaxies would form in ribbons or filaments around the string, and the universe would resemble a spongy array of intertwining ribbon formations wrapped around great bubbles of void. Some astronomers believe the spongy structure is a better fit to observations than is a soapsuds pattern.

For in the wide wombe of the world there lyes,
In hatefule darkness and in deep horrore,
An huge eternal Chaos, . . .
All things from thence doe their first being fetch,
And borrow matter, whereof they are made, . . .

—EDMUND SPENSER,
"The Faerie Queen," 1590

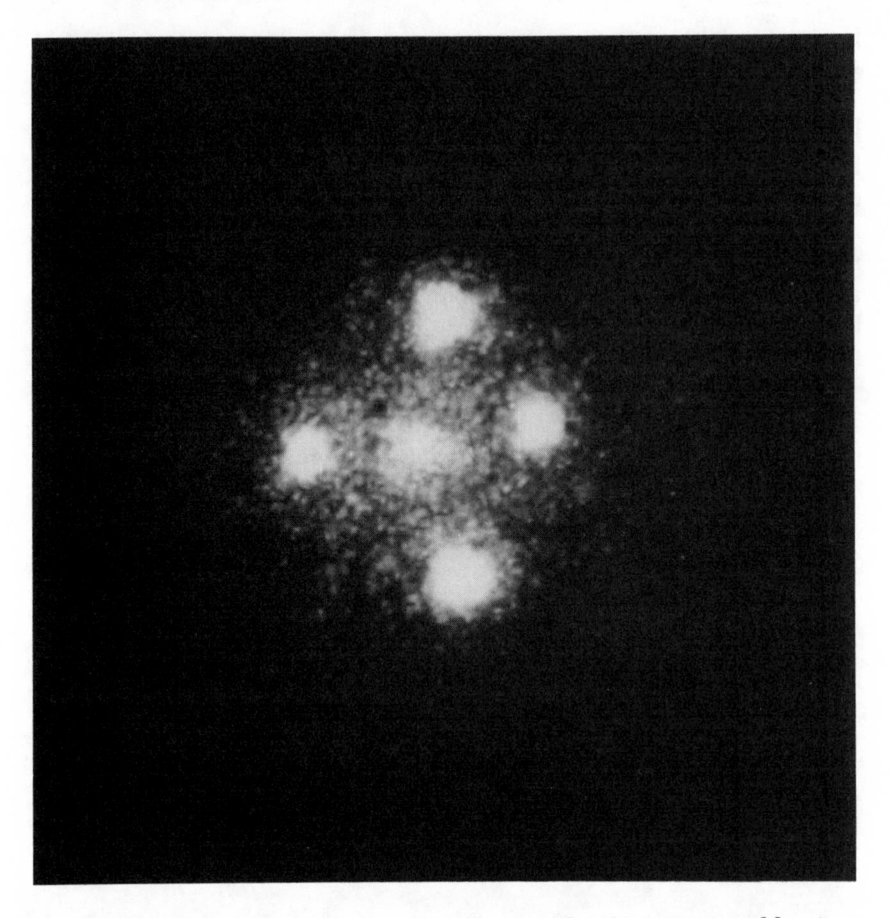

FIGURE 85. *The Einstein Cross is a fine example of gravitational lensing. The image source is a single distant quasar that appears on the four corners of the cross, while the lensing object is the central bulge of a spiral galaxy about 400 million light years from us. The quasar is approximately 8 billion light years distant. John Huchra first observed the Einstein Cross at the Whipple Observatory in Arizona. Even before its first servicing mission, the* HST *imaged it with much greater clarity.*

CHAPTER 8

The Cosmic Perspective

An overwhelming impression of the black depths of space and the immensity of the universe has always boggled the mind. Wrote French philosopher Bernard de Fontenelle (1657–1757):

> *Behold a universe so immense that I am lost in it.*
> *I no longer know where I am,*
> *I am just nothing at all.*
>
> —"Conversations with a Lady on the Plurality of Worlds," 1686

Fontenelle lived to be 100 years old but couldn't find himself. The search for identity in the vast cosmos is too baffling for most persons, who turn to mysticism and theology for answers. Not so the modern breed of cosmologists, who feel confident they are closing in on the solution to the great mysteries of how everything began and how it will all end. About the new concepts, Lev Landau, the brilliant Soviet physicist, once commented, "Cosmology is often wrong but never in doubt." Landau's words reflect the enthusiasm that always accompanies new conceptual "breakthroughs" and the later sober realization of how difficult it is to establish proof.

The Dark of the Night

The seemingly childish question, "Why is the sky dark at night?" has intrigued philosophers and poets and challenged scientists for centuries. It requires twentieth-century cosmology to answer the question satisfactorily.

Johannes Kepler was opposed to the growing popularity of the seventeenth-century notion that the universe could be uniform and infinitely large. Countless stars would fill every realm of space. Kepler argued as follows: If the universe were filled with eternal stars, then no matter where we looked the field of view should be solidly blanketed with stars and as bright overall as the surface of the Sun. There would be no gaps through which to see the dark vault beyond. It is a simple matter to calculate that to collect enough starlight for a bright night sky, we would have to receive radiation from a uniform distribution of stars out to a distance of at least 1 trillion trillion (10^{24}) light-years. Since the sky is black, Kepler reasoned that the universe could not be infinite and filled with eternal stars.

Kepler's analysis failed to recognize the limitations set by the velocity of light and the mortality of stars. The argument based on the horizon limit set by a finite velocity of light was well expressed by poet Edgar Allen Poe in his essay "Eureka" in 1847:

> Were the succession of stars endless, then the background of the sky would present us a uniform luminosity, like that displayed by the galaxy—since there could be absolutely no point, in all that background, at which would not exist a star. The only mode, therefore, in which, under such a state of affairs, we could comprehend the voids which our telescopes find in innumerable directions, would be by supposing the distance of the invisible background so immense that no ray from it has yet been able to reach us at all. (p. 273)

The second problem of the finite life of stars was beyond the state of knowledge in Kepler's time. Because stars burn out after about 10 billion years on the average, we can never receive more than a trickle of starlight, at most 10^{-14} of the combined luminosity required to make the sky blaze with light.

In 1744 Swiss astronomer Jean Phillipe de Cheseau had the brilliant insight to suggest that the space between the stars could contain absorbing matter, in the form of gas and dust, that would diminish the brightness of the night sky. His concept of the existence of interstellar matter was not confirmed until nearly 100 years later, when Heinrich Olbers, a German astronomer whose specialty was comets, revived the puzzle and rediscovered Cheseau's answer. He, too, proposed that the sky is dark at night because dust in space absorbs enough starlight to blot out the distant stars. Soon thereafter John Herschel, son of William, rejected Olbers's hypothesis by showing that the dust would heat up to equilibrium with the energy absorbed from starlight until it became as bright as the stars. In spite of such arguments against Olbers's idea, the problem of the dark night sky came to be known as Olbers's Paradox.

When the present century brought us the concept of the expanding universe, astronomers thought that at long last they had the correct answer to the riddle. As the universe expands, the wavelength of light stretches with it. The more distant the star, the greater the redshift and the weaker the energy that we receive. The explanation seems so pat and simple that few astronomers considered it critically until Edward Harrison of the University of Massachusetts showed that the total effect of the redshift would be to weaken starlight by only a factor of 2.

Finally we come to the Big Bang concept that has dominated cosmology since the discovery of the 3°K microwave background radiation in 1964. It leads immediately to a satisfactory explanation. Since the universe has been expanding from an almost infinitesmal beginning to its present size for only 15 billion years, we cannot see anything more remote than 15 billion light-years, certainly not the required 10^{24} light-years, regardless of how long stars continue to shine.

Harrison also introduced another cogent argument for the darkness of the night sky from straightforward consideration of the energy content of the universe. The average density of the cosmos is only about three atoms of hydrogen per cubic meter. He calculated that if all this mass were converted to energy, the temperature of the universe would not exceed 20°K. The inescapable conclusion is that the universe does not contain enough energy to create a bright sky.

Finally, even though the sky between the stars is not bright, neither is it completely black. The universal microwave background radiation now has a temperature of slightly less than 3°K, but when the universe was about 500,000 years old this background was indeed visible and filled the entire sky with the brightness of the Sun. Expansion of the universe has redshifted the relic of the early fireball to merely a faint, invisible afterglow.

The Shape and Scale of the Universe

Nothing puzzles me more than time and space, and yet nothing troubles me less.

—CHARLES LAMB, 1810 letter

Two-thousand years ago Euclid postulated that through a given point only one straight line can be drawn parallel to another straight line. Popular expressions of the axiom are, "A straight line continues forever"; and, "parallel lines never meet." Surprisingly, mathematicians have never been able to prove the axiom, and it remains one of the most tantalizing unsolved problems of mathematics. In Euclidian geometry space is flat and the sum of the angles of a triangle is 180 degrees. All our normal experience conforms to this concept. In the past 2 centuries, however, mathematicians have postulated alternatives to Euclidian geometry to which the terms *open* and *closed* are applied.

A non-Euclidian geometry of positively "curved space" was constructed by Bernhard Riemann of Göttingen University in 1854, in which two parallel light rays must eventually converge, and no line can be drawn through a given point parallel to a given straight line. In Riemann's geometry we can visualize a two-dimensional analogy of *closed* space on the surface of a three-dimensional sphere. Great circles correspond to straight lines; that is, they are the shortest distances between any two points on a globe. Any two great circles must intersect just as lines of longitude on the Earth's surface intersect at the poles, and the sum of the angles of a triangle on a sphere is always greater than 180 degrees.

In the 1820s the Russian mathematician Nikolai Ivanovich Lobachevski formulated a space geometry of negative curvature. The geometry of three-dimensional negatively curved space can be visualized in two dimensions as a saddle surface, or the throat of a horn, on which the sum of the angles of a triangle is always less than 180 degrees. In this geometry two light rays that start out parallel must eventually diverge, and there are an unlimited number of lines through any point that do not intersect a given line. Negatively curved, or "hyperbolic," space is popularly called *open.*

These non-Euclidian geometries remained mathematical abstractions until Einstein developed his theory of general relativity, a geometrical theory of four-dimensional spacetime in which gravitation is a "warp" in the geometry, introduced by the presence of mass. In Chapter 6 we saw that it takes a very large mass to deform space in its neighborhood, so that even the Sun produces only a small distortion and the deviation from Newtonian gravity is difficult to detect. On the scale of the universe at large, all the stars in all the galaxies, all the diffuse and invisible matter between the galaxies, and possibly the concentrations of matter in black holes as well, combine to determine if cosmic space bends positively or negatively into a closed or open universe.

Aldous Huxley, a novelist and a contemporary of Einstein, expressed the popular bafflement with concepts of curved space in his essay "Views of Holland": "We have learned that nothing is simple and rational except what we ourselves have invented: that God thinks neither in terms of Euclid or Riemann; that science has explained nothing; that the more we know the more fantastic the world becomes and the profounder the surrounding darkness" (*Essays New and Old,* 1927). The scientific outlook of German physicist Karl Schwarzschild was more cheerful. In a 1900 paper presented at a meeting of the Deutsche Astronomische Gesellschaft "Concerning the Possible Measure of the Curvature of Space," he commented, "We find ourselves—if we want to—in a geometrical fairyland, but it is the beauty of the fairyland that we do not know whether it may not in the end be real after all" (pp. 337–338).

The Expanding Universe

We live in an expanding universe, but surely gravity must be slowing it down. Will the expansion ever halt and revert to collapse, or will the rate of deceleration be so small that the headlong outward rush of all the galaxies will never stop? Einstein's equations showed that if space is Lobatchevskian—open—the universe will expand forever; if space is Riemannian—closed—the universe will eventually stop expanding and collapse. Einstein's intuition strongly favored the closed universe. For most of the past 2 decades the evidence has favored an open universe, but recent theories of an "inflationary" universe demand that it be precisely flat. Accordingly, the expansion of the universe will steadily slow down but never come to a halt. In infinite time the universe would coast to infinite size.

Over the range of distances out to 1 or 2 billion light-years the velocities of recession derived from redshifts are closely proportional to distance and the universe appears to be flat within the uncertainties of measurement. Although much larger redshifts are measurable with high accuracy, the corresponding distance determinations become relatively uncertain. Until it becomes possible to measure the distances to galaxies many billions of light-years away with high precision, no subtle bending of space will be clearly revealed.

When it was first recognized in 1929 that nebulous patches thought to lie within the Milky Way were in fact distant galaxies, the universe was judged to be only about one-tenth the size that we now estimate. Since then, the task of sizing the universe has fallen to a small priesthood of dedicated astronomers at the world's greatest observatories. Scaling the distances to celestial objects is a bootstrap operation that begins with the nearby universe and steps farther and farther outward from one rung of the distance ladder to the next. An error in any one of the steps affects all subsequent steps, so that by the time the astronomer's outreach extends to the most remote objects the cumulative error becomes quite substantial.

The story of scaling the universe begins more than 70 years ago, when the American astronomer Vesto Melvin Slipher, at the Lowell Observatory in Arizona, discovered the redshift phenomenon in galaxies. At the time it was still generally thought that the Milky Way contained the entire universe and that the faint nebulous patches that were observed lay within it. His first target was the Andromeda nebula, which showed a blueshift in its spectrum, but the large majority of the patches he next observed were redshifted. Stars in the Milky Way were both blueshifted and redshifted, which indicated that some were moving toward the Earth and others away. But the largest redshifts of his sample of galaxies were about 10 times as great as those of any stars. As Slipher continued his work he reached fainter and fainter nebulosities that showed larger and larger redshifts. Astronomers soon learned that the Milky Way is a member of a small local cluster of galaxies, including Andromeda, that are gravitationally bound together and have relative motions toward each other. The more-distant galaxies are grouped in clusters that partake of a general expansion and are moving away. The situation was now ripe for Hubble to develop his evidence for the expanding universe.

By the end of the 1920s Hubble had teamed with his assistant, Milton Humason, to analyze the spectra of many more galaxies with the 100-inch telescope on Mount Wilson. Humason started as a mule driver, transporting material up the mountain when the Mount Wilson Observatory was being built, and subsequently was promoted to janitor; he soon began to assist with spectroscopic observations. He had only an eighth-

FIGURE 86. *Two-dimensional analogs of: (a) Three-dimensional flat space of zero curvature—an infinite plane; (b) spherical space of constant positive curvature—a sphere; and (c) hyperbolic space of constant negative curvature—a saddle surface.*

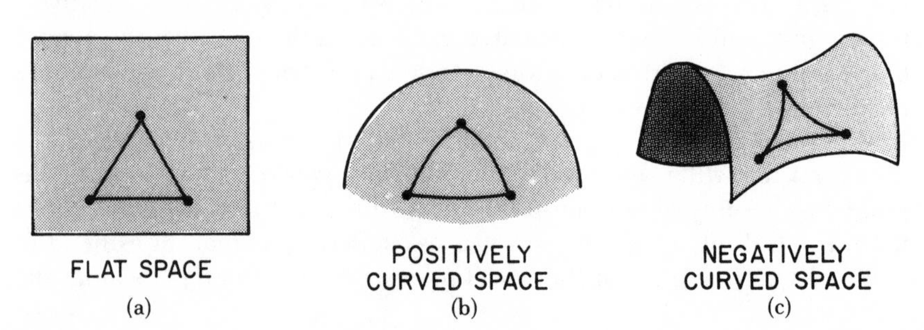

grade education but quickly advanced to fulltime operator. His skills became legendary and his 28-year devoted collaboration with Hubble was remarkably productive.

In today's electronically managed environment, it is difficult to appreciate Humason's highly specialized skills. A spectrum photograph is far more difficult to obtain than a photographic image of a star field. After searching out a faint nebulosity amid myriad bright stars, Humason had to position the fine slit of his spectrograph on the nebula and hold it steady from dusk to dawn. Every few minutes he had to finely reset the center of the nebular image on the narrow slit to correct for the minute wandering of the tracking motion of the telescope. The fainter the galaxy, the longer the exposure, often extending from one night to the next and sometimes over a span of several nights. To stand the chill and fatigue of night-after-night vigils on the mountain top took a rugged endurance and remarkable devotion to the task.

By 1936 data were accumulated on more than 100 galaxies and some ten clusters. When Hubble plotted the redshift versus distance, his galaxies fell on a straight line, showing that redshift was directly proportional to distance. If a galaxy at 10 million light-years were receding at 100 miles per second, one at 1 billion light-years would rush away at 10,000 miles per second. Hubble made the great leap from these preliminary data to the conclusion that every galaxy beyond our local group was racing away with a velocity proportional to its distance, and that the universe was expanding.

Hubble's revelation gave the impression that we are located at the center of the expanding universe, but it is only an illusion. It helps to think of the analogy of a raisin yeast cake. If each raisin represents a cluster of galaxies and imaginary observers are located on one raisin, as the cake bakes and swells they see each raisin-galaxy move apart from every other raisin-galaxy. If the observers were located on any other raisin their impressions would be the same, and if the cake were infinitely large they would not be able to find its center.

> The whole visible world is only an imperceptible atom in the ample bosom of nature. . . . We may enlarge our conceptions beyond all imaginable space. . . . It is an infinite sphere, the center of which is everywhere, the circumference nowhere.
>
> —BLAISE PASCAL,
> *Pensées,* Sec. II paragraph 72, 1670

Climbing the Cosmic Distance Ladder

> . . . of the parallax or parallactic drift of so called fixed stars, in reality ever moving wanderers from

> immeasurably remote eons to infinitely remote futures in comparison with which the years threescore and ten, of allotted human life formed a parenthesis of infinitesimal brevity.
>
> —JAMES JOYCE,
> *Ulysses*, Random House, 1986, p. 579

When an object is comparatively close to the Solar System, astronomers use the method of parallax just as surveyors on the ground resort to simple trigonometry. To determine the distance to a star, the shift in its apparent position is observed twice, at 6-month intervals and from opposite extremes of the Earth's orbit about the Sun.

All that needs to be known is an accurate value for the diameter of the Earth's orbit. The distance determined by parallax is given in parsecs (parallax arc seconds). One parsec equals 3.26 light-years. In 1838 the German astronomer–mathematician Friedrich Bessel made the first determination of the distance to a star beyond the Sun, 61 Cygni, by measuring a parallax of 0.31 arc second. Hence the stellar distance was 3.4 parsecs. At distances greater than about 30 parsecs the parallax method becomes uncertain.

Beyond the range of parallax, astronomers need *standard candles*, celestial objects of characteristic luminosity whose distances can be compared with a nearby object by the inverse square law of brightness versus distance; that is, if a light is twice as far away, it appears only one-fourth as bright. The first rung on the distance ladder is the Hyades, a cluster of about 100,000 stars. Every cluster is a statistical assortment of all the types of stars, from giants to dwarfs, that are burning hydrogen on the main sequence. The color, brightness, and temperature of a star depend only on its consuming hydrogen for its energy. In a cluster such as the Hyades, nearly all of the stars are about the same age even though their masses differ widely. When the apparent brightnesses of stars in the Hyades are plotted against color, they fit the normal Hertzsprung–Russell diagram; that is, the curved line of the main sequence.

At a distance of 46 parsecs there are some bright stars in the Hyades that can still be measured by parallax. With this distance as a benchmark, some 100 stars on the main sequence in the cluster can be used as a combined standard candle. When the apparent brightnesses of a similar collection of stars in a more distant cluster are measured, it can be determined how far the Hyades stars would have to be moved to match their brightness.

Clusters of stars often contain one or more variable stars of a class of supergiants called *Cepheids*, which vary in luminosity as regularly as clockwork. A Cepheid variable expands and contracts like a coiled spring. When it overshoots its average size, gravity regains control and stops the expansion. As it swings back again into compression, past its equilibrium position, the internal pressure sends it rebounding outward again. In each cycle expansion cools the surface and lowers the brightness; then compression heats the surface and increases the brightness.

The variability results from internal instabilities that cause the star to pulsate, expanding and contracting over a period of from as short as 2 to as long as 40 days. For example, the familiar North Star, Polaris, is a Cepheid variable with a period of about 4 days. Cepheids are hotter and more luminous than red supergiants and have maximum radii that can exceed 200 million kilometers. Some Cepheids reach 10,000 times the brightness of the Sun and can be observed to extragalactic distances.

Early in this century Henrietta Swan Leavitt of the Harvard College Observatory made a study of Cepheid variables in the Small Magellanic Cloud. For several decades the Harvard College Observatory station in the mountains of Peru had archived photographs of the two Magellanic Clouds, first noted by Ferdinand Magellan's crew members when they sailed around Cape Horn. The Magellanic Clouds are now known to be satellite galaxies of the Milky Way about 200,000 light-years away. Their stars are at such great distance compared to those in the Milky Way that to a close approximation all the Cepheids in the Small Magellanic Cloud could be taken to be at the same distance from us. Leavitt discovered a total of 1777 variable stars of which only 17 were regularly variable Delta Cephei stars. She found that the dim Cepheids pulsed faster and the brighter ones more slowly. The shortest period was 1.25 days and the longest was 127 days. The apparent luminosities could be translated directly into absolute luminosities and distances simply by measuring the periods. When a Cepheid in the Hyades cluster was next matched to the distance determined by the main-sequence brightness method, astronomers had their standard star.

Once astronomers discover a Cepheid variable they observe it for 1 week or more and plot the variation in brightness while they clock the rhythm. The luminosity is determined from a chart of the period–luminosity relationship. This true brightness is compared with the apparent brightness of the star as it is seen by telescope and dimmed by distance. Using the inverse-square law of brightness versus distance, it is a simple matter to calculate the distance of the star.

Cepheids are so bright that they are easily observed in nearby galaxies out to about 15 million light-years. When Hubble found a Cepheid in Andromeda on the night of 6 October 1923, he was able to resolve the debate over whether the nebulosity was within the Milky Way or was an external "island universe." His calculated distance for Andromeda was over 1 million light-years, well beyond the bounds of the Milky Way. Before long he found a dozen Cepheids in other nearby galaxies and was able to use the distance–redshift relationship to estimate the age of the universe. The result turned out to be embarrassingly small, somewhat less than 2 billion years. Geologists by that time could prove from radio-

active dating that the Earth was more than 4 billion years old. It was not until 1952 that Walter Baade, at Mount Palomar, resolved the dilemma when he discovered that there were two kinds of Cepheids, each with its own characteristic brightness–luminosity relationship. Baade's new standard candle made the universe comfortably older than the Earth.

Even Cepheids are not bright enough to be seen in galaxies more distant than about 5 megaparsecs. Supernovae are so much brighter that they can be observed to distances an order of magnitude greater. The idea of using supernovae as standard candles would be fine if all supernovae had the same outbursts of luminosity. Unfortunately, they do not. Hale Observatory's Allan Sandage believes that all Type I supernovae do indeed have very similar light curves, that they are "well-machined bombs" that produce nearly identical explosions. Other astrophysicists believe that the spectrum of the expanding shell of a supernova explosion resembles a black body, so that the radiation temperature can be used to calculate the intrinsic brightness and hence the distance of the star. Preliminary tests of this approach have been encouraging.

Recently a purely geometrical approach to finding the distance to a supernova has been proposed and tested successfully by a team of astronomers from the Harvard–Smithsonian Astrophysical Observatory, the Max Planck Institute for Radio Astronomy, the Haystack Observatory, and the National Science Foundation. The shell of exploding material from a supernova expands very rapidly, and its light is Doppler-shifted accordingly, so that the rate of expansion is readily derived. It then becomes a simple matter to calculate the size of the shell as it grows. If the apparent angular size could be measured independently to a first approximation, astronomers would have a straightforward determination of distance.

This method has been applied to a brilliant supernova outburst in the

FIGURE *87.* Henrietta Swan Leavitt (1868–1921). *Henrietta Swan Leavitt joined Harvard College Observatory in 1902. She soon became head of the photographic photometry department, which was devoted to accurate determinations of stellar brightnesses. The Harvard plate archives contained thousands of sky photographs, including photographs of the Magellanic Clouds that were taken in Peru. She searched out variable stars in the clouds, and by 1908 had discovered 1777 of them. Only 17 were pulsating stars of the Delta Cephei variety, for which she noted a direct relationship between period and luminosity. Slow was brighter and fast was dimmer. The slowest variable had a period of 127 days, and the fastest period was only 1.25 days. When the distance to one of these was determined it became possible to convert the apparent luminosities to absolute luminosities and thus derive standard candles based on observations of periodicities. The relationship has provided a major tool for the determination of stellar distances in the Milky Way and other nearby galaxies.*

galaxy M-100 in the Virgo cluster. The angular size was less than one-millionth of 1 degree, but the remarkable accuracy of very long baseline radio interferometry made study feasible. The distance to the supernova was determined to be 19 million parsecs, and led to a Hubble constant of about 65 kilometers per second per megaparsec; that fit comfortably in the middle range of the most credible determinations by other methods.

The greatest drawback to this new supernova method is the long wait between supernovae. The event in M-100 occurred in 1979 and there has not been a comparable opportunity since then. An optimistic estimate would predict about one suitable event per year, so perhaps the past several years have just been statistically uneventful.

In spite of the dedicated efforts of Hubble's successors, most notably Allan Sandage, the distance scale remains reliable only to about 30% in the farthest reaches of the universe. When quasars were discovered, it was thought that they might serve as standard candles at very great distances because they are so bright compared to ordinary galaxies. That promise has not been fulfilled because we understand so little of the internal evolution of quasars and cannot separate the cosmological indications from their intrinsic differences. Our knowledge of the evolution of so-called normal galaxies is far better. If we could observe them to the epoch of formation out of the intergalactic gas, we might begin to understand some of the outstanding mysteries of the early universe. The *HST* should take us much closer to that goal.

Breaking the Redshift Record

The largest quasar redshifts known in early 1986 were about 3.8 (wavelength-shifted 380%), and that seemed to mark the earliest epoch of their formation. In August 1986 British astronomers used the 48-inch Schmidt telescope in Australia to set a new quasar-distance record by detecting a redshift of 4.01. The key to detecting an object at a very large distance is to search for the hydrogen Lyman Alpha line in its spectrum, shifted from its far ultraviolet wavelength of 1216 angstroms into the visible range, or better still, into the near infrared. The work of the British astronomers was facilitated by a new instrument called the Automatic Plate Monitoring Facility (APM), which sifts through all the information on the infrared sensitive photographic plate with a laser-scanning device and, with the aid of a computer, identifies classes of objects. Out of a field of some 0.25 million starlike objects, the APM found the highly redshifted quasar at a wavelength of about 6000 angstroms. Its magnitude was 19.8. The light that we now receive was emitted when the universe had evolved to less than 1% of its present volume. But the redshift record has since then been broken several times. In September 1987 a new record of 4.43 was announced by Cambridge University astronomers.

Normal galaxies are typically about 100 times fainter than quasars, and until recently the most distant galaxies observed were at redshifts of somewhat less than 2. In 1986 Hyron Spinrad and his student Stanislaus Djorgovski detected a galaxy at $z = 3.218$ (z stands for redshift). This redshift corresponds to an age roughly 10% of that of the Milky Way. It is generally believed that galaxies were formed at values of z around 4 or 5. Spinrad's observations therefore take us back to a time closer to when we may observe the process of galaxy formation and perhaps even the connection with the special case of the origin of quasars. The magnitude of Spinrad's galaxy was only 24.5, which corresponds to the brightness of a candle at a distance of about 10,000 miles. It should be clear now why astronomers expect so much from the HST.

As the search accelerates, dramatic new evidence is appearing of very young, newly forming galaxies at redshifts larger than 6, or more than 95% closer to the time of the Big Bang. These new observations come from a team of University of Arizona astronomers led by George Rieke. Since the prevailing ideas in cosmology call for the epoch of galaxy formation to have begun no earlier than a redshift of 5, the larger redshifts come as a great surprise.

The Arizona astronomers have taken advantage of new long-wavelength CCDs of the type described in Chapter 2 for optical astronomy. They were developed for military use and were only recently declassified for civilian applications. Instead of scanning the sky point by point, a full infrared image of a region of the sky can now be taken, like an optical photograph. George Rieke comments, "If I had started this research with the detectors available when I started doing astronomy sixteen years ago it literally would have taken the age of the universe to complete it."

Gravity, the Cosmological Glue

> All things . . . linked are, that thou canst not stir
> a flower without troubling of a star.
>
> —Francis Thompson,
> "Mistress of Vision," 1897

The law of expansion is expressed in terms of Hubble's constant, velocity per unit of distance. The most commonly used unit is kilometers per second per megaparsec. There is general agreement today that the constant is somewhere between 50 and 100, but it must be determined much more precisely to satisfy the needs of cosmologists.

The time that has passed since the Big Bang set the universe on the course of its expansion is the reciprocal of Hubble's constant: from 10 billion years for $H = 100$ kilometers per second per megaparsec to 20 billion years for $H = 50$ kilometers per second per megaparsec. The lower value for the age of the universe is uncomfortably close to the best estimates of the age of the oldest stars and galaxies. The Hubble age is always too large because gravity must slow down the expansion as time goes on, so that H is smaller today than it must have been in the past. Exactly how much the expansion has slowed depends on the total mass of the universe. The greater the mass, the greater the retarding influence of gravity on the expansion. Every particle of mass in the universe acts on every other particle to bind the universe together.

The curvature of the universe depends entirely on its density. There is a critical density of about 10^{-29} gram per cubic centimeter (about three hydrogen atoms per cubic meter) that defines a flat universe that will just barely escape collapse and coast outward forever. Higher density leads to a closed universe; lower density leads to a more rapidly expanding open universe. The ratio of the energy density of the universe to the critical density is a quantity that cosmologists designate by the Greek letter omega (Ω). If $\Omega = 1$, the universe is flat; $\Omega > 1$ is a closed universe; and $\Omega < 1$ is an open universe. All of the visible stars in all the galaxies add up to only a minor fraction—less than 10%—of the critical mass of the universe. An intensive search is now being undertaken by astronomers everywhere to assess the amount of invisible matter in order to arrive at a more accurate mass budget for the universe and thus find the key to its shape: closed, open, or flat.

In 1979 Robert H. Dicke and P. James E. Peebles of Princeton University proposed what is known as the *flatness theorem.* They pointed out that if Ω deviated ever so slightly from 1 just an instant after the Big Bang, the difference would grow inexorably with time. Had it not been almost precisely flat at the very beginning, the universe would have collapsed almost immediately or would have blown apart too fast for galaxies of stars to have formed. Within the wide uncertainty of present evidence, Ω could be anything from 0.1 to as high as 2. To be even this narrowly constrained today, it must have been within 1 part in 1000 trillion of being exactly 1 at 1 second after the Big Bang. If Ω were originally precisely 1, it would remain equal to 1 forever.

THE INFINITE UNIVERSE

Classical scholars struggled with the concept of an edge to the universe. In the first century B.C. Roman poet Lucretius argued as follows:

> If for the moment all existing space be held to be bounded, supposing a man runs forward to its outside borders, and stands

> on the utmost verge and then throws a winged javelin, do you choose that when hurled with vigorous force it shall advance to the point to which it has been sent and fly to a distance, or do you decide that something can get in its way and stop it? For you must admit and adopt one of the two suppositions; either of which shuts you out from all escape and compels you to grant that the universe stretches without end. For whether there is something to get in its way and prevent its coming whither it was sent and placing itself in the point intended, or whether it is carried forward, in either case it has not started from the end. In this way I will go on and, wherever you have placed the outside borders, I will ask what then becomes of the javelin. The result will be that an end can nowhere be fixed, and that the room given for flight will prolong the power of flight. (*On the Nature of Things*, 50 B.C., Book I, paragraph 958)

When astronomers talk about the size of the universe they mean the "observable universe." If the age of the universe is about 15 billion years, the oldest light that reaches our telescopes can have traveled no more than 15 billion light-years. That is the distance to the edge of the observable universe no matter how large our telescopes and no matter that the universe may be infinite. An edge of the universe therefore exists in time, if not in space. If we could see right to the edge, we would witness the birth of the universe. In the standard Big Bang theory, there was an interval after the universe started expanding and before galaxies of radiant stars began to form. Prior to the advent of starlight, except for the flash of the Big Bang fireball and its relic afterglow, the universe was dark. Theory sets the time of darkness only very approximately, perhaps as little as 10 million years or as long as 2 billion years. If the first stars or quasars appeared 15 billion years ago, then 15 billion light-years is the limit of the visible universe. Looking farther there is only darkness, because there was nothing visible at the earliest times.

The Dark Universe of Invisible Matter

In 1974 four of the younger generation of astronomers, J. Richard Gott III and James Gunn of the California Institute of Technology and David Schramm and Beatrice Tinsley of the University of Texas, tried to nail down the case for an open universe. They drew on the evidence gathered from all the available publications. When they totaled all the visible mass in the universe it came to only from 0.01 to 0.1 of the closure density, and they confidently concluded that the case for a universe that would expand forever was overwhelming. Instead of shutting off all further controversy, however, almost all subsequent theory and observation has been driven by the expectation that the universe is more likely to prove to be flat ($\Omega = 1$). In the remainder of this chapter we shall examine some of the reasons why so many theorists today believe that the universe is flat and why so many astronomers are searching for proof of the existence of dark matter that they believe could make up 90% of the universe.

Early in the century astronomers thought that the distribution of light revealed the location of mass in the universe. Today we suspect that only a few percentages of the mass in the universe is radiating at any wavelength that can be detected from the ground or in space. The invisible matter could be in the form of extremely small, dim stars, Jupiter-like planets, black holes, neutrinos, or new hypothetical particles such as photinos, axions, and gravitinos, none of which has yet been detected.

Galaxies in clusters swarm about a center of gravity because of the mutual gravitational attraction of all the galaxies. Otherwise, the clusters would evaporate. But almost 50 years ago, Fritz Zwicky and Sinclair Smith at the California Institute of Technology observed that the relative velocities of the galaxies in a large cluster didn't fit the simple concept of gravitational binding. The individual members were moving so fast that they should be flying apart and the cluster dispersing. They concluded that within the cluster there must be more dark matter, "missing mass" as Zwicky called it, than visible star mass.

Since the pioneering work of Zwicky and Sinclair, other astronomers have followed their lead and looked for the gravitational evidence of nonluminous mass in spiral galaxies, in pairs of galaxies, and in clusters of galaxies. All the evidence points to a large component of ubiquitous, invisible mass. In recent years careful studies of the rotation of the giant spiral Andromeda and other galaxies have clearly shown evidence for large amounts of dark mass well beyond the visible outer limits.

In the Solar System, where the planets are held in orbit by the gravitational pull of the Sun, the orbital velocities decrease outward as the reciprocal of the radial distance squared. Closest-in Mercury races around the Sun at a speed of 108,000 miles per hour; farthest-out Pluto cruises sedately at only 10,500 miles per hour. This behavior derives simply from the balance between Newtonian gravitation and centrifugal force first expressed by Kepler in his laws of planetary motion. In a spiral galaxy, the stars, gas, and dust parade around the galactic center according to the same laws. At any distance from the center, all the mass within the radius of the orbit exerts its gravitational force as though it were concentrated at the center. Mass outside a particle's orbit has no effect.

The optical brightness of a galaxy typically peaks strongly at the nucleus and falls off very rapidly with radial distance. Ten years ago, to everyone's surprise, it was discovered that the mass is not distributed in the same way as the luminosity. Stars more distant from the center did not exhibit smaller Keplerian velocities. Until recent years it was very difficult to measure velocities in the outer regions of galaxies to determine the behavior of rotation with radial distance. When radio measurements were made of neutral hydrogen in Andromeda, the rotation curve showed no tendency to turn down where optical evidence suggested the edge of the galaxy. Gas near the outer limits moved just as fast as gas close to the core of the galaxy. This behavior could only be understood if there were large amounts of mass in the far reaches of some invisible halo surrounding the visible galaxy.

Beginning around 1977, Vera Rubin and her colleagues at the Carnegie Institution began a systematic study of the rotation of galaxies by measuring Doppler velocities of stars. They used the most modern high-resolution spectrographic and electronic imaging techniques employed on large telescopes. If a rotating spiral galaxy is viewed edge-on to the line of site, stars on one side are approaching and Doppler shifts are toward the blue; on the opposite side, stars are rushing away and the Doppler shifts are to the red. Their analysis of observations of more than 60 galaxies gave overwhelming evidence that the rotation curves are nearly flat or are rising out to the visible limits. The inescapable conclusion is that the galactic mass does not follow the light distribution of the galaxy.

To explain this behavior, astronomers generally believe that each spiral galaxy is totally enveloped in a spherical cloud or halo of invisible mass. Theorists emphasize that a flat disk would not be dynamically stable and would break up if not for the strong gravitational influence of a massive halo. This extended volume of dark mass keeps the orbital velocities constant well past the apparent limit of the visible galaxy, the beautiful spiral form of which is merely a skeleton on which the surrounding ghostlike halo is hung.

Rubin wryly comments that when she chose to study the rotation curves of galaxies in the early 1960s, most of the turmoil in astrophysics centered on quasars. She recalls that "I wanted a problem that nobody would bother me about." Her quiet field of study, the search for dark matter, now turns out to be one of the most fashionable subjects in cosmology.

We have no consensus about what the dark matter may be. Cold dark matter concepts have created strong links between cosmology and particle physics. Just when cosmologists were being driven to postulate some form of dark matter, physicists were finding it necessary to propose new exotic particles within the framework of GUT of all the physical forces. Several of the hypothetical particles required by physicists could conceivably satisfy the cosmological need for dark matter. Unfortunately the new particles are only theoretical constructs, and no direct detections have yet been made. Still, there is some virtue in the fact that the physicist's particles were proposed independently of cosmological theory and for entirely unrelated reasons, yet have properties that fit the ideas of dark matter and its behavior.

What Is the Dark Matter?

Hypothetical sources of invisible mass include many candidates from the new physics as well as variations on the theme of black holes. Black holes of stellar mass form from dying stars. Once formed they never vanish and can only grow. There cannot be more of them than the number of supernovae—only a small fraction of all the stars in the universe—that have occurred since galaxies formed. During the chaos of the early moments of the Big Bang, however, extreme pressures could have squeezed primordial matter into ubiquitous baby black holes in enormous numbers.

Neutrinos and antineutrinos from the Big Bang are theoretically as abundant as photons of light, 1 billion times as numerous as nucleons of matter. They fill the entire universe. If each neutrino had as little as one fifty-thousandth of the mass of an electron, their total contribution to Ω would bring its value to unity or greater. Neutrinos would be especially effective in binding together clusters of galaxies, though not individual galaxies themselves. Although intensive laboratory efforts have been underway for many years to determine whether or not the neutrino has mass, the only positive result is claimed by Soviet scientists whose evidence remains controversial.

Supersymmetry is a theoretical framework in which every known particle has a supersymmetric partner similar in nearly all respects except for spin. They could interact with normal matter in much the same way as has been postulated for heavy neutrinos. *Photinos* are conceived to be the supersymmetric partners of photons of light, but would have extremely small interactions with matter through the weak nuclear force and gravity. Theorists believe it possible that photinos with masses from 1 to 50 times the mass of the proton could be sufficiently abundant to close the universe. Some evidence for photinos has been claimed by physicists at the CERN particle accelerator near Geneva.

In the modern theory of supergravity, fluctuations of the gravitational field are carried by particles called *gravitinos* that travel at close to the speed of light and may have small mass.

The theory of axions was created by American physicist Frank Wilczek to get around a mathematical problem with the theory of quarks. They should weigh less than one-millionth of an electron mass and clump readily in galaxies.

Cosmic strings are theoretical thin tubes of spacetime; residual cracks in the universe left over from the earliest epoch of the Big Bang. Their diameters are smaller than an elementary particle, but they can kink and loop through the cosmos for millions of light-years and they weigh as much as 10^7 solar masses per light-year. Their gravitational attraction would be enormous, and galaxies, clusters, and superclusters could grow along a string.

Photinos, axions, and gravitinos are collectively known as *Weakly Interactive Massive Particles (WIMPS).* With that name can they qualify to rule the universe? They are undetectable with any instruments available today, although some very sophisticated devices are now being built. Until WIMPS are experimentally detected they belong to the category of broad speculation.

If brown dwarfs of about 0.07 solar mass and black holes of about 200 solar masses are major contributors to dark matter in galactic halos, gravitational lensing may reveal their presence. Within the Galaxy the average displacement of the image of a star is about one micro-arc second, hence it is referred to as microlensing. In 1993, a collaboration known as MACHO (Massive Compact Halo Object) undertook to monitor about 2 million stars in the Large Magellanic Cloud. Over the first year, results were disappointing, yielding evidence of only 3 microlensings, too few to conclude any substantial source of dark matter. Over the next two years, however, 8 lensings were detected, from which a mass contribution to the halo from MACHOs of 200 billion suns was deduced. In 1996, 41 events were observed toward the galactic bulge.

Steady-State Cosmology and Continuous Creation

At the end of World War II, cosmology was largely a matter of speculation in the absence of good observational evidence. It is true that in 1946 George Gamow, Ralph Alpher, and Robert Herman were dedicated to the concept of a Big Bang universe that originated in a primordial fireball, and they embarked on calculations of its temperature, evolution, and element synthesis. But many scientists were still confused by conflicting estimates of the age of the universe from Hubble's early observations of the rate of expansion and from the age of the Earth as calculated by geophysicists. Elementary particle physics was certainly elementary and was just beginning to explain the nuclear physics of stellar energy. It was very difficult to model the physics of the extreme conditions of temperature and density in the Big Bang.

Hermann Bondi, Thomas Gold, and Fred Hoyle, who had been wrapped up in technologies connected with the British war effort, were now free to turn their attention to more esoteric research. In 1948 they proposed a bold new approach to cosmology by postulating a steady-state universe that had no beginning and no end and yet satisfied the observational evidence of universal expansion. To make it expand, they proposed that matter was continuously created out of nothing to provide exactly enough pressure for the expansion and to maintain a constant universal density. Only one atom of hydrogen per year in a volume the size of a modern skyscraper would suffice. As galaxies receded from each other, new ones would fill the vacant spaces, and the universe would appear eternally the same.

Although the steady-state–continuous creation cosmology attracted many adherents, it was doomed within a decade. Hoyle held a philosophical distaste for a Big Bang creation and steadfastly sought to preserve a case for continuous creation. I recall how, in 1959, I had occasion to visit with Hoyle in Cambridge and discuss very preliminary observational evidence of a cosmic x-ray background radiation. At that time the steady-state model had been modified to call for the creation of neutrons. With a half-life of about 1000 seconds, the neutrons would decay to protons and electrons. The disintegrations would impart very high energies to the particles, sufficient to raise the temperature of the intergalactic medium to hundreds of millions of degrees. Such a high-temperature cosmic plasma would radiate a background of x-rays in the range that we had observed with our NRL rockets. Hoyle set about enthusiastically comparing the measured x-ray flux with the prediction of his steady-state model, but to his disappointment the x-rays were too weak by a factor of 100 to support his theory.

Soon, studies of radio galaxies and quasars gave very strong evidence of an evolutionary universe. In 1961 Martin Ryle and his Cambridge University colleagues completed a survey of radio sources out to 8 billion light-years and found that the density of faint sources was greater in earlier epochs of the universe. The source counts implied an evolutionary development of the universe of radio galaxies.

In 1966 Dennis Sciama and Martin Rees, also of Cambridge University, compared quasars of large redshift with those of smaller redshifts. They found an excess of quasars at greater distances, implying a more-rapid evolution of quasars in the very early universe. A final nail in the coffin of the steady-state theory was the discovery of the cosmic microwave background radiation and the near-perfect fit of the spectrum with the prediction of relic radiation from the Big Bang.

The Standard Big Bang

Where all that was to be, in all that was,
Whirl'd for a million eons thro' the vast
waste dawn of multitudinous eddying light—

—ALFRED LORD TENNYSON,
"De Profundis," 1852

According to standard Big Bang theory, at the initial moment of time the universe was infinitely hot and infinitesimally small; a physical singularity beyond mathematical description. Creation was considered out of bounds for physics, and any question of what preceded the Big Bang even more so. St. Augustine once asked, "What was God doing before he created Heaven and Earth?" He then answered his own question: "He was preparing a Hell for those who inquire into such high matters." The standard Big Bang is concerned only with the history of what followed the creation of the primordial fireball. Theorists today, however, are not frightened by St. Augustine's caution. They engage in speculation about the nature of things before the onset of the explosion of spacetime and try to connect the first 10^{-35} second with the physicists' newest GUT of the physical forces.

In Chapter 1 we outlined the development of the Big Bang from a starting point of about 100 trillion degrees. Packed inside this incredible inferno was all the material and photon energy of the universe. Collisions and nuclear reactions proceeded so rapidly and violently that every elementary constituent of matter was created and reprocessed over and over again in a bath of high-energy radiation. Within a few minutes the temperature fell below 1 billion degrees. Nuclear collisions became much less energetic, and matter settled into more normal, familiar forms. Out of the holocaust emerged a plasma of protons, electrons, and photons of radiation.

For the next several hundred thousand years the universe consisted of a melange of fast particles and photons rushing everywhere inside the expanding fireball. The average density of electrons and protons was so high that radiation was effectively trapped. As fast as photons were emitted they were absorbed and scattered, so that in spite of all the internal generation of light the universe was opaque. Radiation and matter were so tightly "coupled" that no light could escape.

As expansion continued the universe dropped to a temperature of 3000°K. Protons could now hang on to colliding electrons and form stable neutral hydrogen atoms. Suddenly the new universe of neutral gas grew transparent. Radiation "decoupled" from matter and flooded outward freely. Everywhere space turned bright with a reddish hue. As time went

on, the photons steadily lost energy because of the Doppler redshift. No matter where in the universe hypothetical observers were located, the radiation would reach them from a part of the cosmos that was receding; just as we now see galaxies and quasars rushing away in the expanding universe. The wavelength of light stretched into the microwave spectrum, characterized by a temperature of 2.7°K. In all directions it appears to be almost perfectly uniform; that is, isotropic. It is the most ancient radiation signal from the early universe, preceding even the most-distant quasar light. The microwave relic of the Big Bang fireball is truly a cosmic background radiation.

Convincing support for the connection of the background radiation with the Big Bang is its spectral fit to a black body. (Radiation emitted by a perfect radiator is termed *black body* and has a unique spectral shape. The intensity increases slowly at first, from long toward shorter wavelengths. After peaking at a characteristic maximum wavelength it falls off rapidly to a short wavelength limit.) The microwave background peaks at about 2 millimeters and fits very closely to the shape of a 2.7°K black body.

Although there is a very strong consensus of support for the connection between the Big Bang and the 2.7°K background, there remain a few astronomers who reserve some degree of skepticism. Fred Hoyle, for example, suggests we may be seeing redshifted starlight from other uni-

FIGURE *88.* George Gamow (1904–1969). *George Gamow was born in Odessa, U.S.S.R., and received his doctorate degree from the University of Leningrad in 1928. Immediately afterward he worked with leading physicists in Europe (Neils Bohr and Ernest Rutherford) and discovered the "tunnel effect" that showed how nuclei could be disintegrated by bombardment with particles of a much lower energy than the binding energies in atomic nuclei. Led by Gamow's theory, John D. Cockroft and E. T. S. Walton used an 800-kilovolt particle accelerator in Rutherford's laboratory to achieve the first artificial disintegration of an atom (lithium).*

Upon returning to the Soviet Union, Gamow's freedom to travel was severely restricted. In an attempt to leave permanently, Gamow and his wife set out from a holiday in Odessa to cross the Black Sea for Turkey in a rubber boat. A storm carried them back and washed them up on the shores of Odessa. In 1933 they defected while attending the Solvay Conference in Brussels. Gamow's most productive years were spent in the United States, at George Washington University.

Gamow was one of the earliest proponents of the concept of the Big Bang, and with two students, Ralph A. Alpher and Robert Herman, he set out to develop a theory of primordial nucleosynthesis of all the elements. That effort failed, but Alpher and Herman recognized early that a relic background microwave radiation must have survived the Big Bang. Their calculation of a 5°K background 17 years before Penzias and Wilson measured the 3°K temperature was remarkably close to correct.

verses bordering on ours. Martin Rees speculates that the universe could have been filled with superluminous stars when it was very young, perhaps between 1 and 10 million years old. Their blue light would have been converted to heat inside thick clouds of gas and dust that could have attained a temperature of a few hundred degrees. In the universal expansion, infrared heat radiation would redshift to microwaves in much the same way as the background radiation of a Big Bang.

The microwave background permeates the entire universe. The question is often asked, If the background radiation is a relic of the Big Bang, where is its center? Answer: There is no center. The photons reach observers from the ancient universe that existed 15 billion years ago and arrive from all directions in equal numbers. Those photons that were produced here at the time of decoupling are now 15 billion light-years away. When we observe the background we see the "center" of the universe all over the sky. Radiation appears from all directions as though we are

FIGURE *89.* Fred Hoyle (1915–). *In 1948 British astrophysicists Hermann Bondi and Thomas Gold put forward a theory of a "steady state," or "continuous creation," universe. Their premise was that if the universe is homogeneous in space it must also be homogeneous in time, and that it must have looked the same in the past, and will look in the future, as it does now. Their theory was largely intuitive. At about the same time, Fred Hoyle had set out to develop a rigorous mathematical theory of the steady state consistent with Einstein's theory of general relativity. He arrived at a result, later refuted, that described how the rate of origin of matter and the rate of expansion regulate each other to maintain a constant mean density in the universe. The creation of local matter was related to the expansion energy of the entire universe.*

Hoyle originally coined the name "Big Bang" in derision. Later he and R. J. Tayler demonstrated theoretically that the large abundance of helium in the present universe could not have been produced in stars but would have been created in a Big Bang. It is interesting to note that Hoyle himself adopted this line of reasoning that contradicted the steady-state model of the universe.

In 1957 Hoyle, with E. Margaret Burbridge, Geoffrey Burbridge, and William A. Fowler, proposed an explanation for the existence of elements heavier than the helium and lithium produced in a Big Bang by processes of heavy element nucleosynthesis in stars. They worked out a comprehensive explanation for nucleosynthesis of all the heavy elements in the universe by neutron capture processes in supernovae.

Hoyle has made prolific contributions to a diversity of subjects. In 1955 he and Martin Schwarzhild made the first use of computer modeling to follow the evolution of a normal star to the red giant phase. He has offered new insights into possible origins of life. His extracurricular successes include a musical comedy and works of science fiction.

surrounded by a homogeneous "photosphere," the shell in which decoupling occurred.

Can we detect any motion of the Solar System through the background? Because of the Doppler shift, the radiation in our direction of motion would appear ever so slightly hotter than 3°K; in the opposite direction it should appear cooler. In the mid-1960s observations were improved to the extent that an upper limit of only 1 part in 1000 could be placed on any deviation from uniformity. The accuracy was improved still further when scientists in the 1970s sent their telescopes above the atmosphere on balloons and U-2 aircraft. Evidence was established of a slight anisotropy, about 4 parts in 100,000. Our solar system appeared to be moving through the background at about 600 kilometers per second.

> *With an awful, dreadful list*
> *Towards other galaxies unknown*
> *Ponderously turns the Milky Way . . .*
>
> —Boris Pasternak,
> "Night," from *When It Clears Up*,
> 1945–1957

When corrections are applied for the orbital motion of the Earth and the motion of the Sun around the center of the Milky Way, our entire galaxy seems to be moving toward the constellation Serpens Caput. It is not clear how many galaxies beyond the local group partake of this general motion.

Taking the evidence of motion of the Local Group of galaxies with respect to the background radiation, infrared astronomers have employed the Infrared Astronomy Satellite (IRAS) survey to extract a new clue to the abundance of dark matter. Michael Rowan-Robinson and David Walker of Queen Mary College, London, and Amos Yahill at Stony Brook State University of New York have taken the 60-micron infrared observations that can be attributed almost entirely to galaxies and used them to calculate the gravitational attraction of each galaxy. They make the reasonable assumption that the infrared luminosity is a measure of the mass of the Galaxy.

When they combine the gravitational effects, the net pull on the Local Group is in approximately the same direction as its motion relative to the microwave background. The volume that IRAS sampled is about 600 million light-years in radius. Rowan-Robinson and his colleagues then assumed that the IRAS sample was typical of the universe at large and calculated how much mass there must be, distributed like that in the IRAS sample, to produce the velocity of 600 kilometers per second of the Local Group relative to the microwave background. Their answer came out to be $\Omega = 1$ with an estimated uncertainty of 20%. The infrared universe therefore appears to contain 90% or more of its galactic mass in invisible form.

Big Bang Nucleosynthesis

In the Introduction we outlined the course of the rapid fall-off of temperature in the early moments of the Big Bang. At 1 second after Time Zero, the temperature had dropped to 10 billion degrees kelvin, and a few minutes later it had cooled to a few hundred million degrees kelvin, the temperature regime of nuclear fusion in the cores of stars. Before 1 second, elementary particles raced about with such violent energy that pairs of particles could not fuse. George Gamow referred to this primordial soup as *ylem*, a word traceable to Aristotle and defined by Webster's Dictionary as "the first substance from which the elements were supposed to be formed."

Once the temperature fell below a few billion degrees kelvin, protons and neutrons could begin to stick together, forming mainly helium nuclei and much smaller trace amounts of deuterium. When Gamow, Alpher, and Herman attempted to model the build-up of heavier elements in the early minutes of the Big Bang by successive capture of neutrons, they discovered that there was an intrinsic barrier at atomic number five. It seemed impossible to get past helium to lithium in spite of various ingenious proposals by Enrico Fermi and Anthony Turkevich at the University of Chicago, and Eugene Wigner at Princeton University. Fred Hoyle, still strongly committed to the steady-state universe, pointed the way out of Gamow's dilemma with the Big Bang by suggesting that nucleosynthesis of the heavy elements occurred in the evolution of stars, as was discussed in Chapter 6.

The hot Big Bang model correctly predicts a cosmic abundance of 75% hydrogen, 25% helium, and a small fraction of 1% of deuterium. It turns out that the deuterium abundance is a critical indicator of the density of the universe at the age of a few minutes.

Deuterium was the first step on the road to nucleosynthesis of the heavier elements. Most of the deuterium in the universe must have been made about 3 minutes after the Big Bang. In those early minutes, the frequency of energetic capture collisions that fused protons with neutrons to form deuterons was very dependent on the particle density. Only a small trace of deuterium was left over from the helium fusion epoch to survive in the present universe. If the primordial density were high enough for a closed universe, the trace of leftover deuterium would be correspondingly low; an open universe would contain relatively more deuterium.

Prior to 1972 there was no evidence for deuterium anywhere in the universe except in the Earth. To detect deuterium in interstellar space requires observation of its ultraviolet absorption spectrum above the

NEW GENESIS

In the beginning God created radiation and ylem. And ylem was without shape or number, and the nucleons were rushing madly over the face of the deep.

And God said: 'Let there be mass two.' And there was mass two. And God saw deuterium and it was good.

And God said: 'Let there be mass three.' And there was mass three. And God saw tritium and tralphium, and they were good. And God continued to call number after number until he came to the transuranium elements. But when He looked back on his work He found that it was not good. In the excitement of counting, He missed calling for mass five and so, naturally, no heavier elements could have been formed.

God was very disappointed, and wanted first to contract the universe again, and to start all over from the beginning. But it would be much too simple. Thus, being almighty, God decided to correct his mistake in a most impossible way.

And God said: 'Let there by Hoyle.' And there was Hoyle. And God looked at Hoyle . . . and told him to make heavy elements in any way he pleased.

And Hoyle decided to make heavy elements in stars, and to spread them around by supernovae explosions. But in doing so he had to obtain the same abundance curve which would have resulted from nucleosynthesis in ylem, if God would not have forgotten to call for mass five.

And so, with the help of God, Hoyle made heavy elements in this way, but it was so complicated that nowadays neither Hoyle, nor God, nor anybody else can figure out exactly how it was done.

Amen

From *My Word Line* by GEORGE GAMOW.

atmosphere. That feat was accomplished with the NASA *Copernicus* satellite by Princeton astronomers who observed the deuterium absorption lines of the interstellar medium against the spectrum of the star Beta Centauri. They found an abundance of 1.5 parts in 100,000, a concentration that translates to a cosmic density of about 10^{-30} grams per cubic centimeter, or less than 10% of closure density. It follows that if all of the observed deuterium is primordial and the universe is made of normal matter, the universe must be open. A closed universe would require a deuterium abundance of no more than 1 part in 1 million. It must be proven beyond a doubt, however, that all the observed deuterium is primordial; that is, that no significant amount was made in stars and passed out to the interstellar medium, even though deuterium produced by fusion in the core of a star converts to helium almost instantly. If the deuterium evidence is correct and the universe is flat or closed, 90% or more of its mass must be dark matter.

In summary, three important predictions in support of the Big Bang model are confirmed by observation. First, the model is consistent with Hubble's evidence of the expanding universe. Second, it predicts the cosmic microwave background almost exactly as observed. Last, Big Bang nucleosynthesis predicts correctly the cosmic abundances of helium and deuterium produced before stellar evolution began to seed the universe with the heavy elements.

These successful predictions that follow from the first microsecond after the start of the Big Bang are counterbalanced by problems traceable to earlier moments. Foremost is the difficulty of explaining the remarkable uniformity of the microwave background. In the standard Big Bang, the universe emerges too fast for thermal equilibrium to attain the observed degree of homogeneity.

The Horizon Problem: Cosmic Communication at the Speed of Light

The difficulty with the homogeneity arises from the finite speed of light, which sets the limit on how fast information about any physical process can propagate. At any instant of the Big Bang, there is a maximum length, the *horizon distance*, that a light signal could have traveled since Time Zero.

In the chaos of the superdense fireball from which the universe emerged, we would expect microscopic random fluctuations to have been present. Some regions would have been slightly hotter or cooler than others. When the microwave background radiation originated, sources in opposite directions of the sky were already separated from each other by perhaps 100 times the horizon distance. If these regions could not communicate, how could they have evolved in such near-perfect equilibrium?

On the scale of the Big Bang, slightly different temperature regimes should have retained their separate characteristics with expansion. Their imprint would now be detectable in the microwave background as slightly different radiation temperatures from widely separated directions of space. But we have seen that the microwave background is very uniform.

The Big Bang model simply sets aside the horizon problem with the assumption that the uniformity is intrinsic to the initial conditions. The universe would then evolve uniformly, but not because of any physical property derived theoretically from the model.

Inflation Cosmology

> I could be bounded in a nutshell and count myself a king of infinite space were it not that I have bad dreams.
>
> —WILLIAM SHAKESPEARE, *Hamlet*, Act II, Scene II

Theorists today dream of infinite universes bounded in microscopic bubbles of space. The most fashionable cosmologies are versions of an inflation scenario inspired by Alan Guth at Cornell University in 1979. At a very early time in its cooling, the universe would have undergone a phase change in which cooling and condensing suddenly reverted to violent expansion. A popular analogy is the expansion of water when it freezes. The expansion pressure is so great that it will shatter a container.

Compared to the expansion of ice, Guth's inflation was on a super-explosive scale. The entire observable universe is supposed to have grown from an almost infinitesimal bubble of space, only one-trillionth the size of a proton. The horizon distance was then much larger than the size of the bubble, so that it had to be almost perfectly uniform. No part of the bubble could have been hotter or colder than any other part. Once in this condition, the bubble universe would remain homogeneous no matter how much it subsequently expanded. Here was an answer to the uniformity dilemma of the Big Bang.

Behind Guth's concept of inflation are the new GUTs of particle physics that link the strong and weak nuclear forces and electromagnetism. They embrace all the known physical forces except gravity. At the earliest instant of time none of the discrete structures that characterize matter today—neither atoms nor molecules—could have existed in the inferno of super-high temperature. The infant cosmos must have resembled a formless soup of essentially pure energy that subsequently congealed as it cooled into the states we now recognize as matter. The GUT formulas deal with energies greater than 10^{24} electron volts (10^{15} gigaelectron volts) and interactions across distances of less than 10^{-29} centimeters. Physicists had started to build an accelerator with a track 50 miles in circumference, the Superconducting Super-Collider (SSC), projected to cost over $8 billion, but it was cancelled in favor of the space station; however, to reach energies of 10^{15} gigaelectron volts would require accel-

THE PHYSICAL FORCES OF NATURE

Physicists identify four fundamental forces in nature. Most familiar is the weakest force, gravity, which holds the planets in orbit about the Sun and keeps us from falling off the Earth. Second is the electromagnetic force that governs the flow of electric charges and their behavior in magnetic fields. Third is the strong force that holds the atomic nucleus together. Last is the weak force that operates at even shorter distances than the strong force and is responsible for radioactive decay. This is the force that governs neutrino physics. Unification of the electromagnetic and weak forces has been demonstrated in experiments with high-energy particle accelerators.

Gravity is 1 trillion trillion billion times weaker than the electromagnetic force. A small toy magnet easily lifts an iron nail even though the gravitational pull of all the Earth's mass is working against it. When a massive astronomical body collapses to a black hole, however, gravity overwhelms all other forces.

The late Richard P. Feynman at the California Institute of Technology used the following example to impress beginning science students with the strength of the electromagnetic force: If you were standing at arms length from someone and each of you had 1% more protons than electrons in your body . . . the repelling force would be enough to lift a weight equal to that of the Earth! In ordinary matter protons and electrons exactly balance and no stress occurs.

The strong nuclear force is the "glue" that binds protons and neutrons in a nucleus. It is about 100 times as powerful as electromagnetic force, enough to prevent the electrical repulsion of positively charged protons in a nucleus from blowing it apart. Although far stronger than gravity, the weak nuclear force is about 1000 times weaker than electromagnetism and therefore is effective only at extremely short range in the nucleus.

Back toward the beginning of time, the temperature of the universe was so high that these forces, which now appear so different, were merged into one "unified" force. Theories that define the merger are called Grand Unified Theories, or GUTs. At one-trillion-trillion-trillionth of a microsecond (10^{-42} second) gravity broke free, leaving the strong, weak, and electromagnetic forces still unified.

eration over a distance of light-years. There is little likelihood that man-made accelerators much larger than the SSC will ever be built. The laboratory of the universe will always be the physicist's super-accelerator.

The energy regimes and interaction ranges to which GUTs apply are very appropriate to the universe at times less than 10^{-35} seconds from the instant of creation. To complete the dream of grand unification of all physical forces, a theory of quantum gravity is needed that describes the universe at 10^{-43} second after Time Zero. Thus far gravity has eluded all attempts to incorporate it into the GUT design. Otherwise, progress in particle physics has been remarkable, and the next generation of accelerators should probe much deeper into unification regimes that only cosmology can now test. As Leon Lederman, director of the great accelerator program at Fermi laboratory, described the physicists' goal: "We hope to explain the entire universe in a single simple formula that you can wear on your T-shirt."

Bubble Universes

> Nothing shall come of nothing.
>
> —WILLIAM SHAKESPEARE,
> *King Lear,* Act I, Scene I

> The universe may be the ultimate free lunch.
>
> —ALAN GUTH, 1982

Alan Guth set the scene of the inflationary episode in the early stages of the standard Big Bang. (Other contributors to inflation theories include Andrei D. Linde and A. A. Starobinsky in the U.S.S.R., Alexander Vilenkin at Tufts University, and Stephen Hawking in England.) In one option, he proposed a primordial soup of matter in a condition of random chaos. Some localized regions would be hot and expanding, others cold and contracting. In a particular volume, which was momentarily superheated and expanding, inflation would get its start. Alternatively he proposed that the universe could inflate from nothing. What physicists refer to as "nothing" is a hypothetical quantum mechanical vacuum state. It is conceived as a sea of "virtual" subatomic particles that exist only ephemerally. These elementary particles pop out of the vacuum for the briefest flash of reality and then almost instantly disappear by annihilation. Such instants of particle appearance and disappearance are called *vacuum fluctuations.* Guth proposed that the entire universe might have emerged from the vacuum just as elementary particles pop out in fluctuations of the vacuum.

The new inflationary cosmology tracks the universe from 10^{-39} second, when its temperature was about 10^{30} degrees. A single unified force, except for gravity, acted on all the elementary constituents of matter, and the high temperature forced the universe to expand in all directions in a smooth manner. For every one-hundredfold advance in time, the volume of the observable universe grew by a factor of 10. As the universe underwent expansion it cooled in a well-behaved way: From 10^{-39} second to 10^{-35} second the temperature fell to 10^{28} degrees.

GUTs identify 10^{28} degrees as a critical temperature below which the particles that carry the unified force disappear and the strong and the *electroweak* forces part company. At the end of GUTs, cosmologists arrive at an unexpected development. The universe had stored a tremendous amount of energy not in particles, but in space itself. This is called the *false vacuum*, and the energy could be released abruptly to the *true vacuum*. When that transition occurred, space was filled with small bubbles and the energy release produced a phase of exponential growth of each bubble. These regions underwent 100 doublings in size in the next 10^{-30} second. Our special universe evolved as one of these bubbles. It grew inside a much larger bubble that contained an unlimited number of bubble universes like ours.

Some theorists suggest that the symmetry of grand unification broke unevenly, in a way that resembles the patchy forming of ice on a winter pond, rather than in a smoother sheet. The irregular patches may have subsequently evolved as the universe expanded, until they seeded the formation of galaxies. Theorists speculate that cosmic strings could have developed along the boundaries. When grand unification broke down, heavy particles that are the mediators of the force may have been left behind. Perhaps these remain in today's universe as dark matter.

At the end of inflation our observable universe was as big as a baseball. Each bubble then resumed its expansion at the rate normal for the standard Big Bang before inflation, a modest doubling every one-thousand-billion-billionth of 1 second. As the universe continued to cool, the electroweak force broke into two forces at one-hundred-trillionths (10^{-10}) second. The four forces that we recognize in the universe today were then independent. A little later on, at 1 microsecond, quarks disappeared inside protons and neutrons and, finally, at about 2 minutes, atomic nuclei fused from protons and neutrons according to familiar nuclear physics.

Many theorists have joined Guth in pursuing the idea of inflation in the early universe, but attractive as the concept may be it is still almost pure hypothesis; there is no direct experimental evidence of the validity of the GUT regime that precedes inflation. Because inflation would proceed so rapidly, the universe would almost instantly become perfectly flat and, accordingly, should remain flat to this time. To test this aspect of inflation theory astronomers have made the search for sufficient invisible mass to make $\Omega = 1$ a high priority.

. . . the great Architect
Did wisely to conceal, and not divulge
His secrets to be scanned by them who ought
Rather admire; or if they list to try
Conjecture, he his Fabric of the Heav'ns
Hath left to their disputes, perhaps to move
His laughter at their quaint Opinions wide
Hereafter, when they come to model Heav'n
And calculate the Stars, how they will wield
The mighty frame, how build, unbuild, contrive
To save appearances, . . .

—JOHN MILTON,
"Paradise Lost," Book VIII, 1667

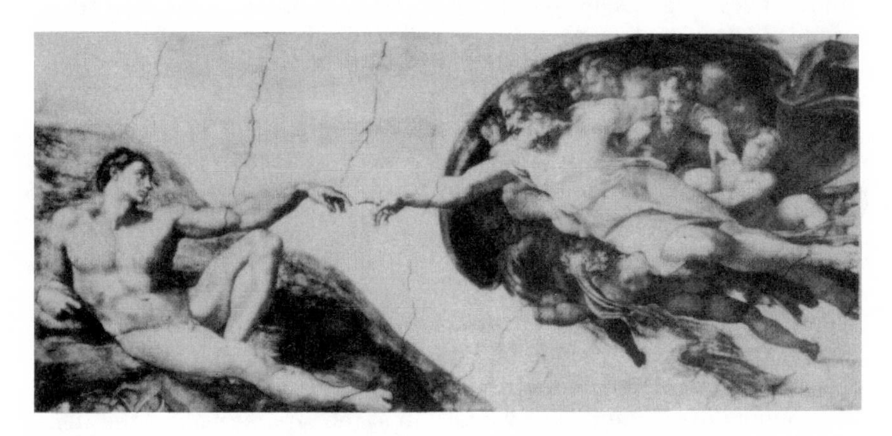

FIGURE *90. "God Creating Adam," by Michelangelo. (Bettmann Archive.)*

CHAPTER 9

The Search for Life in the Universe

For in fact what is man in nature?
A Nothing in comparison with the Infinite,
an All in comparison with the Nothing,
a mean between nothing and everything.

—Blaise Pascal
Pensées, Sec. II paragraph 72, 1670

Philosophical contemplation of an infinite universe is invariably linked to intimations of the proliferation of life throughout the cosmos. Lucretius (98–55 B.C.) wrote in *De Rerum Natura* "nothing in the universe is the only one of its kind . . . there must be countless worlds and inhabitants thereof." In John Milton's "Paradise Lost" (1667), the angel Raphael explains to Adam how and why the world was created:

Witness this new-made World, another Heav'n
. . . .
Of amplitude almost immense, with Starrs
Numerous and every Starr perhaps a World
of destined habitation;

Then Raphael counsels Adam:

Dream not of other Worlds, what creatures there
Live, in what state, condition or degree. . . .

Although the human race on Earth has dreamed and speculated about extraterrestrial life, the only creatures in the heavens have been mythological, and all the remarkable instruments described in this book have failed to provide a positive clue about any alien civilization on any distant planet. "Are we alone?" is a question that still defies scientific investigation.

Much more is known about life's tenacious hold on our watery planet. It is some 3 billion years since the most primitive forms of life first appeared on Earth. Evolution to intelligent beings has been a long, slow process. Two-and-one-half billion years elapsed before creatures resembling the marine life of today appeared in the oceans. Hundreds of millions of years later plant forms and animals emerged from the sea to spread over the land. About 60 million years ago the dinosaurs and other giant reptiles became extinct and mammals were left to roam the Earth. A few million years ago marked the beginning of the Stone Age, when tree dwellers came down and fashioned tools out of rocks. Hunting and farming skills came within the last few tens of thousands of years. People learned to replant seeds only about 10 thousand years ago, and settled civilizations began to replace nomadic tribes.

Until the eighteenth century mechanical power came primarily from the use of animals and human muscles, to be replaced only in fairly recent times by the invention of steam engines and electrical power. The first industrial revolution, just a few generations back, brought the material benefits of mass production and ease of transportation. In the past half-century we have witnessed a second technological revolution, out of which have emerged the miracles of electronic communications, computers, and nuclear power. When we try to assess the probabilities of the existence of alien civilizations elsewhere in the cosmos, we must recognize that it took our civilization a large fraction of the age of the Galaxy to develop and that, during our lifetime, technological capability has arisen with startling rapidity.

We must not think of evolution as a sequence that inevitably leads to intelligent life once the replication of living material has started. Evolution is rather a chance process based on random mutations stimulated by the environment and subject to environmental selection. The end form of life is that best suited to survival in the environment. It may not necessarily involve memory or intelligence. Freeman Dyson comments,

> I do not believe we yet know enough about stars, planets, life and mind to give us a firm basis for deciding whether the presence of intelligence in the universe is probable or improbable. Many biologists and chemists have concluded from inadequate evidence that the development of intelligent life should be a frequent occurrence in our galaxy. Having examined their evidence and heard their arguments, I consider it just as likely

> that no intelligent species other than our own has ever existed. The question can only be answered by unprejudiced observation. ("Intelligence in the Universe," *Mercury*, November–December 1972, p. 9)

Life on Earth depends on the chemistry of carbon and the availability of water. The energy of life is captured from the Sun through the process of plant photosynthesis. Conditions on several of the planets of the Solar System resemble those on Earth sufficiently that one might expect at least some primitive forms of life, but none has been detected. Our efforts thus far have failed to find so much as a microbe or even a fossil remnant of life on any of the planets explored with space probes.

The surface of Venus is hot enough to melt lead, and its atmosphere consists almost entirely of carbon dioxide topped by clouds that contain liberal amounts of sulfuric acid, an environment very hostile to life. Mercury is also too hot and is almost devoid of any sensible atmosphere. Speculation continues about the possibility of living organisms in the atmosphere of Jupiter, but the ecological zone in which conditions might be favorable is very narrow. At the cloud tops the atmosphere is too cold. Though it increases inward to a level at which temperate conditions must prevail, further down the temperature rapidly becomes too high. Between cooking and freezing, the tolerable regime would be very constricted, and circulation up or down would be destructive of life. Unlike Earth, Jupiter shows no evidence of having a solid or liquid surface to which life might cling.

Among the Jovian moons only Io has the warmth and atmosphere to sustain life as we know it, but it is rife with volcanic activity that spews lava over the surface and belches sulfurous fumes into the atmosphere.

Mars, a Failed Life Experiment

Of all the planets Mars has been the subject of most popular speculation as an abode of intelligent beings having the attributes of Earthlings. Today Mars resembles a dried-out relic of a once-living planet. Astronomers of the nineteenth and most of the twentieth century observed seasonal changes on the Martian surface that they connected with the advance and retreat of polar icecaps. At lower latitudes the Martian surface appeared to be broken up into bright and dark areas that they characterized as deserts and maria (seas). In the late spring and summer the maria look dark bluish green; in the fall and winter the color fades to a dusty brownish tone and then turns dark again in the following spring. The seasonal patterns on Mars suggested that the color is related to the growth of vegetation in spring and summer and its disappearance in fall and winter.

The surface of Mars is criss-crossed by an apparent network of sharp, straight lines running for hundreds of miles. The lines vary in color seasonally, as do the broader surface features. Giovanni Schiaparelli speculated in the early 1900s that they could be the work of intelligent Martians, and named them *canali* (channels).

Percival Lowell, a member of a distinguished New England family (his brother became president of Harvard University) and endowed with considerable wealth, was so persuaded by Schiaparelli's idea of canals made by Martians that he established the Lowell Observatory in Flagstaff, Arizona, and dedicated his fortune to the pursuit of evidence of life on Mars. He was a tireless writer and lecturer about his convictions of the existence of Martian life and did much to stimulate the general public's interest in Mars.

Lowell recognized that, apart from the polar caps, Mars was very dry, yet he interpreted the seasonal oscillation in color between the two polar caps as evidence of the transfer of water from one pole to the other. As for the maria, he agreed that the absence of reflected sunlight ruled out the presence of liquid water. But Lowell was convinced that the semiannual transfer of water was accompanied by a darkening of the dark areas that could only be explained as a wave of vegetation traveling with the moisture. From that observation he concluded that the canali were a gigantic artificial irrigation system that made the planet habitable. In Lowell's view, the Martians were a race superior to us, carrying on a struggle for survival in a hostile environment on a dying planet.

Lowell's idea of Martian civilization, in spite of its appeal to the general public, was thoroughly discounted by the scientific community before he died. The canals were recognized as an illusion resulting from observational difficulties, but his ideas of Martian meterology and the coupling of the transfer of water with the seasonal variation in vegetation seemed convincing. These Lowellian concepts were finally discredited scientifically by close-up explorations when the *Mariners* visited the planet.

Space Age exploration of Mars began in 1965, when the NASA *Mariner 4* spacecraft flew by Mars and measured the atmospheric density. As the trajectory of the spacecraft carried it behind the planet, its radio beams to Earth were blocked for nearly 1 hour. Upon entering and emerging from eclipse, *Mariner's* radio signal was bent by atmospheric refraction as it grazed the surface of the planet. From the amount of refraction the pressure was found to be only 6 millibars (Earth's atmospheric pressure at ground is about 1000 millibars), rather than the 85 millibars that had previously been estimated from Earth-based observations. Coupled with spectroscopic measurements from Earth that gave the carbon dioxide pressure as 4 millibars, the new data indicated that carbon dioxide was the major constituent of the Martian atmosphere.

Mariners 6 and *7* arrived at Mars in 1969. Measurements of temperature over the polar caps were consistent with carbon dioxide snow rather than ice. The evidence that polar caps were composed of carbon dioxide—present at all times in the atmosphere—threw into doubt the model of water migrating back and forth between the poles on a seasonal basis. But without water to nurture vegetation, how could one explain the seasonal color variations? It is now believed that the annual changes in the dark areas result from the redistribution of dust transported by seasonal high winds. When the dust is blown away, darker bedrock is exposed.

The most positive rejection of the possibility of life on Mars was deduced from the two NASA *Viking* missions that reached Mars in 1976. Upon arrival at the planet, each spacecraft separated into a lander and an orbiter. As the landers descended to the surface, the orbiters photographed the planet from on high and made global measurements of temperature and atmospheric composition. Information from the landers was transmitted to the orbiters and relayed to Earth. The landers were unmanned biochemical laboratories, the most sophisticated robots ever built.

Testing the soil is the best way to find microorganisms indicative of the presence of life. Soil on Earth is inhabited by bacteria, yeast, and molds that are present in great abundance; they are very hardy survivors even in extreme environments. The logic of the *Viking* exploration called for three instruments designed to detect metabolic activities of soil microorganisms. When these organisms act on food to produce energy, waste gases must be released. A mechanical arm on the *Viking* lander was commanded to reach out and scoop up a sample of soil to be treated with nutrient chemicals. The experiments looked for evolution of tell-tale gas as evidence of living metabolism. In the judgment of the principal investigators, the experiments revealed no evidence of organic matter.

In theory the hydroxyl (OH) produced by the ultraviolet dissociation of any water vapor in the Martian atmosphere would be a strong oxidant capable of destroying organic matter. Its presence could also explain the red color of Mars as the pigmentation of oxides of iron. Still another global oxidation process was subsequently shown to operate in a powerful way. In the presence of ultraviolet light and titanium oxide (a surprising 0.5% of Martian soil), atmospheric oxygen becomes a very efficient reactant with organic matter. On the other hand, it has been argued that enough oxygen to color the planet red could only have come from photosynthetic organisms.

The Martian atmosphere, although now cold and dry, could have been wet and warmer 1 billion years ago. At that time water may have been precipitated in large amounts and run off in what now appear to be dry river beds. If life existed in that distant past, deep sample returns from well-chosen sites may eventually reveal a fossil life-history. Planetary scientists propose a sequence of exploration to uncover the planet's history that would deliver landers to emplace a network of geophysical stations, let loose rovers for sampling a variety of geological sites, and gather and return sample borings to Earth.

The failure to find any evidence of life on Mars was keenly disappointing to many members of the biological sciences community, and some have disputed the negative conclusion. If there is no evidence of life on Mars, the implications are enormous. Earth is the only life-supporting planet orbiting the Sun. We are alone in the Solar System and must extend the search for life far beyond its limits.

Distant Stars and Their Planets

The difficulty of estimating the possible number of civilizations in the Galaxy can be easily appreciated when we admit that we have no positive evidence of the existence of a single planetary system other than our own.

In a simplistic view, it would seem that the universe must be teeming with stars that have planetary systems like the Sun's. At one time it was speculated that Solar System planets were produced by a grazing collision between the Sun and a passing star. In that case, planets would be extremely rare because such collisions are highly unlikely. Now the prevailing view is that planets are the normal accompaniment of the condensation of Sun-like stars from great clouds of interstellar gas. If so, the Galaxy should indeed contain planets in great numbers.

One connection that astronomers take to be significant is that between the mass of a star and its speed of rotation. Massive stars spin faster; small stars such as the Sun rotate more slowly. For stars much smaller than the Sun, the rotation becomes difficult even to detect. As the Sun slowed down, it must have transferred angular momentum to material that had been ejected from it and left in orbit. Today, less than 1% of the Solar System's angular momentum resides in the Sun, and all the rest is associated with the planets. If somehow the planets could be drawn into the Sun, the Sun's spin would increase by a factor of about 200 and would then compare very closely with that of more-massive stars. This relationship suggests that all stars less massive than about 2 suns have been slowed down in the process of formation by the transfer of angular momentum to the orbiting debris out of which planets form. Accordingly, planets should accompany the large majority of Sun-like stars.

Detecting a planet adjacent to a nearby star is a daunting problem for even our most powerful telescopes. If we reverse the situation and imagine observers on a planet of the nearest star, Alpha Centauri at a distance of 4.3 light-years, the Sun would appear to be a rather bright star, but Earth would be 1.6 billion times fainter and so close as to be lost in the glare. Even a Jupiter-sized object in isolation at a distance of several light-years from Earth would be almost impossible to detect.

There is an indirect way of detecting a planet that is orbiting another star. As the planet circles the parent star, it tugs the massive body ever so slightly back and forth across the line of sight so that the track on a photographic plate will wiggle about the direction of motion of the center of gravity of the star and its planet. But the expected motion is very small at best. For example, a planet the size of Jupiter orbiting a Sun-like star at a distance of 25 light-years would produce a wiggle of only 1 milliarcsecond (about the angular diameter of a human hair at a distance of 5 kilometers).

The best-studied example of this dynamic effect is Barnard's Star, a red dwarf 5.4 light-years away. For more than 40 years Peter van de Kamp observed it with the Sproul telescope at Swarthmore College. In some years he exposed as many as 6000 plates by using the telescope every clear night. He persistently claimed that there is a wiggle in the star track with a period of 11.5 years that could be attributed to a pair of planets that are 0.7 and 0.5 times the mass of Jupiter, respectively.

If the planets of Barnard's Star were real, they would be the closest to our Solar System and a likely target for our first interstellar space probes. Van de Kamp worked diligently to perfect his astronomy and counter criticisms about technical shortcomings, but substantial doubt persists about his detection of a planetary mass. From observations with the same Sproul telescope in 1973, John Hershey published a study of 12 stars. He found wiggles similar to those seen in Barnard's Star in all of them. One especially large jump appeared in 1949, when a new cell for the 24-inch lens was installed. Again in 1957, a smaller glitch appeared when adjustments were made to the objective lens. Clearly, there were troubles with the stability of the Sproul telescope, but van de Kamp stubbornly defended his results. George Gatewood and Heinrich Eichhorn published an analysis of 241 plates from other observations in 1973 and found no evidence for a planetary companion of Barnard's Star. The capability of the HST will far exceed that of any ground-based imaging telescope, and for the next decade it will be the primary hope for finding direct evidence of a nearby planetary system. But van de Kamp estimated the visual brightness of his hypothetical planets at thirtieth magnitude, which is probably just beyond the grasp of the HST.

New instruments have been designed specifically for detection of planet-induced wiggles in star positions and promise to do far better than traditional high-resolution telescopes. One such device is the multichannel astrometric photometer that is mounted on the 30-inch refractor of the Allegheny Observatory in Pittsburgh. It looks simultaneously at 12 stars in its field through separate fiber-optic light pipes while a precisely ruled grid of 4 lines per millimeter is translated across the field with a very smoothly controlled movement. The times of the resultant intensity variations are computer analyzed to reveal relative positions with a precision of about 0.003 arcs second for each night of observation. Two weeks of data on the star Procyon revealed not only its proper motion but also the wiggle induced by its orbiting white dwarf companion, which has a 40-year period. Such accuracy far exceeds any previous astrometry. It has been proposed that a dedicated astrometric telescope of this type, designed specifically for the detection of planets, be flown on a spacecraft or mounted on a space station. In principle it could achieve one-hundred-thousandth of 1 arc second resolution in distinguishing the wiggle effect of a Jupiter-sized planet orbiting a bright star. But it would need to function for several decades to give an unambiguous detection.

Star Trek: Science Fiction or Visionary Future?

Sick of our circling round and round the sun
Something about the trouble will be done.
Now that we've found the secret out of weight,
So we can cancel it however great.
. . . .
(Our gravity has been our major curse)
We'll cast off hauser for the universe
. . . .
Taking along the whole race for a ride. . . .

—Robert Frost,
"The Prophets Really Prophesy as Mystics . . .
The Commentators Merely by Statistics," 1962

The possibility of direct contact through space travel, or even of remote reconnaissance with unmanned messenger craft, seems to lie in the far-distant future, but this does not deter enthusiasts from conceiving exotic forms of propulsion for the journeys.

In Chapter 3 we described the progress of *Pioneer 10* past the orbit of the outermost planet, on its way to the stars. Engineers at the Jet Propulsion Laboratory (JPL) calculate that it will make a remote flyby of a red dwarf star at a distance of about 3 light-years from the Earth in about 32,600 years. Also moving away from the Solar System are *Pioneer 11* and *Voyagers 1* and *2*. The JPL computers indicate that *Voyager 2* will come closer than 1 light-year to Sirius, the Dog Star, in A.D. 359,000.

All four spacecraft are powered by radioisotope thermoelectric generators (RTG) that may die before the spacecraft reach the helisophere, the boundary between the region dominated by the solar wind and the edge of interstellar space. As they cruise toward the stars the spacecraft will be speechless and senseless, no longer communicating in any way with Earth or broadcasting a signal that would attract the attention of intelligent extraterrestrial life forms. The chance that any of the spacecraft will ever be examined by intelligent creatures is incredibly slim. Parenthetically, even if the RTGs could function for a much longer time, the signal strength from any of the spacecraft would rapidly fade with increasing distance. Signals received from *Pioneer 10* by the large 64-meter NASA tracking dish are already reduced to one-billion-billionth of 1 watt.

Starship Daedalus

Scientists of the British Interplanetary Society have conceived an imaginary starship *Daedalus* that would use nuclear propulsion and refuel on its way out of the Solar System. Instead of deuterium and tritium, *Daedalus* would use deuterium and helium-3. The required 30,000 tons of helium-3 would be extracted from the atmosphere of Jupiter by an enormous isotope separation operation carried out by more than 100 unmanned spaceship factories operated by robots. Manned fuel shuttles would carry the helium-3 to a storage depot orbiting one of Jupiter's moons.

The hypothetical destination of *Daedalus* is Barnard's Star. To power its engine, frozen pellets of deuterium and helium-3 would be released into the engine chamber. As the pellets entered the chamber they would be hit by a beam of high-energy electrons that trigger nuclear fusion. Each pellet would generate the energy of several tons of TNT to provide enormous thrust. After 2 years of continuous burning of deuterium and helium-3, the starship would reach 7% of the speed of light and then jettison the massive first stage. Over the following 21 months, a second stage of nuclear propulsion would drive the starship up to 12% of the speed of light, 36,000 kilometers per second. Then *Daedalus* would cruise for 47 years until it reached Barnard's Star.

Upon arriving at its target, *Daedalus's* telescopes would image everything in sight while robot probes sortied close in to the area where planets might be. Still rushing onward, *Daedalus* would process its precious observations and radio the refined scientific data back to Earth, where it would be received about 56 years after the launch.

Project Orion: Bombs Away

The concept of nuclear rocket propulsion for deep exploration of the Solar System was seriously studied between 1958 and 1965. Although 7 years of strenuous effort by a team of highly talented physicists went into design studies, engineering tests of components, and laboratory physics experiments, the Orion propulsion concept was finally abandoned for political reasons. At the time of the project's demise, the development team was convinced of the feasibility of a vehicle that could carry eight men and a payload of 100 tons on a fast round-trip to Mars.

The proposed Orion propulsion system involved small nuclear bombs of about 30 kiloton yield, called *pulse units,* carried in external canisters. The bombs would be ejected sequentially and exploded behind the vehicle. Each explosion would transmit an impulse to the spacecraft in a fraction of 1 millisecond through a pusher plate. Several shock-absorbing columns would connect the plate to the main vehicle.

In the original design concept, propulsion to Earth orbit was to be accomplished by Saturn V rockets making perhaps as many as three deliveries to bring up Orion, 2000 pulse units, and payload. For a 500-day Mars landing mission, the required weight upon departure from Earth orbit was 640,000 pounds. In principle there was little difficulty in raising the payload and bomb propellant loading to a departure weight from Earth orbit as high as 2 million pounds, sufficient to carry out a Mars surface expedition and return in 200 to 250 days. Ultimately, a 1-mile-diameter Super Orion was envisioned that could pulse away to a nearby star carrying 1 million bomb propellants.

Although Orion was conceived as a mode of propulsion for detailed exploration of the Solar System, many spaceflight enthusiasts have cherished much grander ambitions of flights to the stars. But our fastest rockets now travel at about 0.01% of the speed of light and would take about 100,000 years to reach the nearest stars. To reach distances of a few hundred light-years would require trips as long as 10 million years.

For interstellar travel to be contemplated with any seriousness, propulsion systems must achieve velocities close to the speed of light. The most efficient energy source would be matter–antimatter annihilation. In contrast to nuclear fusion, which converts only 0.7% of the mass of fuel to energy, matter–antimatter collisions convert 100% of the particle mass to energy. The energy from 35 milligrams of antimatter would suffice to put the NASA space shuttle into orbit. In principle it is possible to manufacture and store antimatter, but applications are far beyond our horizon.

The production and storage of antiparticles is a marvel of accelerator technology. In the accelerators at CERN near Geneva, Switzerland, and at Fermilab in Illinois, a proton bunch of some 10 trillion particles is directed into a target to produce a shower of a few million antiprotons out the other side. The antiparticles are then shunted into a storage ring, the antiproton accumulator, where they fill a few cubic meters. The particles are next cooled down and concentrated into a very fine filament. While the first bunch is cooling, the following bunch arrives. Bunch after bunch is "stacked," up to tens of thousands of strands, by which time hundreds of billions of antiprotons are circulating in the storage ring. That amount is still less than one-trillionth of 1 gram. As one physicist at CERN put it, "If dropped into a glass of water to annihilate with protons in the water, enough energy would be generated to raise the temperature of the water by no more than 1 degree."

For practical use as a fuel, antimatter would need to be stored in a small volume. Physicists at the University of Washington have constructed a container that confines the particles in a very small vacuum region by means of electric and magnetic fields that prevent the antimatter particles from touching the walls. Thus far they have achieved as much as 100 seconds of confinement. Next it will be necessary to freeze the antiprotons into tiny crystals of antihydrogen ice suspended in a vacuum.

At the technology level achieved in modern-day accelerators, it would take thousands of years to produce one-millionth of 1 gram of antimatter. Even if it should become technologically feasible to use matter–antimatter annihilation for propulsion, it would require thousands of times as much fuel as payload to travel interstellar distances at close to the speed of light. Were it possible to approach the speed of light, however, the law of general relativity might provide a bonus. At 99.5% of the speed of light, a 100-year trip as timed from Earth would take less than 10 years by the clock of the traveler aboard the space ship.

To make antimatter fuel a more interesting prospect, the efficiency of production of antiprotons in particle accelerators would have to be improved 100,000 times. It has been speculated that antimatter could eventually be produced for a cost of perhaps $10 million per milligram. That amount could produce the energy of 20 tons of liquid oxygen–H_2 (LOX–H_2) the most powerful fuel of present rockets. Furthermore, the 20 tons of LOX–H_2 would cost an additional $100 million or more for delivery to a launch station in low Earth orbit.

In the rest frame of a starship the bombardment by ambient interstellar gas molecules could boost radiation to lethal levels. Spacecraft velocities more than 1000 times those reached by the most powerful of today's rockets come close to a natural limit based on radiation dosage considerations. At these higher velocities interplanetary or interstellar gas bombards the rocket as hard as the rocket hits the gas, with energies that exceed 1 million electron volts and at a rate greater than 1 billion per square centimeter. The shielding required to protect the astronauts from the hazards of solar flares and galactic cosmic rays now becomes itself a source of deadly radiation. The particles excite nuclear reactions that, in turn, create penetrating nuclear gamma rays and neutrons in such abundance that the dose rate becomes lethal.

Since our most powerful rockets at present fail the requirements of interstellar travel by a factor of 10,000, talk of starships sounds like nothing more than science fiction. But what of civilizations far more technically advanced than ours? I find it utterly fantastic to view the giant propulsion system of the space shuttle against my background of work with the small (6-inch diameter) solid-propellant rockets of only 30 years ago. If life in the universe is a statistical occurrence, there must be thousands of civilizations in the Galaxy that passed our point of technological development millions of years ago. How far might their capability for space travel have brought them? Could they not be roaming the Galaxy far and wide?

Ronald Bracewell of Stanford University has suggested that more-advanced civilizations may already have dispatched automated spacecraft to thousands of targets in the Galaxy. After arriving at a distant star and beginning their orbit, the messenger spacecraft could draw power from the star's light. Radio broadcasts from the probe to the star's planets would be automatically initiated and could hardly fail to attract the attention of intelligent listeners. Why have we not had a visit from one of these interstellar voyagers?

Michael H. Hart of Trinity University and Frank J. Tipler of Tulane University go further than Bracewell. They make the case that within a few hundred million years of the emergence of the first spacefaring intelligence in the Galaxy, the Milky Way should be completely colonized. Since the age of the Galaxy is some 10 billion years and it has taken life on Earth "only" 3 billion years to emerge, their scenario has had ample time to develop. But again, if these space travelers exist, why have they not found us?

Some scientists believe the urge to explore is so deeply ingrained in intelligence that our lack of extraterrestrial contact must mean we are the only intelligent life in the Galaxy.

Iosif Shklovsky, the Soviet astrophysicist, was impressed with this argument. He had been one of the pioneers in exploring the question of extraterrestrial life. On the occasion of the meeting of the International Astronomical Union in Montreal in 1980, he was invited to chair a symposium on the subject of life in the universe. Unfortunately, his visa was not forthcoming until the day before his scheduled symposium, by which time he had abandoned all hope of getting to Montreal. When permission to travel came, he raced off to Montreal, arriving just barely in time. At the end of the session, one of the participants asked, "Dr. Shklovsky, what is your own best personal assessment of the existence of extraterrestrial intelligence?" He replied, "Yesterday, I thought there was no possibility whatsoever that I could be here for this meeting, yet here I am." That evening, at dinner with me, he rued his glib remark, commenting that his black humor was always getting him into trouble with the bureaucrats at home and jeopardizing his contacts with intelligent life in the Western world.

Interstellar travel is the stuff that dreams are made of. For visionaries such as Konstantin Tsiolkovsky, Hermann Oberth, and Robert H. Goddard, it was only a matter of time until the gap between reality and dreams began to close. In a diary, 14 January 1918, Goddard speculated about the "last migration" of the human race aboard fleets of nuclear propelled rockets. "When the sun grows colder rockets should be sent to all parts of the Milky Way. . . . With each expedition should be taken as much as possible of all human knowledge in as . . . condensed and indestructible a form as possible so that the new civilization can begin where the old ended" (*The Last Migration*, manuscript at Goddard Library, Clark University). To introduce a biography of Goddard, Charles A. Lindbergh quoted from the Chinese:

FIGURE *91. Robert B. Goddard at Auburn, Massachusetts, 16 March 1926, before the first flight of a liquid-fueled rocket.*

Am I a man who dreamed of being a butterfly,
Or am I a butterfly dreaming myself to be a man?

—in MILTON LEHMAN,
This High Man, 1963

Eavesdropping on the Galaxy

To reach the nearest star at close to the speed of light would take 4.3 years and could consume the equivalent of all the electrical power produced in the United States in 1 million years. By contrast, it would take only a few watts of power to beam easily detectable radio messages between stars at the speed of light at distances up to a few thousand light-years. Could the universe be filled with interstellar communications? Are we ready to join a cosmic network?

We have been leaking radio photons into space for more than 50 years, since the introduction of radio communications. With the coming of television soap operas as well as civilian and military radar, the radio pollution of space has amplified enormously, and the photons proceed on their almost unobstructed way forever. But the spectrum of radio noise covers a very wide range of frequencies, and the signal that a cosmic listener could tune in to on any wavelength would be a very small fraction of all the escaping radio noise. Furthermore, the radio leakage has been diluted by spreading in all directions. Only if there is a deliberate attempt to beam a message toward a specific star can there be much hope of it being recognized. If everybody is listening, however, and nobody is beaming, eavesdropping remains the only possibility for detection.

The most desirable frequency range for radio penetration lies between 1000 and 10,000 megahertz. Higher frequencies are absorbed by water vapor in the atmosphere; lower frequencies are masked by the noisy background of radio emission from the Milky Way. At 1420 megahertz, atomic hydrogen radiates strongly throughout the Galaxy. A similar strong emission at 1620 megahertz comes from hydroxyl molecules (OH). The frequency gap between these two prominent radiations is called the "water hole" that interstellar radio communicators would most likely choose to establish friendly contact.

The hydrogen frequency fills the Galaxy like a cosmic beacon, but by listening to frequencies displaced only a few megahertz, the natural noise can be avoided. As far as we know, all forms of life use water as a solvent. H and OH combine to form water, and every civilization in the universe would recognize its special meaning for life. If any particular frequency is to be selected for the search, the water hole is the most likely range of choice. But even this narrow-frequency band allows for a great number of channels.

The domain of the water hole has a bandwidth of 200 megahertz. Any discrete radio-frequency propagating through the interstellar medium will broaden slightly by atomic processes, but not by more than about 0.015 hertz. If 200 megahertz is to be searched in 0.1-hertz steps, several hundred million frequency channels must be scanned. Fortunately, radio technology may soon provide the capability of scanning at the rate of hundreds of millions of channels per second.

In 1960 Frank Drake, then at Cornell University, bravely undertook the first search for radio signals beamed by an extraterrestrial intelligence. As head of the National Radio Astronomy Observatory, he scheduled about 200 hours on an 80-foot radio telescope to listen for signals from two candidate stars. He called his brief look "Project Ozma," after the queen of the storybook land of Oz. By all odds Drake and his partners should not have succeeded, and they didn't. Theirs was hardly more than a symbolic gesture toward finding the needle in the haystack, but they had made a start at sifting the noisy radio background for an artificial radio contact.

Drake and Carl Sagan, also of Cornell University, then arranged to use the largest antenna in the world, the 1000-foot Arecibo dish that had been fitted into a large sinkhole in the landscape of Puerto Rico. The listening was largely bootlegged on other projects' time because there was no money specifically allocated for an effort deemed so unlikely to succeed.

Besides listening, Drake and his associates beamed a coded message on 16 November 1974 that could be deciphered by intelligent listeners into a broad description of us. The radio beam was toward the globular star cluster in Hercules, 25,000 light-years from Earth. Of course since the message will not reach its destination for 25,000 years, we cannot expect to receive an answer before 50,000 years have passed, almost an eternity for humankind, but still a small interval in the life of the Galaxy.

Even this shoestring operation at Arecibo was vehemently criticized, and when NASA formally proposed in 1978 to spend about $2 million dollars per year on the Search for Extraterrestrial Intelligence (SETI), former Senator William Proxmire presented the agency with his Golden Fleece award, a monthly recognition of an extravagant, foolish waste of public money.

For a time Proxmire succeeded in preventing any NASA funding of SETI. When he discovered that NASA was spending about $1 million per year on technology related to SETI, he persuaded Congress to shut off even that paltry effort. Sagan addressed Proxmire's political challenge with a determined counterattack backed by a large number of distinguished scientists from many nations. Supporters of SETI lobbied Washington, D.C., politicians, including Senator Proxmire, in an effort to persuade them of the intellectual imperative of SETI and the valuable technological benefits that would accompany a substantial effort. Martin

Rees of Cambridge University was often quoted as saying, "The absence of evidence is not evidence of absence."

In 1981 Paul Horowitz, a young radio scientist at Harvard University, received a modest grant from the Planetary Society, a grass-roots organization of about 100,000 space exploration enthusiasts, presided over by Sagan. Horowitz set out to design the most efficient electronic-search device that current technology could produce to scan the radio background spectrum. He developed what he then called "Suitcase SETI" and brought it to Arecibo for testing in 1983. When he took his suitcase back to Harvard University it was installed on an 85-foot dish dedicated to what became known as Project Sentinel. Designed originally to scan 250,000 radio-frequency channels per second, Sentinel could process more signals in 1 minute than Drake's Project Ozma could in 100 years.

SETI scientists have recognized the importance of Doppler shifts of any very narrow-band frequency beamed toward the Earth. The orbital motion of the Earth about the Sun, as well as its spin, both produce significant Doppler shifts in a signal received from a distant transmitter. Another source is the motion of the local cluster of galaxies relative to the universe as a whole as revealed by the asymmetry of the cosmic background radiation. If an alien civilization tried to contact us precisely at the hydrogen frequency, we would need to scan about 10 million channels to compensate for galactic Doppler shift relative to a cosmic rest-frame.

Senator Proxmire eventually relented, and NASA is being allowed to spend about $12.5 million over a few years to improve the technology for the radio search by a large step. With the additional aid of a private grant from Steven Spielberg, producer of the movie *E.T.*, Project Sentinel was upgraded in 1985 to scan 8.4 million channels simultaneously.

The great advance in radio spectrum scanning is still only a small stab at searching the millions of stars necessary to any chance of success. NASA's newest approach will scan millions of channels per second with very much higher sensitivity. The advanced electronics that are being developed will almost certainly have substantial benefits outside SETI that should satisfy even former Senator Proxmire.

Freeman Dyson regrets that the search for artificial signals of intelligent life is concentrated almost exclusively in the radio spectrum. He urges that every channel of the spectrum be a candidate source of evidence; a long-range search for life is therefore hardly different from a reasonable program of general astronomical exploration. He particularly favors the use of the infrared, where progress has been extremely rapid, culminating in the recent IRAS mission. All of the infrared sources thus far discovered can be explained as natural astronomical objects of luminosities far exceeding that of the Sun. But Dyson points out that any technologically advanced species must radiate waste heat copiously in the far infrared. His model of an artificial source would be a star comparable to the Sun, enclosed in a shell of habitable cometesimals. Future infrared astronomy will be sensitive enough to search out such faint sources. The philosophy of the search is to find a celestial object that cannot be explained by any of our normal experience.

"Are we alone?" is one of the most profound questions that human beings can ask. Philip Morrison once remarked that "the discovery of life on another planet would transform life from a miracle into a statistic." Thus far the miracle has been preserved. The evidence is strong that life exists on no other planet in the Solar System. We must focus our search on much more distant realms. And if we conclude that our small life-supporting planet really is unique, will humanity resolve to avoid self-destruction, choosing instead to preserve our delicately balanced environment and to shun the ultimate folly of nuclear War?

Of man, what see we but his station here,
From which to reason, or to which refer?
Thro' worlds unnumber'd tho' the God be known,
'Tis ours to trace him only in our own.
He who thro' vast immensity can pierce,
See worlds on worlds compose one universe,
Observe how system into system runs,
What other planets circle other suns,
What vary'd Being peoples every star,
May tell why Heav'n has made us as we are. . . .

—Alexander Pope,
"Essay on Man," Epistle I, 1733

Epilogue:

Reflections on Mind and Cosmos

I do not know what I may appear to the world, but to myself I seem to have been only like a boy playing on the seashore, and diverting myself in now and then finding a smoother pebble or a prettier shell than ordinary, whilst the great ocean of truth lay all undiscovered before me.

—Sir Isaac Newton,
in *Brewster's Memoirs of Newton, Vol. 2,* 1855

Why should I mince my words? The truth of Nature which I had rejected and chased away, returned by stealth through the back door, disguising itself to be accepted. . . . I thought and searched, until I went nearly mad, for a reason, why the planet preferred an elliptical orbit. . . .

—Johannes Kepler, on his discovery in 1609 that the orbit of Mars is an ellipse, after astronomers had believed for a thousand years that all planetary orbits were circular.

ASTRONOMICAL DISCOVERY

Other desires perish in their gratification, but the desire of knowledge never: the eye is not satisfied with seeing nor the ear filled with hearing. . . . The sum of things to be known is inexhaustible, and however long we read we shall never come to the end of our storybook.

—A. E. HOUSMAN,
in John Carter, ed., *Selected Prose,* 1961

The most beautiful experience we can have is the mysterious. It is the fundamental emotion which stands at the cradle of true arts and science.

—ALBERT EINSTEIN,
The World As I See It, 1934

I support the cosmological hypothesis that states that the development of the universe is repeated in its basic characteristics an infinite number of times. Further, other civilizations, including more "successful" ones should exist an infinite number of times on the "preceding" and "subsequent" pages of the book of universe. Nevertheless, this *weltanschauung* cannot in the least devalue our sacred aspirations in this world, into which, like a gleam in the darkness, we have appeared for an instant from the black nothingness of the ever-unconscious matter, in order to make good the demands of Reason and create a life worthy of ourselves and of the Goal we only simply perceive.

—ANDRÉ SAKHAROV, at the conclusion
of his 1975 Nobel Peace Prize lecture

In my entire scientific life, extending over forty-five years, the most shattering experience has been that an exact solution of Einstein's equations of general relativity . . . provides the *absolutely exact representation* of untold numbers of massive black holes that populate the universe. This 'shuddering before the beautiful,' this incredible fact that a discovery motivated by a search after the beautiful in mathematics should find its exact replica in Nature, persuades me to say that beauty is that to which the human mind responds at its deepest and most profound.

—S. CHANDRASEKHAR (upon encountering a theory
of rotating black holes by New Zealander Roy Kerr),
Truth and Beauty, 1983

In no other discipline . . . do men confront mystery and challenge of the order of that which looms down on the astronomers in the long watches of the night. The astronomer knows at first hand . . . how slight is our earth, how slight and fleeting are mankind. . . . But more than that, he senses . . . the majesty which resides in the mind of man because that mind seeks in all its slightness to see, to learn, to understand at least some part of the mysterious majesty of the universe.

—VANNEVAR BUSH,
in *Science Digest*, June 1978

I have approximate answers and possible beliefs and different degrees of certainty about different things, but I am not absolutely sure of anything. There are many things I don't know anything about, such as whether it even means anything to ask why we are here. But I don't have to know an answer. I don't feel frightened by not knowing things, by being lost in a mysterious universe without any purpose, which is the way it really is, so far as I can tell. It doesn't frighten me.

—RICHARD FEYNMAN,
from a 1981 interview

APPENDIX A:

The Full Spectral Range of Electromagnetic Radiation

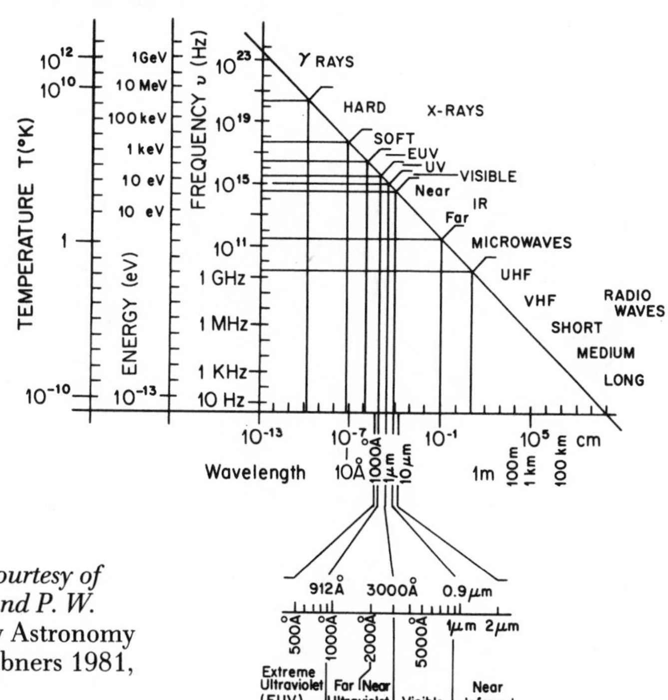

FIGURE 92. *Courtesy of J. L. Culhane and P. W. Sanford,* X-Ray Astronomy New York: Scribners 1981, p. 30.

Vertical scales give frequencies in hertz, the equivalent quantum energy in electron volts, and characteristic temperature in degrees kelvin. On the horizontal scales, wavelengths are given in metric units and in angstroms (1 angstrom = 10^{-8} centimeter). Radio wavelengths are expressed in units of centimeters, meters, and kilometers; the optical and x-ray range in angstroms and nanometers (1 nanometer = 10^{-9} meter); and the infrared in microns or micrometers (1 micron = 10^{-6} meter). The optical range is expanded below. Frequency ν, in hertz (cycles per sec), and wavelength λ, in centimeters, are related to the velocity of light c in centimeters per second by,

$$\lambda\nu = c$$

(wavelength × frequency = velocity of light).

APPENDIX B:

The Doppler Effect

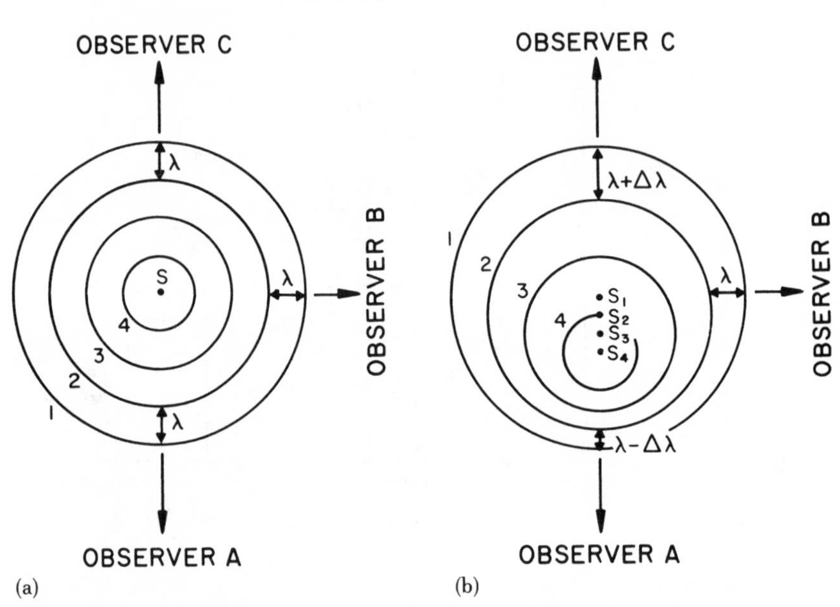

FIGURE *93. The Doppler effect. (a) The light source, s, is stationary with respect to the observer. (b) The light source, s, is moving relative to the observer.*

In 1842 Christian Doppler pointed out that if a light source is approaching an observer, the light waves are bunched more closely; if the source is receding, the waves are stretched out. The effect is similar in sound waves as well. As a police car races toward you the siren's pitch is higher than when the car has passed and is traveling away. In Figure 93a the light source is stationary with respect to the observer. Like ripples from a regular drip in a tub of water, successive wave crests spread out uniformly in all directions and pass an observer with a constant separation (λ), the wavelength. If, however, the source is moving relative to the observer, as in Figure 93b, the successive waves are emitted with the source at different positions: S_1, S_2, S_3, S_4. To observer A, the passing waves are separated by a distance less than λ, whereas observer C sees a separation greater than λ. The wavelength of radiation seen at A is shortened or blueshifted; at C it is lengthened or redshifted. To an observer at B, the wavelength is unchanged. Motion toward or away from an observer is termed *radial velocity*. Observers positioned between A and B or between B and C will see a lesser shortening or lengthening of wavelength in proportion to the component of velocity in the line of sight.

The simpler Doppler shift formula for speeds much less than the speed of light, denoted by c, is:

$$\text{Redshift } z = \frac{\textit{Change in wavelength } (\Delta\lambda)}{\text{wavelength } (\lambda)} = \frac{\textit{velocity } (v).}{\text{c}}$$

If a star is receding its relative velocity is obtained from the redshift.

$$v = c\Delta\lambda/\lambda.$$

For redshifts greater than about 0.1 a relativistic Doppler formula must be used. The relationship between redshift and velocity of recession at large redshifts is given in the following table:

Redshift	Velocity of Recession (% c)
0.1	10
1.0	60
10.0	98.5
100.0	99.95

APPENDIX C:

The Keck Telescopes

The largest ground-based optical telescopes currently in service are the twin pair, Keck I and Keck II, atop Mauna Kea, Hawaii, at a height of 4200 meters (13,600 feet). They belong to a new generation of advanced technology telescopes. Each 10-meter primary mirror is a mosaic of 36 hexagonal segments 3 inches thick and 1.8 meters across, fitted together in a honeycomb pattern. The composite mirror and telescope structure weighs only 158 tons, about one-third as much as the Hale telescope on Mount Palomar even though the mirror area is four times as large. The grinding and polishing was accomplished by an ingenious procedure developed by Jerry Nelson and his associates at the Lawrence Berkeley Laboratory. Instead of polishing a set of shallow off-axis paraboloids, the procedure was reduced to the relatively easy task of polishing to a concave spherical surface. Nelson used a technique of "stressed mirror polishing." Each 1.8-meter mirror was warped in a special jig by a precisely computed amount and ground to a spherical figure. After polishing, the jig was removed and the mirror segment sprang back into exactly the right shape. The segment was then cut around the perimeter to hexagonal shape. Because the fully assembled mirror distorts under gravity as well as under the influence of temperature variations and wind, sensors along the rims of the segments tell a computer to signal actuator pistons on the underside of each segment to correct its position relative to adjoining segments twice each second to an accuracy of less than 1 wavelength of visible light.

At the height of 4200 meters, water vapor freezes out of the air and turbulence is minimal, making Mauna Kea an ideal site for infrared and submillimeter as well as optical astronomy. The physical stress of working atop Mauna Kea is not trivial. Buildings at the top are supplemented with oxygen to protect workers against oxygen starvation, and personnel sleep at a lower level of 2800 meters.

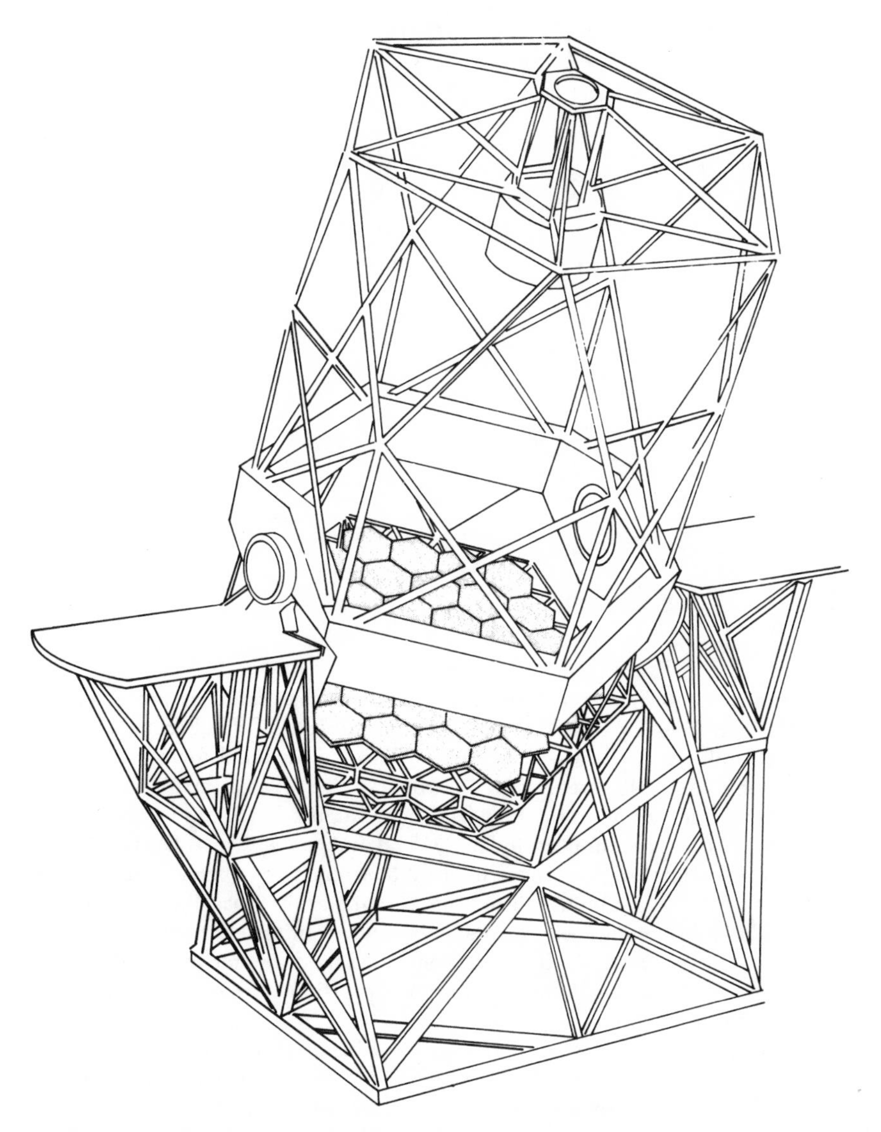

FIGURE 94. *Diagram of the Keck telescope. Thirty-six hexagonal mirror segments fit together to equal the aperture of a single hyperbolic mirror, 10 meters in diameter. (Courtesy of the California Association for Research in Astronomy.)*

APPENDIX D:

The Very Large Telescope

A consortium known as the European Southern Observatory is proceeding to build the largest optical telescope of the next century atop Cerro Paranal in Chile. The $255 million Very Large Telescope (VLT) is scheduled for completion in 2000. It will comprise four 8-meter mirrors spaced 98 feet apart on separate mountings. In combination the four telescopes can act as a multiple mirror telescope (MMT) to superimpose the individual images with the light-gathering power of a single 16-meter (more than 52 feet) diameter mirror. Alternatively, the four telescopes can be used as an interferometer on a baseline of up to 150 meters. The thin "meniscus" mirrors, only a few centimeters thick, are flexible and backed by arrays of small pistons controlled by computer that continuously adjust 350 contact points to correct for any distortions due to gravitational sag and temperature variations.

The VLT will operate in open air instead of inside a dome or boxlike shelter. Smaller thermal distortions are thus expected. An aerodynamically designed wind screen will maintain a smooth air flow over the telescope.

Appendix E:

New Technology Telescopes

On La Silla Mountain in Chile, an 8-meter telescope is operated remotely by astronomers sitting at computers in Garching, Germany. The isolation of astronomers from their telescopes and their dependence on computerized operations is a revolutionary change. Powerful computers execute the commands to track the stars, control the shape of flexible mirrors, and manage the collection of enormous amounts of data. Adaptive optics refers to what are popularly called "rubber" mirrors that change shape to compensate for distortions in the wavefront of light by inhomogeneities in the atmosphere. Some adaptive optics systems create an artificial star by means of a laser beam. The beam is directed at a position in the sky close to the target star being observed and excites atmospheric luminescence. The distortions produced in the artificial star by the shimmering atmosphere are continuously followed by a sensor that signals a computer to drive the flexible mirror to fix the image of the artificial star and at the same time the star being observed. The result is nearly perfect "seeing."

Gemini, a project of AURA (Associated Universities for Research in Astronomy), is building a pair of 8.1-meter telescopes, one on Mauna Kea, Hawaii, and the other at Cerro Paranal in Chile. Project Columbus is producing a pair of large spin-cast mirrors with the combined light-gathering power of an 11.8-meter dish.

FIGURE 95. *The Very Large Telescope (VLT). (Courtesy of the European Southern Observatory.)*

Further Reading

Asimov, I. *The Exploding Suns.* E. P. Dutton, New York, 1985.
Bartusiak, M. *Thursday's Universe.* Times Books, New York, 1986.
Begelman, Mitchell C., and Rees, Martin J. *Gravity's Fatal Attraction.* W. H. Freeman, Scientific American Library, New York, 1996.
Bok, B., and Bok, P. F., *The Milky Way.* Harvard Univ. Press, Cambridge, Mass., 1979.
Ferris, T. *The Red Limit.* William Morrow, New York, 1977.
Field, G., and Chaisson, E. *The Invisible Universe.* Birkhauser, Boston, 1985.
Friedman, H. *Sun and Earth.* W. H. Freeman. New York, 1986.
——— *The Amazing Universe.* National Geographic Society, Washington, D.C., 1975.
Gamow, G. *A Star Called the Sun.* Viking Press. New York, 1964.
Goldsmith, D. *The Evolving Universe.* Benjamin Cummings, 1985.
Greenstein, G. *Frozen Star.* Scribner, New York, 1983.
Harrison, E. R. *Cosmology.* Cambridge Univ. Press, Cambridge, England, 1981.
Hawking, Stephan W. *A Brief History of Time.* Bantam Books, New York, 1988.
Henbest, N., and Marten, M. *The New Astronomy.* Cambridge Univ. Press, Cambridge, England, 1983.
Hodge, P. W. *Galaxies.* Harvard Univ. Press, Cambridge, Mass., 1968.
Horowitz, N. H. *To Utopia and Back.* W. H. Freeman, New York, 1986.

FURTHER READING

Kippenhahn, R. *Light from the Depths of Time.* Springer-Verlag, Berlin, 1987.
——— *100 Billion Suns.* Basic Books, New York, 1983.
Noyes, R. *The Sun, Our Star.* Harvard Univ. Press, Cambridge, Mass., 1983.
Pagels, H. R. *Perfect Symmetry.* Simon and Schuster, New York, 1985.
Schwinger, J. *Einstein's Legacy.* W. H. Freeman, New York, 1986.
Shipman, H. L. *Black Holes, Quasars, and the Universe.* Houghton Mifflin, Boston, Mass., 1980.
Shklovsky, I. S., and Sagan, C. *Intelligent Life in the Universe.* Holden-Day, San Francisco, 1966.
Silk, J. *The Big Bang.* W. H. Freeman, San Francisco, 1980.
Sullivan, W. *Black Holes.* Doubleday, New York, 1979.
——— *We Are Not Alone.* McGraw-Hill, New York, 1964.
Thorne, Kip S. *Black Holes and Time Warps.* Norton, New York, 1994.
Tucker, W., and Giacconi, R. *The X-Ray Universe.* Harvard Univ. Press, Cambridge, Mass., 1985.
Weinberg, S. *The First Three Minutes.* Basic Books, New York, 1977.

GLOSSARY

A

absolute temperature Temperature measured on the kelvin scale (K), which begins at absolute zero and increases by centigrade units so that water freezes at 273.16°K and boils at 373.16°K.

absolute zero Temperature at which a gas has no thermal energy. This is at 0°K or at −459.69°F.

absorption line A narrow region of a spectrum of photons within which the intensity of radiation is lower than the adjoining regions.

accretion An infall of matter on an object.

accretion disk An accumulation of matter in a spiraling, flattened pattern around a compact object.

alpha particle Nucleus of helium-4 consisting of two protons plus two neutrons.

angstrom A unit of wavelength equal to 10^{-8} centimeter and often abbreviated Å.

angular momentum A measure of the amount of spin possessed by an object.

angular size The part of a circle of 360 degrees over which an object appears to extend; measured in degrees, minutes of arc, and seconds of arc.

annihilation Interaction of a matter particle and an antimatter particle in which the mass is converted to gamma radiation.

antiparticle The antimatter complement of an elementary particle having the same mass and spin but with equal and opposite charge.

antimatter Matter composed of antiparticles.
aperture The diameter of the lens or mirror in a telescope.
apogee For an object in an elliptical orbit about another object, the point in its orbit at the greatest elongation.
apparent brightness The brightness that a celestial object appears to have to a distant observer, hence a measure that depends on the distance between object and observer.
apparent magnitude Apparent brightness measured in magnitude units, in which 5 magnitudes corresponds to a brightness ratio of 100, and each magnitude to a ratio of 2.512. The higher the magnitude, the lower the brightness.
astrometry A branch of astronomy concerned with the precise positioning of celestial objects.
astronomical unit The average distance from the Sun to the Earth, equal to 149,597,900 kilometers. Sometimes abbreviated A.U.
astrophysics The study of the composition and other physical properties of celestial objects.

B

Balmer Series Spectral lines of the hydrogen atom associated with the second energy level. The strong visible red line at 6563 angstroms involves a transition from the third level to the second level.
Big Bang The primordial explosion about 15 billion years ago that launched the expansion of the universe to its present size.
binary star Two stars in orbit about their common center of mass.
bit The smallest unit of information in a computer equivalent to a zero or one; that is, one binary decision. Contraction of *binary digit.*
black body An object that absorbs all radiant energy that strikes it and therefore appears black.
black body radiation The spectrum characteristic of a "perfect radiator," namely one that radiates the maximum amount of energy per unit surface area at a given temperature and is perfectly black at 0°K. Radiation in a state of thermal equilibrium.
black dwarf A burned-out star that has no energy left to radiate light.
black hole A spherical volume of space surrounding an object so compressed that its gravity completely closes spacetime around itself and neither photons nor particles can escape.
black hole radius Equal to 3 kilometers times the mass in units of the Sun's mass.
blue giant A massive star of high surface temperature.
bremsstrahlung Continuum radiation emitted when two charged particles pass in near collision. (German for *braking radiation.*)
byte A sequence of bits, usually eight, operated on as a unit in a computer.

C

cD galaxy An extremely massive galaxy centered within a rich cluster of galaxies.
celestial equator The circle on the celestial sphere that lies above the celestial equator and is everywhere equidistant from the celestial poles.
celestial poles Two points on the celestial sphere that lie directly above the north and south poles of the Earth.
celestial sphere The imaginary sphere of the sky that is centered on the Earth.
center of mass The point within a body or a collection of bodies that makes the product of mass times distance the same in any direction from that point.
Cepheid Variable A giant pulsating star whose period of variation is related to its absolute luminosity.

charge-coupled devices (CCD) An array of tiny electromagnetic radiation detectors (pixels) made of semiconductor material that can be read out to provide an image.

chromatic aberration Failure of a lens to focus different colors to the same focal point.

chromosphere The portion of the atmosphere of the Sun and other stars—from 10,000 to 100,000 kilometers immediately above the photosphere—that merges into the outer corona.

constellation A grouping of stars named after a mythological character or animal. Some 88 such groups are recognized by astronomers.

continuous spectrum A spectrum in which the emission of radiation is smoothly continuous over all wavelengths.

corona The outermost portions of the solar atmosphere and of other stars reaching millions of kilometers into space at temperatures in the range of millions of degrees.

cosmic microwave background A relic radiation of the Big Bang fireball. It now pervades the entire universe with a characteristic temperature of 2.7°K.

cosmic rays Mostly electrons, protons, and helium nuclei moving through the Galaxy at close to the speed of light.

cosmology The science of the structure of the universe as a whole and of its evolution.

critical density The average density of matter in the universe that determines whether the universe will expand forever or collapse. The critical value is 10^{-29} grams per cubic centimeter. Higher density leads to reversal of the present expansion; lower density leads to infinite expansion.

D

deuterium An isotope of hydrogen in which the nucleus is composed of one proton and one neutron.

Doppler effect A wavelength shift in which wavelengths appear to shorten as a source approaches an observer and to lengthen as a source recedes.

dwarf galaxy A very small galaxy, usually elliptical, containing fewer than 100 million stars.

E

eclipsing binary A binary star system in which one star passes in front of the other once each revolution, obscuring its view from the Earth.

electron An elementary particle with one unit of negative charge and a mass of 9.1×10^{-28} gram.

emission line A spike of photon intensity confined to a narrow range of wavelength.

erg A metric unit of energy equal to the work done by a force of 1 dyne acting over a distance of 1 centimeter.

escape velocity The minimum velocity required by a rocket to leave its place of launching, never to return.

event horizon The spherical surface surrounding the center of a black hole at its radius. Any event inside that radius cannot be observed from outside because no signal from inside the black hole can pass through the event horizon.

excited state A state of an atom in which at least one electron occupies an orbit larger than the minimum energy orbit allowed. The atom can return to its ground state by the transfer of an electron from the excited state to the smaller orbit with the emission of a photon.

exclusion principle The quantum-mechanical rule that for certain kinds of elementary particles (electrons, protons, and neutrons) no two particles of the same type can occupy the same position and have the same velocity.

F

flare A local region of the Sun or other star that suddenly brightens 10 times or more in x-rays and radio waves as well as in visible light.

flux The flow of energy per square centimeter received from a radiation source.

frame of reference A coordinate system for the purpose of assigning positions and times to events.

fusion Nuclear process that builds up heavier nuclei by joining protons and neutrons with the release of enormous amounts of energy.

G

galactic cannibalism The capture of a smaller galaxy by a larger one as a result of gravitational forces.

galactic coordinates Objects seen from the Earth are specified with respect to the galactic equator by galactic latitude and with respect to the direction to the galactic center by galactic longitude. Zero degrees is the direction to the galactic center and 180 degrees is the opposite direction.

galaxy A large collection of stars, gas, and dust held together by their mutual gravitational attraction. Most galaxies contain from a few million to trillions of stars.

galaxy cluster A group of galaxies gravitationally bound together from their time of formation. A cluster may contain as few as a dozen galaxies or as many as thousands.

gamma rays Photons with the highest energies (in excess of 1 million electron volts) and highest frequencies (above 10^{20} hertz).

giant molecular cloud A localized region of interstellar space containing a much higher density of cool gas and dust, typically about 1 million solar masses of material at a density of about 10^{-19} grams per cubic centimeter within a diameter of several light-years to dozens of light-years.

globular star cluster A spherical condensation of as many as 100,000 stars in a localized region of a galaxy often well outside the galactic plane. The concentration is so great that several stars may be found within a volume of only a single cubic light-year. The stars are typically as old as the galaxy.

Grand Unified Theory A theory of elementary forces that unifies the electroweak theory of electromagnetic and weak nuclear forces with the quantum chromodynamic theory of strong nuclear forces. It still does not include gravitation. Sometimes abbreviated GUT.

gravitation Force of attraction that matter exerts on other matter, proportional to the product of the masses of two objects and inversely proportional to the square of their distance of separation.

gravitational lens A massive object that exerts enough gravitational force to bend light rays from a distant source to a focus.

gravitational redshift A decrease in the energy of photons as they leave the surface of a massive object. The shift of wavelength to the red is proportional to the mass of the object divided by the radius of its surface.

gravitational waves Waves in the gravitational field that are analogous to light waves in the electromagnetic field. Gravitational waves travel at the speed of light.

graviton The quantum of gravitational radiation analogous to the photon of light.

H

heliocentric Centered on the Sun.

heliosphere The volume of space surrounding the Sun out to the distance where the solar wind encounters the interstellar gas.

hertz A unit of frequency equal to

one cycle per second. Sometimes abbreviated Hz.
Herzsprung–Russell Diagram A plot of the absolute magnitudes of stars against their temperatures.
Hubble's Constant. The constant of proportionality that relates the velocities of recession of galaxies to their distances. The constant is not precisely established but is in the range of from 50 to 100 kilometers per second per megaparsec.

I

interferometer A pair of telescopes that combine their signals so that the interference of waves arriving at the telescopes permits the determination of direction and image resolution with the definition of a single telescope of diameter equal to the separation of the individual telescopes.
intergalactic medium Gas in intergalactic space outside the bounds of galaxies.
inverse-square law The apparent brightness of a source decreases in proportion to the square of its distance. A star 100 light-years away looks one-fourth as bright as the same star at a distance of 50 light-years.
ionization Process by which an atom loses one or more electrons to gain a positive electrical charge.
ionosphere The region of the upper atmosphere from about 50-miles to 200-miles altitude that is electrified by solar ultraviolet and x-ray radiation.
isotope A subspecies of a chemical element that is characterized by the number of neutrons in the nucleus. The number of protons defines the element.
isotropic The property of being uniform in all directions. The radio microwave background and the diffuse x-ray background are highly isotropic.

K

kilogram Unit of mass in the metric system equal to 1000 grams, sometimes abbreviated kg. One kilogram equals 2.2 pounds.
kilometer One kilometer equals 1000 meters, sometimes abbreviated km. (1 kilometer = 0.61237 mile)
kinetic energy Energy of motion. For an object moving at much less than the speed of light, kinetic energy equals one-half the product of its mass and the square of its velocity.

L

light-year The distance light travels in 1 year, equal to 9.46×10^{17} centimeters.
Local Group The cluster of about 20 galaxies in the neighborhood of the Milky Way.
luminosity Total energy emitted per second in the form of photons.
Lyman series The spectral lines of a hydrogen atom that are emitted when an electron jumps from an excited state to the first energy level or ground state. The Lyman-Alpha line at 1216 angstroms is the strongest line in the hydrogen spectrum. Absorption lines arise at the same wavelengths when an electron absorbs radiant energy in the ground state and moves to an excited state.

M

Magellanic Clouds Two small, irregular satellite galaxies of the Milky Way at a distance of about 48 kiloparsecs from the Sun.
magnetic field A field of force in space created by a magnet or by the flow of electric current that deflects the trajectories of charged particles.
main sequence The band of the Hertzsprung–Russell diagram that stars

occupy during the phase of their evolution in which they derive energy from fusion of hydrogen to helium in their cores.

meridian An imaginary circle that passes overhead from an observer's due-north point through the zenith to the due-south point of the horizon.

meteroroid A comparatively small stony or metallic object in orbit around the Sun that may impact the Earth.

meter The fundamental unit of length in the metric system, equal to 39.37 inches.

microwaves Radio photons with wavelengths between 1 millimeter and a few centimeters.

momentum The product of an object's mass and velocity that tends to hold its motion in a straight line at constant speed unless acted upon by an external force.

Mossbauer Effect Emission of gamma radiation of very precise frequency in radioactive decay.

N

nebula A diffuse cloud of interstellar gas and dust often excited to luminescence by the ultraviolet light of embedded stars. In some contexts, a galaxy.

neutrino An elementary particle with zero or very small rest mass and no electric charge that travels close to the speed of light; emitted or absorbed in particle interactions governed by the weak nuclear force.

neutron An elementary particle with a mass almost identical to that of the proton (1.647×10^{-24} gram) and no electric charge. The neutron is believed to be made of three quarks; it is stable within a nucleus but decays rapidly into a proton, electron, and antineutrino once outside the nucleus.

neutron star An object formed in the collapse of the core of a massive star to a density of 1 billion tons per cubic inch and a diameter of between 10 and 15 miles.

nonthermal radiation Photon emission that does not follow the spectral shape of a black body.

nova Thought to be one of the stars of a binary pair that flashes to sudden brilliance when it receives a critical infall of protons from its companion capable of igniting nuclear fusion.

nucleus The central region of an atom composed of one or more protons and none or more neutrons.

O

occultation The passage of one celestial object in front of another as seen by an observer at a distance.

P

parallax shift The apparent displacement of a star position on the celestial sphere that is caused by the Earth's motion around the Sun. The parallax method of determining stellar distances is accurate out to only a few hundred light-years.

parsec A unit of length equal to 3.262 light-years. It is the distance of an object whose parallax shift is 1 arc second.

photoionization Ionization produced by absorption of a photon.

photon A quantum of electromagnetic radiation with no mass and no electric charge, traveling at the speed of light.

photosphere The visible surface of a star from which most of its radiation escapes into space.

planetary nebula A shell of gas ejected from a star late in its evolution to the red giant stage and illuminated by ultraviolet radiation from the hot core.

plasma An ionized gas in which the

electrons that have been removed from atoms are free to move about.

polarization Aligning of the vibrations of a light wave to one plane.

Population I stars A younger generation of stars with ages from a few million years to about 1 billion years and with a relatively large fractional abundance (about 1% of mass) of elements heavier than helium.

Population II stars An older generation of stars with ages up to 15 billion years and a relatively low percentage (less than 1% by mass) of elements heavier than helium.

positron Particle with electron mass and unit positive charge.

precession The wobble of a spinning rigid body; for example, a toy top or a rotating star.

prominence A looping region of hot, glowing gas arching above the solar photosphere to great heights.

proper motion A star's apparent motion against the background of very distant stars.

proton The nucleus of a hydrogen atom; a fundamental particle of unit positive charge and mass of 1.6724×10^{-24} grams composed of three quarks.

protostar A star in the process of formation from a collapsing gas cloud within which nuclear fusion has not yet been ignited.

pulsar A spinning neutron star radiating oppositely directed beams of radio waves and sometimes visible light, x-rays, and gamma rays as well.

Q

quark An elementary particle that comes in several forms. Three quarks make a proton or a neutron.

quasar An enormously dense galactic nucleus, so small as to appear starlike, yet equal in brightness to hundreds of normal galaxies. Spectrum shows large redshift.

R

radiation pressure Pressure exerted by photon collisions.

radio waves The lowest energy, longest-wavelength form of electromagnetic radiation, with photon energies less than 10^{-15} erg and wavelengths greater than about 1 millimeter.

radio galaxy A galaxy that produces as much energy in radio waves as in visible light.

recombination The capture of an electron by an ion.

red giant A star that has left the main-sequence track of evolution and expanded to as much as 100 or more times its original size.

relativistic Moving with a velocity close to that of light.

resolving power The ability of a telescope to reveal two closely spaced stars as individual points rather than as a single point. Resolution is proportional to telescope diameter and inversely proportional to wavelength.

rich cluster of galaxies A cluster of hundreds to thousands of galaxies.

S

Schwarzschild radius The radius of a black hole.

Seyfert galaxy A spiral galaxy with an extremely bright compact nucleus whose spectrum indicates violent internal movement.

shock wave A discontinuous pressure disturbance moving through a gas at high speed.

solar eclipse The blocking of the view of the Sun's disk by the passage of the Moon directly between the Sun and the Earth.

solar nebula The original gas cloud thought to have assumed the shape of a flattened, spinning disk within which the Sun, the planets, and their satellites, asteroids, and comets formed.

solar mass The mass of the Sun, 2×10^{33} grams.
solar wind Streams of electrons, protons, and other ions that escape the solar corona into interplanetary space.
spacetime According to Einstein's theory of general relativity, events in the physical universe are viewed as taking place in a four-dimensional continuum, three dimensions of space and one of time.
spectral class A designated type of stellar spectrum based on the strength of hydrogen absorption lines in the visible region of the spectrum and molecular absorption lines in cooler stars.
spectroscopy Observation and analysis of the spectrum of light from a laboratory or celestial source.
spectrum The distribution in wavelength or frequency of the number of photons emitted or absorbed by a source.
spicule A localized jet of gas rising about 10,000 kilometers in the chromosphere of the sun.
spin The intrinsic angular momentum of an elementary particle such as a proton, neutron, or electron. It can have only integer or one-half integer values of Planck's constant; $(6.626 \times 10^{-27}$ erg per second)/2.
spiral galaxy A flattened disk of billions of stars with a central bulge and a large galactic corona. Within the disk, spiral arms are outlined by young, luminous stars.
standard candle A celestial source of standard luminosity useful in the determination of stellar distances from observation of the apparent brightness.
steady-state theory A cosmological theory in which the overall appearance of the universe does not change with time. As the universe expands new matter is continuously created to fill the new space.
strong force One of the four fundamental forces. It is always attractive and operates only at distances less than 10^{-13} centimeter to hold together the nucleons inside an atomic nucleus.
sunspot A dark, cool region in the solar photosphere whose temperature is 1000 or 2000 degrees lower than the surrounding photosphere.
supercluster of galaxies A very large association of galaxies up to hundreds of millions of light-years across and bound together by their self-gravitation.
supernova An exploding star that briefly reaches the luminosity of an entire galaxy of stars and fades in a matter of weeks or months.
supernova remnant An expanding cloud of glowing gas left over from a supernova explosion.
synchrotron radiation Photons emitted when electrically charged particles moving at nearly the speed of light are accelerated in a magnetic field.

T

thermal radiation The emission of a continuum of photons as a result of the heat motion of the radiating particles.
totality The phase of a solar eclipse in which the Moon completely obscures the disk of the Sun.
transient A short-lived emission of radiation.
tritium An isotope of hydrogen in which the nucleus contains one proton and two neutrons.

U

ultraviolet Usually defined as photons of higher frequencies than visible light in the range from 10^{15} to 10^{16} hertz.

V

Van Allen Belts Regions high above the surface of the Earth in which energetic particles are trapped by the Earth's magnetic field.

W

wavelength The distance between two successive wave crests or troughs in a wave of light.

weak force One of the four fundamental forces that acts only between certain elementary particles in a nucleus; responsible for radioactive decay.

white dwarf A star about the size of the Earth that has fused helium to carbon and has no further reserve of nuclear fusion energy. It is supported by electron degeneracy pressure in its core.

X

x-rays Photons of energies greater than those of ultraviolet and less than those of gamma rays, with frequencies from 10^{17} to 10^{20} hertz.

Page numbers in *italics* refer to captions.

INDEX

INDEX

INDEX

INDEX

INDEX

INDEX

INDEX

INDEX

INDEX